中国农业科学院
兰州畜牧与兽药研究所科技论文集
（2011）

中国农业科学院兰州畜牧与兽药研究所　主编

中国农业科学技术出版社

图书在版编目（CIP）数据

中国农业科学院兰州畜牧与兽药研究所科技论文集.2011/ 中国农业科学院兰州畜牧与兽药研究所主编.—北京：中国农业科学技术出版社，2017.10
ISBN 978-7-5116-2951-7

Ⅰ.①中⋯　Ⅱ.①中⋯　Ⅲ.①畜牧学-文集②兽医学-文集　Ⅳ.①S8-53

中国版本图书馆 CIP 数据核字（2017）第 005036 号

责任编辑	闫庆健
文字加工	鲁卫泉
责任校对	贾海霞

出 版 者	中国农业科学技术出版社
	北京市中关村南大街 12 号　邮编：100081
电　　话	（010）82106632（编辑室）　（010）82109704（发行部）
	（010）82109703（读者服务部）
传　　真	（010）82106625
网　　址	http://www.castp.cn
经 销 者	各地新华书店
印 刷 者	北京建宏印刷有限公司
开　　本	880 mm×1 230 mm　1/16
印　　张	21　　彩插　16 面
字　　数	607 千字
版　　次	2017 年 10 月第 1 版　2017 年 10 月第 1 次印刷
定　　价	80.00 元

◆版权所有·翻印必究◆

《中国农业科学院兰州畜牧与兽药研究所科技论文集（2011）》编委会

主 任 委 员：杨志强　张继瑜

副主任委员：刘永明　阎　萍　王学智

委　　　员：高雅琴　梁春年　梁剑平　李建喜
　　　　　　李剑勇　李锦华　刘丽娟　潘　虎
　　　　　　时永杰　田福平　杨博辉　严作廷
　　　　　　杨　晓　曾玉峰　周　磊

主　　　编：杨志强　张继瑜　王学智　周　磊

副　主　编：刘永明　阎　萍　曾玉峰

主要撰稿人：高雅琴　梁春年　梁剑平　李建喜
　　　　　　李剑勇　李锦华　刘丽娟　潘　虎
　　　　　　时永杰　田福平　杨博辉　严作廷
　　　　　　杨　晓

前 言

近年来，在中国农业科学院科技创新工程的引领下，研究所的科研水平快速提升。我所科研人员和管理人员不但有工作上的热情，更有对工作认识上的高度和对学科理解上的深度。他们在紧张繁忙的实践活动中，笔耕不辍，将自己的研究成果写成论文。这不单是科研人员和管理人员的工作总结、过程记录，更是他们智慧的结晶，最终成为研究所的一笔宝贵财富。

为了珍惜这笔财富，加强优秀论文的交流与传播，营造更加浓厚的学术氛围，促进科研水平和管理水平的提升，切实推进研究所的科技创新，科技管理处搜集了 2011 年研究所科研人员公开发表的论文编印成《中国农业科学院兰州畜牧与兽药研究所科技论文集》第三卷，共 109 篇。由于时间仓促，可能还有论文未能收录，敬希鉴谅！

编 者

2017 年 10 月

目 录

Carbohydrate-Functionalized Chitosan Fiber for Influenza Virus Capture
.... LI Xue-bing, WU Pei-xing, GAO George F, CHENG Shui-hong (1)

Optimization of Polymerization Parameters for the Sorption of Oseltamivir onto Molecularly Imprinted Polymers
... YANG Ya-jun, LI Jian-yong, LIU Yu-rong, ZHANG Ji-yu, LI Bing, CAI Xue-peng (17)

Enzyme-Assisted Extraction of Naphthodianthrones from *Hypericum perforatum* L. by $^{12}C^{6+}$-ion Beam-Improved Cellulases
...... LI Zhao-zhou, WANG Xue-hong, SHI Guang-liang, BO Yong-heng, LU Xi-hong,
LI Xue-hu, SHANG Ruo-feng, TAO Lei, LIANG Jian-ping (32)

Lonicera japonica Thunb.: Ethnopharmacology, Phytochemistry and Pharmacology of an Important Traditional Chinese Medicine
.... SHANG Xiao-fei, PAN Hu, LI Mao-xing, MIAO Xiao-lou, DING Hong (47)

Antinociceptive and Anti-inflammatory Activities of *Phlomis umbrosa* Turcz Extract
.... SHANG Xiao-fei, WANG Jin-hui, LI Mao-xing, MIAO Xiao-lou, PAN Hu,
YANG Yao-guang, WANG Yu (93)

A Novel Single Nucleotide Polymorphism in Exon 7 of LPL Gene and its Association with Carcass Traits and Visceral Fat Deposition in Yak (*Bos grunniens*) Steers
.... DING X.Z., LIANG C.N., GUO X., XING C.F., BAO P.J., CHU M.,
PEI J., ZHU X.S., YAN P. (103)

Seasonal and Nutrients Intake Regulation of Lipoprotein Lipase (LPL) Activity in Grazing Yak (Bos grunniens) in the Alpine Regions around Qinghai Lake
.... DING X.Z., GUO X., YAN P., LIANG C.N., BAO P.J., CHU M. (111)

Effect of Several Supplemental Chinese Herbs Additives on Rumen Fermentation, Antioxidant Function and Nutrient Digestibility in Sheep
...... QIAO G.H., ZHOU X.H., LI Y., ZHANG H.S., LI J.H., WANG C.M., LU Y. (122)

Synthesis of Aspirin Eugenol Ester and its Biological Activity
... LI Jian-yong, YU Yuan-guang, WANG Qi-wen, ZHANG Ji-yu, YANG Ya-jun, LI Bing,
ZHOU Xu-zheng, NIU Jian-rong, WEI Xiao-juan, LIU Xi-wang, LIU Zhi-qi (135)

Development of Gastric and Pancreatic Enzyme Activities and Their Relationship with Some Gut Regulatory Peptides in Grazing Sheep
.... LANG Xia, WANG Cai-lian (143)

4-Allyl-2-methoxyphenyl 2-acetoxy-benzoate
.... LIU Xi-Wang, LI Jian-Yong, YANG Ya-Jun, ZHANG Ji-Yu（156）
Supplementary Materials: 4-Allyl-2-methoxyphenyl 2-acetoxybenzoate
.... LIU X.W., LI J.Y., YANG Y.J., ZHANG J.Y.（158）
Simultaneous Identification of *FecB* and *FecX^G* Mutations in Chinese Sheep Using High Resolution Melting Analysis
... YUE Yao-Jing, YANG Bo-Hui, LIANG Xia, LIU Jian-Bin, JIAO Shuo, GUO Jian,
SUN Xiao-Ping, NIU Chun-E, FENG Rui-Lin（165）
Clinical Study on the Treatment of Piroline against Bovine Mastitis
.... LIANG Jian-Ping, HAO Bao-Cheng, WANG Xue-Hong, GUO Zhi-Ting,
GUO Wen-Zhu, SHANG Ruo-Feng, TAO Lei, LIU Yu, LI Zhao-Zhou, HUA Lan-Ying,
WANG Shu-Yang（172）
甘肃地区牛源金黄色葡萄球菌分子鉴定及 RAPD 分型
.... 邓海平，蒲万霞，梁剑平，倪春霞，孟晓琴（178）
犬血浆中塞拉菌素含量的高效液相色谱-荧光检测方法的建立
...... 汪 芳，李 冰，周绪正，张继瑜，李剑勇，李金善，牛建荣，魏小娟，杨亚军（185）
伊维菌素纳米乳注射液的研制与质量安全性评价
...... 刘根新，张继瑜，吴培星，李剑勇，刘 英，周绪正，魏小娟，牛建荣，胡宏伟（192）
奶牛乳房炎金黄色葡萄球菌凝固酶基因型研究
.... 倪春霞，蒲万霞，胡永浩，邓海平（200）
甘肃犬瘟热病毒流行株 N 蛋白和 H 蛋白基因的序列分析
.... 王小辉，王旭荣，郭庆勇，李建喜，李世宏（208）
奶牛乳房炎金黄色葡萄球菌基因多态性分型试验
.... 邓海平，倪春霞，蒲万霞（215）
重金属 Pb^{2+} 的抗原合成与鉴定
.... 张志强，杨志强，李建喜，王学智，陈化琦，张景艳，张凯，孟嘉仁（221）
六茜素新制剂的急性毒性试验
.... 王学红，梁剑平，刘 宇，郭志廷，郭文柱，冯艳丽（228）
塞拉菌素口服与透皮给药对小鼠和大鼠的急性毒性试验
...... 汪 芳，周绪正，李 冰，李金善，李剑勇，牛建荣，魏小娟，杨亚军，张继瑜（228）
特种经济动物纤维超微结构的研究
.... 李维红，郭天芬，席 斌，胡正艳（229）
猪戊型肝炎病毒甘肃分离株衣壳蛋白基因序列分析
.... 郝宝成，梁剑平，兰 喜，刑小勇，项海涛，温峰琴，胡永浩，柳纪省（230）
甘肃青海藏系绵羊微卫星遗传多样性研究
.... 郎 侠（230）
牦牛 *DRB*3.2 基因 PCR-RFLP 研究
.... 包鹏甲，阎萍，梁春年，郭 宪，丁学智，裴 杰，褚 敏，朱新书（232）
骡肠阻塞时舌色舌苔的分析
.... 严作廷，巩忠福，谢家声，杨锐乐，黄虎祖，李绪权，尚清彦（232）

黑曲霉高产纤维素酶菌株的高通量筛选
………………………………… 李兆周，王学红，陈化靓，梁剑平（233）
奶牛乳房炎3种主要病原菌血清学交叉免疫试验研究
……………………… 郁　杰，李宏胜，罗金印，李新圃，徐继英，张礼华（233）
应用响应面法优化金丝桃素的纯化工艺
……………… 胡小艳，刘　宇，王学红，郭　凯，华兰英，石广亮，梁剑平（234）
3种紫花苜蓿草籽蛋白质瘤胃动态降解率的评定
…………………… 乔国华，张怀山，周学辉，路　远，张　茜，王春梅（234）
从生态学观点论青藏高原地区牦牛产业的可持续发展
……………… 朱新书，阎　萍，梁春年，郭　宪，裴　杰，包鹏甲，储　敏（235）
甘肃天祝牦牛乳中水解酶活力测定及分析
………………………………………… 席　斌，李维红，高雅琴，牛春娥（235）
高效液相色谱法测定贯叶金丝桃粉剂中金丝桃素的含量
…………………………… 郭文柱，梁剑平，荆文魁，王学红，尚若峰，郝宝成（236）
贯叶连翘中金丝桃素的提取及含量测定方法
…………………………… 王学红，梁剑平，郭志廷，郭文柱，刘　宇，郝宝成（236）
赖氨酸对绵羊肝脏和背最长肌中GHR和IGF-Ⅰ基因表达调控的影响
…… 程胜利，李建升，冯瑞林，岳耀敬，苗小林，刘建斌，郎　侠，郭天芬，裴　杰（237）
两种家兔体温和部分血常规指标的比较
………… 王东升，严作廷，张世栋，李宏胜，李世宏，龚成珍，李锦宇，荔　霞（237）
六茜素新制剂的药效学试验研究
…………………………… 王学红，冯艳丽，梁剑平，郭志廷，郭文柱，刘　宇（238）
奶牛产奶量和质量的季节性变化规律研究
………………………………… 孙晓萍，刘建斌，杨博辉，朗　侠，李思敏（238）
陶×滩、波×滩一代杂交羔羊生长发育分析
………………………………… 孙晓萍，李思敏，朗　峡，杨博辉，刘建斌（239）
天祝白牦牛被毛形态及纤维类型分析
………………………………………… 牛春娥，张利平，高雅琴，梁春年（239）
兔毛的扫描电镜观察
………………………………… 李维红，席　斌，郭天芬，王宏博，牛春娥（240）
阿司匹林丁香酚酯栓剂的制备及质量控制
………… 于远光，李剑勇，杨亚军，刘希望，张继瑜，牛建荣，周旭正，魏小娟，
　　　　　　　　　　　　　　　　　　　　　　　　　　　李　冰，叶得河（241）
发酵型黄芪提取物对肉仔鸡生产性能及免疫球蛋白的作用研究
…… 张　凯，杨志强，王学智，孟嘉仁，张景艳，秦　哲，王　龙，王　磊，
　　　　　　　　　　　　　　　　　　　　　　　　　　　　　　　　李建喜（241）
宫衣净酊急性毒性研究
………………………………… 王东升，谢家声，李世宏，严作廷，张世栋（242）
黄河首曲——玛曲湿地沼生植物的群系分类研究
……………………… 张怀山，张吉宇，乔国华，王春梅，张　茜，杨　晓（242）

静脉灌注 LPS 对家兔的发热效应及肝蛋白的 SDS—PAGE 比较分析
………… 张世栋，李世宏，王东升，龚成珍，李宏胜，李锦宇，荔　霞，严作廷 (243)

两种不同仪器分析方法对洗净山羊绒含脂率测定的比较
……………………………………… 王宏博，常玉兰，李维红，杜天庆，梁丽娜 (243)

绵羊毛与山羊绒的鉴别
…………………………………………………………… 王晓红，姚　穆，刘守智 (244)

塞拉菌素制剂的豚鼠最大化试验和 Buehler 试验
…… 汪　芳，张继瑜，周绪正，李金善，李剑勇，李　冰，牛建荣，魏小娟，杨亚军 (244)

射干麻黄地龙颗粒治疗鸡呼吸型传染性支气管炎效果观察
………………………………… 谢家声，罗超应，王贵波，罗永江，辛蕊华 (245)

中草药饲料添加剂"禽健宝"对蛋鸡生产性能和防病效果观察
………………………… 严作廷，王东升，荔　霞，谢家声，李世宏，张世栋 (245)

猪戊型肝炎病毒 swCH189 株衣壳蛋白基因 *CP239* 片段原核表达条件的优化
…………… 郝宝成，梁剑平，兰　喜，刑小勇，项海涛，温峰琴，胡永浩，柳纪省 (246)

牦牛 MSTN 基因内含子 2 多态性及与生长性状的相关性
…… 梁春年，阎　萍，刑成峰，裴　杰，郭　宪，包鹏甲，丁学智，褚　敏，朱新书 (246)

高效液相色谱法测定美洛昔康注射液美洛昔康含量
…… 王海为，李剑勇，杨亚军，李　冰，张继瑜，刘希望，周旭正，牛建荣，魏小娟 (247)

环形泰勒虫 *suAT1* 基因的克隆及序列分析
……………………………… 蔺红玲，张继瑜，袁莉刚，魏小娟，李　冰 (247)

益蒲灌注液质量标准研究
……………… 苗小棱，李　芸，潘　虎，杨耀光，苏　鹏，王　瑜，焦增华 (248)

动物福利与毛皮动物处死方法
…………………………………………………………… 郭天芬，李维红，席　斌 (248)

微波辅助合成金丝桃素四磺酸衍生物
………………………………………………… 李兆周，王学红，李雪虎，梁剑平 (249)

不同预处理对羊毛纤维平均直径的影响及其相关性研究
…………………………………………………… 郭天芬，王宏博，牛春娥，席　斌 (249)

制定毛纤维直径成分分析仪检测方法标准重要性的探讨
………………………………………………… 郭天芬，李维红，高雅琴，王宏博 (250)

奥司他韦快速鉴别及含量测定方法的建立
………………………………………………… 刘玉荣，杨亚军，李剑勇，李　冰 (250)

肃北高寒草原不同放牧强度土壤养分变化特征
………………………………………… 杨红善，常根柱，周学辉，路　远，那·巴特尔 (251)

西藏拉萨地区不同施氮量对鸭茅产量的影响
………………………………… 张小甫，李锦华，田福平，余成群，黄秀霞，江　措 (251)

苜蓿引进品种半干旱、半湿润区适应性试验
…………………………………………………… 杨红善，常根柱，周学辉，路　远 (252)

白术提取液急性毒性试验
…………………………… 王东升，李锦宇，罗超应，郑继方，张世栋，胡振英 (252)

犬瘟热病毒甘肃株的分离和 N 蛋白基因遗传进化分析
………………………… 王旭荣，王小辉，韩富杰，李世宏，李建喜，孟嘉仁（253）
常山提取物急性毒性试验研究
………………………………… 雷宏东，梁剑平，郭志廷，王学红，华兰英（254）
电针复合静松灵麻醉对山羊血液指标的影响
………………………… 王贵波，丁明星，郑继方，罗超应，王东升，李锦宇（254）
发酵型黄芪提取物对肉仔鸡生产性能及血液生化指标的影响
………… 张　凯，李建喜，杨志强，王学智，孟嘉仁，张景艳，秦俊杰，王　龙（255）
复合活菌制剂降解虾塘中亚硝酸盐的研究与应用
………………………………………… 陶　蕾，梁剑平，郭　凯，李兆周（255）
贵州地区牛源金黄色葡萄球菌分离株 16S rRNA 分子鉴定及 RAPD 分型研究
………………………………… 邓海平，蒲万霞，梁剑平，倪春霞，孟晓琴（256）
黄白口服液的质量控制研究
………………… 王胜义，齐志明，刘治岐，荔　霞，董书伟，刘世祥，刘永明（256）
六茜素的质量标准研究
………………………… 王学红，刘　宇，郭文柱，郝宝成，郑红星，梁剑平（257）
六茜素注射液的亚慢性毒性试验
………………………… 王学红，梁剑平，郭文柱，冯艳丽，刘　宇，郝宝成（257）
芩连液与白虎汤对脂多糖致家兔肝损伤的疗效比较
…… 张世栋，王东升，李世宏，李宏胜，荔　霞，李锦宇，陈炅然，龚成珍，严作廷（258）
犬瘟热病毒分离株 N、H、F 蛋白基因的变异分析
………………………… 王旭荣，王小辉，张世栋，李世宏，潘　虎，严作廷（258）
天祝白牦牛 Leptin 和 SCD1 基因单核苷酸多态性与肌肉脂肪酸含量的关联性分析
……………………………………………………………………… 刘自增，阎　萍（259）
猪戊型肝炎病毒 ORF2 部分片段原核表达载体的构建
………………………………… 郝宝成，兰　喜，胡永浩，柳纪省，梁剑平（260）
塞拉菌素透皮制剂的眼刺激性试验
…… 汪　芳，周绪正，张继瑜，李金善，李剑勇，李　冰，牛建荣，魏小娟，杨亚军（260）
一起梅花鹿结核病的临床诊断与防治
……………………………………………………… 郭志廷，梁剑平，罗晓琴（261）
喹乙醇单克隆抗体制备中细胞融合条件的筛选
………………… 王　磊，李建喜，张景艳，杨志强，王学智，张　凯，孟嘉仁（261）
喹乙醇免疫抗原的合成及评价
………………… 张景艳，李建喜，王　磊，杨志强，王学智，张　凯，孟　嘉（261）
散养产蛋鸡群感染前殖吸虫的诊治
………………… 郝宝成，王学红，郭文柱，郭志廷，朱银萍，胡永浩，梁剑平（262）
人为干扰对玛曲高寒退化草地的影响
……………………………………………………………………… 田福平，陈子萱（262）
微生物检测中糖量子点的制备及其与凝集素的相互作用
………………………………… 孔潇艺，程水红，刘雪峰，李学兵，吴培星（263）

响应面法优化贯叶连翘中金丝桃素超声提取工艺
…………………………………… 郭 凯，梁剑平，华兰英，王学红，石广亮（263）
大蒜多糖对伴刀豆蛋白A致肝损伤小鼠肝组织抗氧化能力的影响
………………………………………………………… 程富胜，程世红，张 霞（264）
牛源金黄色葡萄球菌16S rRNA分子鉴定与随机多态性扩增分型
…………………………………… 邓海平，蒲万霞，梁剑平，倪春霞，孟晓琴（265）
几种中草药对绵羊瘤胃发酵、抗氧化功能和营养物质消化率的影响
………………………… 乔国华，杨 晓，周学辉，张怀山，王晓力，李锦华（265）
不同苜蓿品种叶面积与抗旱性的关联性研究
…………………………………… 张怀山，代立兰，乔国华，王春梅，杨 晓（266）
流产母马血液流变学研究
………………………………………… 严作廷，荔 霞，王东升，张世栋（266）
黄花矶松栽培驯化试验
…………………………………… 常根柱，路 远，周学辉，杨红善，屈建民（267）
12个引进苜蓿品种适应性栽培试验
………………………………………… 杨红善，常根柱，周学辉，路 远（267）
牧区放牧管理传统乡土知识挖掘
………………………………………… 杨红善，周学辉，苗小林，常根柱（268）
沙拐属植物种质资源描述规范和数据标准的研究
…………………………………… 田福平，杨志强，时永杰，路 远，张 茜（268）
我国毛皮产业风险预警系统建设
………………………………………………………… 郭天芬，王宏博，李维红（269）
旱獭、麝鼠、兔狲、青鼬、石貂毛绒纤维超微结构比较
………………………………………………………………… 李维红，吴建平（269）
阿拉善沙拐枣引种驯化栽培试验
…………………………………… 常根柱，张 茜，路 远，杨红善，周学辉（270）
兰州大尾羊微卫星DNA多态性研究
………………………………………………………………… 郎 侠，吕潇潇（270）
鸡传染性喉气管炎及其防制
………………………………………………………… 严作廷，刘家彪，严建鹏（271）
阿司匹林丁香酚酯的高效解热作用及作用机制
…… 叶得河，于远光，李剑勇，杨亚军，张继瑜，周旭正，牛建荣，魏小娟，李 冰（272）
阿司匹林丁香酚酯的抗炎作用及其可能的作用机制
…… 李剑勇，王棋文，于远光，杨亚军，牛建荣，周旭正，张继瑜，魏小娟，李 冰（273）
非抗生素疗法在防治奶牛子宫内膜炎上的研究概况
…………………………………… 严作廷，荔 霞，王东升，陈炅然，张世栋（274）
西门塔尔牛在甘肃省的利用与发展
………………………………………… 郭 宪，梁春年，丁学智，阎 萍（274）
针刺麻醉机制研究进展
…………………………………… 王贵波，罗超应，郑继方，李锦宇，王东升，丁明星（275）

现代生物技术在动物营养中的应用
.... 郭天芬，王宏博，李维红（275）

羊毛纤维的结构及影响羊毛品质的因素
.... 郭天芬，李维红，牛春娥 王宏博（276）

非甾体抗炎药物研究进展
.... 王海为，李剑勇，杨亚军，于远光（276）

预防围产期奶牛酮病的营养策略
.... 乔国华，李锦华，周学辉，杨 晓，张 茜，王春梅（277）

STATs家族基因多态性在奶牛育种中的研究进展
.... 褚 敏，阎 萍，裴 杰，梁春年，郭 宪，曾玉峰，包鹏甲（277）

大环内酯类抗寄生虫药物的研究进展
.... 汪 芳，周绪正，李 冰，张继瑜（278）

环形泰勒虫主要膜蛋白的研究进展
.... 蔺红玲，张继瑜，魏小娟，李 冰（278）

家畜H-FABP基因的研究进展
.... 曾玉峰，阎 萍，梁春年，郭 宪，包鹏甲，裴 杰，褚 敏（279）

牦牛数量性状基因的研究进展
.... 包鹏甲，阎 萍，梁春年，郭 宪，裴 杰，褚 敏，朱新书（279）

饲料添加剂喹胺醇的研究进展
.... 郭文柱，梁剑平，王学红，郭志廷，郝宝成，尚若锋，华兰英（280）

抑制素免疫及其在肉牛繁殖中的应用
.... 郭 宪，阎 萍，梁春年，包鹏甲，朱新书，褚 敏，裴 杰（280）

青蒿琥酯治疗泰勒焦虫病的研究进展
.... 张 杰，张继瑜，李 冰，周绪正（281）

我国乳品质量标准在安全管理中存在的问题及对策
.... 王宏博，高雅琴，牛春娥，郭天芬，李维红（281）

药动学-药效学结合模型及其在兽用抗菌药物中的应用
.... 杨亚军，李剑勇，李 冰（282）

中药鸭跖草的研究进展
.... 王兴业，李剑勇，李 冰，杨亚军（282）

非生物学药物筛选方法及其应用
.... 刘玉荣，李剑勇，杨亚军（283）

黄花补血草醇提物的抗炎作用研究
.... 刘 宇，蒲秀英，梁剑平，华兰英，王学红，尚若锋（283）

中兽药防治动物疫病研究现状
.... 郝宝成，王学红，郭志廷，郭文柱，胡永浩，梁剑平（284）

我国主要毛皮市场及其规范对策
.... 郭天芬，席 斌，李维红，常玉兰（284）

牛繁殖性状候选基因的研究进展
.... 梁春年，阎 萍，郭 宪，包鹏甲，裴 杰，褚 敏（285）

山羊痘的流行及防治措施
.... 郝宝成，梁剑平，王学红，郭志廷，郭文柱，胡永浩（285）

基因组学和蛋白组学在纳米药物毒理研究中的应用
.... 高昭辉，董书伟，薛慧文，荔　霞，申小云，刘永明，王胜义，刘世祥，
齐志明（286）

如何进行奶牛乳房炎金黄色葡萄球菌疫苗的合理评估
.... 王旭荣，李建喜，王小辉，李宏胜，杨志强，孟嘉仁（286）

我国奶牛普通病防治现状和发展对策
.... 严作廷，杨志强，荔　霞（287）

治疗奶牛卵巢疾病性不孕症中药制剂研究进展
.... 王东升，严作廷，李世宏，张世栋，谢家声（287）

基因芯片技术及其在兽医学中的应用
.... 龚成珍，胡永浩，王东升，张世栋，严作廷（288）

植物精油对奶牛和肉牛瘤胃发酵的影响研究进展
.... 乔国华，张怀山，周学辉，常根柱，张　茜，路　远（288）

羔羊腹泻综合防治技术
.... 严作廷，王　东（289）

抗血小板药物研究及临床应用状况
.... 张泉州，于远光，李剑勇，杨亚军（289）

高速公路绿化工程系统构建概述
.... 常根柱，周学辉，杨红善，路　远，屈建民，石天欢（290）

紫草的药理研究及其在兽医临床上的应用概况
.... 严作廷，杨国林，巩忠福（290）

Effects of Arecoline Hydrobromide on Taeniasis and Cysticercosis in Domestic Animals
.... ZHOU Xu-zheng, ZHANG Ji-yu, LI Jian-yong, LI Jin-shan, WEI Xiao-juan,
NIU Jian-rong, LI Bing（291）

Study on the Correlation Between Leaf Area and Drought Resistance in Different Alfalfa Varieties
...... ZHAN GHuai-shan, DAI Li-lan, QIAO Guo-hua, WANG Chun-mei, YANG Xiao（291）

Progress in Determination Methods of Degradation Rate in Ruminants
.... WANG Hong-bo*, GAO Ya-qin, NIU Chun-e, LI Wei-hong, GUO Tian-fen（292）

Comparison on Two Different Instrumental Methods for Determination of Fat Content in Washed Cashmere
...... WANG Hong-bo, CHANG Yu-lan, LI Wei-hong, DU Tian-qing, LIANG Li-na（293）

4个美国标准秋眠型苜蓿对自然光周期的生长反应
.... 冯长松，李锦华，李绍钰，卢欣石（293）

不同断奶日龄对羔羊育肥效果的影响
.... 张鹏俊，郎　侠（294）

复方杨黄灌注液对奶牛子宫内膜炎的药效试验
.... 倪鸿韬，王　瑜（294）

甘肃省绒毛生产销售情况调查研究
………… 牛春娥，杨博辉，曹藏虎，岳耀敬，张　力，郭　健，孙晓萍，郎　侠，
刘建斌，程胜利，焦　硕，冯瑞林（295）

狼尾草属牧草种质资源的开发利用
……………………………………………………………………………… 张怀山（295）

青海省湟源县奶牛养殖存在问题与对策
………… 严作廷，荔　霞，齐志明，王胜义，刘永明，王国仓，付明善（296）

我国奶牛主要疾病研究进展
………………………… 严作廷，王东升，王旭荣，张世栋，李宏胜（296）

西藏达孜箭筈豌豆西牧324播种期试验
………………………… 李锦华，张小甫，田福平，乔国华，黄秀霞，苗小林（297）

西藏山南地区微肥对苜蓿种子产量影响的研究
………………………… 马海兰，刘学录，李锦华，江措，田福平，乔国华（297）

应用不对称竞争性PCR技术确定藏羊Agouti基因拷贝重复数
………………………… 杨树猛，岳耀敬，杨博辉，郭　健，孙晓萍，牛春娥，冯瑞林，
郭婷婷（298）

中草药防治奶牛子宫内膜炎的现状及其药理作用机制
………………………… 严作廷，王东升，荔　霞，张世栋，李世宏（298）

鸡大肠杆菌病及其中草药防治的概况
………………………… 严作廷，王东升，陈炅然，张世栋（299）

甘南牧区牦牛产业发展与生态环境建设浅议
………………………… 丁学智，郭　宪，梁春年，阎　萍（299）

甘肃省牛种业现状与"十二五"科技发展方向
………… 梁春年；阎　萍；丁学智；郭　宪；包鹏甲；裴　杰；朱新书（300）

高寒牧区牦牛繁育综合技术措施
………………………… 郭　宪，丁学智，梁春年，包鹏甲，阎　萍（300）

高寒牧区牦牛冷季暖棚饲养技术要点
………………………… 包鹏甲，阎　萍，梁春年，郭　宪，丁学智（301）

牦牛ATP6和ATP8基因生物信息学分析
………………………… 吴晓云，阎　萍，梁春年，郭　宪（301）

GPA对小鼠肝损伤血清及肝组织中GOT、GPT的影响
……………………………………………………………… 张　霞，程富胜（302）

动物疫病发展与防控趋势分析
……………………………………………………………… 倪鸿韬，王　瑜（302）

附红清治疗小白鼠人工感染猪附红细胞体试验观察
………………………… 苗小楼，李　芸，李宏胜，潘　虎，尚小飞（303）

复方茜草灌注液质量标准研究
………………………… 王学红，王作信，郭文柱，刘　宇，郝宝成，梁剑平（303）

黑曲霉发酵对黄柏中游离态小檗碱含量的影响
………………………… 陶　蕾，王宏伟，许冠英，殷劲松，梁剑平，郭　凯（304）

家畜脉诊研究进展
………………………………………… 严作廷，王东升，万玉林，荔　霞（304）

六茜素稳定性影响因素研究
………………………… 王学红，郭文柱，刘　宇，郝宝成，郑红星，梁剑平（305）

鸦胆子苦木素对小鼠血清胆碱酯酶和总胆红素的影响
………………………………………… 程富胜，程世红，张　霞，王华东（305）

射干麻黄地龙散对小白鼠的止咳祛痰作用
………………………… 罗永江，谢家声，辛蕊华，郑继方，胡振英，邓素平（306）

紫花苜蓿航天诱变田间变异观察研究
………………………… 杨红善，常根柱，李红民，柴小琴，周学辉，路　远（306）

甘南藏系绵羊饲养管理现状及发展对策
………………………… 王宏博，阎　萍，梁春年，郎　侠，丁学智（307）

饲草青贮系统中的乳酸菌及其添加剂研究进展
………………………… 王晓力，张慧杰，孙启忠，玉　柱，郭艳萍（307）

母牦牛的繁殖特性与人工授精
………………………… 郭　宪，裴　杰，包鹏甲，梁春年，丁学智，褚　敏，阎　萍（308）

动物用抗球虫药-常山酮
………………………… 王学红，梁剑平，郭文柱，刘　宇，郝宝成（308）

一例雏鸭曲霉菌病的临床诊治
………………………………………… 郭志廷，梁剑平，罗晓琴（309）

奶牛乳房炎元乳链球菌抗生素耐药性研究
………………………… 李宏胜，罗金印，李新圃，王　玲，苗小楼，王旭荣（309）

我国奶牛乳房炎病原菌区系调查及抗生素耐药性检测
………………………… 李宏胜，罗金印，李新圃，王　玲，王旭荣，李建喜，杨　峰（310）

细胞因子在奶牛乳房炎生物学防治中的应用
………………………… 王　玲，李宏胜，陈炅然，苗小楼，王正兵，尚若锋（311）

马属动物感冒的辨证施治
………………………………………………………… 谢家声，严作廷（311）

牦牛 ATP6 和 Cytb 基因的分子进化研究
………… 吴晓云，阎　萍，梁春年，郭　宪，包鹏甲，裴　杰，丁学智，褚　敏，
　　　　　　　　　　　　　　　　　　　　　　　　焦　斐，刘　建（312）

牛 FAS 基因的生物信息学分析
………… 焦　斐，阎　萍，梁春年，郭　宪，包鹏甲，裴　杰，丁学智，褚　敏，
　　　　　　　　　　　　　　　　　　　　　　　　吴晓云，刘　建（312）

牛 FTO 基因功能结构生物信息学分析
………… 刘　建，阎　萍，梁春年，郭　宪，包鹏甲，裴　杰，丁学智，褚　敏，
　　　　　　　　　　　　　　　　　　　　　　　　吴晓云，焦　斐（313）

牛 PCR 性别鉴定研究进展
………… 裴　杰，阎　萍，程胜利，梁春年，郭　宪，包鹏甲，褚　敏（313）

奶牛乳腺的生理防御及乳房炎病理学研究

.... 李新圃，罗金印，李宏胜（314）

奶牛乳房炎以疫苗预防为主的综合防控措施应用效果研究
.... 李宏胜，罗金印，李新圃，苗小楼，王　玲，王旭荣，王小辉（314）

奶牛乳房炎综合防控技术研究进展
.... 李宏胜，李新圃，罗金印，苗小楼，王　玲，李建喜（315）

奶牛隐性子宫内膜炎的生物学诊疗想法
.... 张世栋，严作廷，王东升，李世宏（315）

清宫助孕液治疗奶牛子宫内膜炎临床试验
...... 严作廷，王东升，李世宏，张世栋，谢家升，王雪郦，杨明成，朱新荣，陈道顺（316）

如何进行奶牛乳房炎金黄色葡萄球菌疫苗的合理评估
.... 王旭荣，李建喜，王小辉，李宏胜，杨志强，孟嘉仁（316）

中兽药在奶牛健康养殖中的作用
.... 李建喜（317）

中药子宫灌注剂治疗奶牛不孕症综述
.... 王东升，严作廷，张世栋，谢家声，李世宏（317）

蛋白质组学研究概况及其在中兽医学研究中的应用探讨
.... 董书伟，荔　霞，刘永明，王胜义，王旭荣，刘世祥，齐志明（318）

微量元素硒在奶牛上的研究进展
.... 王胜义，刘永明，齐志明，荔　霞，董书伟，刘世祥（318）

动物性食品中兽药残留分析检测研究进展
.... 李　冰，李剑勇，周绪正，杨亚军，牛建荣，魏小娟，李金善，张继瑜（319）

规模化肉牛养殖场苍蝇的防控策略
.... 周绪正，张继瑜，李　冰，魏小娟，牛建荣，李金善，李剑勇，杨亚军（319）

银翘蓝芩注射液中绿原酸、黄芩苷和连翘苷的含量测定
.... 杨亚军，王兴业，李剑勇，李　冰，周绪正，刘希望（320）

Carbohydrate-Functionalized Chitosan Fiber for Influenza Virus Capture

LI Xue-bing[1,2]*, WU Pei-xing[1,2], GAO George F[1], CHENG Shui-hong[1]

(1. CAS Key Laboratory of Pathogenic Microbiology and Immunology, Institute of Microbiology, Chinese Academy of Sciences, Beijing 100101, China; 2. Lanzhou Institute of Animal Science and Veterinary Pharmaceutics, Chinese Academy of Agricultural Science, Lanzhou 730050, China. ❍ *Supporting Information*)

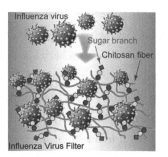

Abstract: The high transmissibility and genetic variability of the influenza virus have made the design of effective approaches to control the infection particularly challenging. The virus surface hemagglutinin (HA) protein is responsible for the viral attachment to the host cell surface via the binding with its glycoligands, such as sialyllactose (SL), and thereby is an attractive target for antiviral designs. Herein we present the facile construction and development of two SL-incorporated chitosan-based materials, either as a water-soluble polymer or as a functional fiber, to demonstrate their abilities for viral adhesion inhibition and decontamination. The syntheses were accomplished by grafting a lactoside bearing an aldehyde-functionalized aglycone to the amino groups of chitosan or chitosan fiber followed by the enzymatic sialylation with sialyltransferase. The obtained water-soluble SL-chitosan conjugate bound HA with high affinity and inhibited effectively the viral attachment to host erythrocytes. Moreover, the SLfunctionalized chitosan fiber efficiently removed the virus from an aqueous medium. The results collectively demonstrate that these potential new materials may function as the virus adsorbents for prevention and control of influenza. Importantly, these materials represent an appealing approach for presenting a protein ligand on a chitosan backbone, which is a versatile molecular platform for biofunctionalization and, thereby, can be used for not only antiviral designs, but also extensive medical development such as diagnosis and drug delivery.

---

* Received: July 14, 2011; Revised: October 4, 2011; Published: October 6, 2011

Corresponding Author. Tel: +86-10-62526982.Fax: +86-10-62526982.E-mail: lixb@ im.ac.cn.Li and Wu contributed equally to this work.

INTRODUCTION

Influenza is a severe viral disease of the respiratory tract that causes over 300000 human deaths annually.[1] Recent outbreaks of avian H5N1 and 2009 A/H1N1 influenza have caused public panic and economic losses,[2,3] emphasizing the urgent need for effective measures for virus prevention and treatment. Antivirals and vaccinations are critical in the fight against influenza; however, the effectiveness of the currently available drugs (rimantidine, amantadine, oseltamivir, and zanamivir) is increasingly diminished by the viral resistance,[4-6] and the current process of constructing a new vaccine strain based on the newly circulating virus is quite time-consuming (over six months[7]), which can hardly meet the requirements of rapid mass vaccination for curtailing the pandemic. Developments of alternatives to the present drugs therefore must be a high priority to meet challenges of the emerging drug-resistant and novel pandemic viruses. Furthermore, the effective approaches for endowing fibrous materials with a virus-capture capability are strongly demanded as their potential applications, such as the virus filters for face mask, air conditioner, air cleaner, and so on, would significantly limit the spread of the virus infection.[8]

It is well established that many pathogens use host cell surface glycans as anchors for attachments, which subsequently lead to infection.[9,10] Prominent examples of such pathogens include influenza virus, Escherichia coli O157, and Shiga toxin. Hemagglutinin (HA), the major surface antigen of the influenza virus, is responsible for the viral attachment to host cells via binding to its glycoligands, such as sialyllactose [SL, Neu5Acα-(2,3)Galβ(1,4)Glc].[11,12] The structural study on HA proteins complexed with SL or its analogues reveals that the specific binding is driven by multiple interactions consisting of hydrogen bonds, hydrophobic interactions, and van der Waals contacts.[13] Similarly, *E. coli* O157 and its virulence determinant, Shiga toxin, recognize a cell surface trisaccharide, globotriose, to facilitate the bacterial colonization and toxin entry into the cells.[14,15] The binding complex of Shiga toxin with globotriose is also well-defined by the crystal structure.[16] These examples illustrate that carbohydrates can play an indispensable role in infectious diseases and suggest that pathogen.carbohydrate interactions can be an excellent target for medical development. Indeed, coupling of glycoligands with scaffolds (molecules or materials) may generate appealing multivalent systems that not only contribute to anti-infective drug designs, but also advance the field applications. The artificial multivalent glyco-materials have been shown to be powerful tools for biomedical research and development.[17-20]

Herein, we demonstrate two novel carbohydrate-functionalized chitosan-based materials (Figure 1). The first is a water-soluble SL-chitosan conjugate (SLCC 1) designed as a highly active influenza virus HA inhibitor. The second is a simple filtration device featured by a sheet of SL-modified chitosan fiber (SLCF 2) to effectively remove the virus from the medium. Sugar-incorporated systems reported in the past are diverse with various functions, which include conjugated polymers and dendrimers, nanoparticles, vesicles, magnetic beads, quantum dots, membrane, and microarrays.[21-28] To the best of our knowledge, however, this work is the first to functionalize a fibrous material to confer the molecular recognition capability with the sugar. Importantly, the use of polysaccharide chitosan, which has been widely used in the pharmaceutical and medical areas due to its unique properties, such as biocompatibility, biodegradability, and antimicrobial activity,[29-33] as the scaffold enables not only the improved safety of SLCC 1 as a potential anti-influenza agent, but also attractive applications of SLCF 2 as a filter material in the control of the virus transmission.

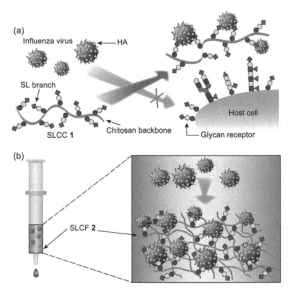

Fig. 1 Schematic illustrations of (a) inhibition of influenza virus adhesion by soluble SLCC 1 and (b) capture of influenza virus by fiber
Note: SLCF 2.Abbreviations: HA, hemagglutinin; SL, sialyllactose.

EXPERIMENTAL SECTION

Materials. Chitosan flakes were purchased from Yaizu Suisankagaku Industry Co.Ltd.The molecular weight (M_w) and polydispersity index (PDI: M_w/M_n) were, respectively, 5000 kDa and 1.8, as measured by a gel permeation chromatography (GPC) system equipped with a laser light scattering (LLS) detector.The deacetylation degree was 80%, as determined by the ^{1}H NMR spectrum.Alkene 3 was synthesized according to procedures described previously.[34,35] α2,3-Sialyltransferase (α2, 3-SiaT, rat recombinant, *Spodoptera frugiperda*, M_w 3.9 kDa) was purchased from Calbiochem Co.Sialyllactose and cytidine-5′-monophospho-N-acetylneuraminic acid sodium salt (CMP-Neu5Ac) was purchased from Sigma Co.Recombinant influenza virus hemagglutinin (HA, A/Equine/La Plata/93, H3N8) was purchased from Katakura Industries Co., Ltd.Human influenza A/PR/8/34 (H1N1) virus was propagated in the allantoic cavity of 10 day old embryonated chicken eggs.Infected allantoic fluid was pooled and clarified by sucrose-density gradient centrifugation.The hemagglutination unit (HAU) of the purified virus was determined by hemagglutination assay.[36]

Measurements. ^{1}H NMR spectra were recorded on a Bruker DRX 600 spectrometer at 600 MHz.Elemental analyses were performed using a CHN elemental analyzer (CHN corder MT-6, Yanaco Co.).IR spectra were recorded on an IR spectrometer (FT/IR-4200, Jasco Co.) by the KBr pellet method.Surface morphology of the fibers was observed using a scanning electron microscope

(SEM; S-2250, Hitachi, Ltd.) operated at 25 kV. Prior to analysis, the fibers were dried in vacuo and coated with gold-palladium alloy according to a standard procedure.[37] The M_w and PDI of the starting chitosan and all soluble conjugates were measured by a GPC system (CTO-20A, Shimadzu Co., Ltd.) equipped with a multiangle LLS detector (DAWN HELEOS-II, Wyatt Technology Corp.) in NaOAc buffer (pH 4.5). Dialysis was conducted using a cellulose ester (CE) membrane tube with the MWCO of 300 kDa (Thermo Fisher Scientific Inc.). Surface plasmon resonance (SPR) analysis was performed by a biosensor system (BIAcore 3000, Biacore, Inc.).

Synthesis of lactose-chitosan conjugate 5. O_3 was successively bubbled through a stirred solution of alkene 3 (398 mg, 0.94 mmol) in MeOH (40 mL) at -78 ℃ until TLC (9:4:2 ethyl acetate-isopropanol-water) showed no starting material left. N_2 was subsequently bubbled through the solution for 10 min to remove the remaining O_3. After adding Me_2S (139 μL, 1.88 mmol), the reaction mixture was stirred at room temperature overnight followed by evaporation to quantitatively yield 4 (400 mg, white solid), which was directly subjected to the next coupling reaction with chitosan. Thus, to a stirred aqueous 5 wt% AcOH solution (10 mL) of chitosan (100 mg, 0.59 mmol, 0.47 mmol of GlcN) was added dropwise a solution of the above aldehyde 4 in methanol (6 mL). The mixture was stirred for 1 h at room temperature, and then $NaCNBH_3$ (118 mg, 1.88 mmol) was added. After stirring the mixture overnight, the acidic solution was neutralized with 1 N aqueous solution of NaOH. The mixture was subjected to dialysis. After filtration to remove the precipitate, lyophilization gave 214 mg of conjugate 5 as a white amorphous solid. The DS_{Lac} is 0.80, as determined from the ^1H NMR spectrum and elemental analysis. The yield is 73% on the basis of the 0.8 DS_{Lac}. The M_w is 15000 kDa as measured by GPC. ^1H NMR (D_2O, 27 ℃): δ4.42 (1H, d, J=7.8 Hz, H-1 of Gal), 4.38 (1H, d, J=7.8 Hz, H-1 of Glc), 3.93-3.45 (19.5H, m, protons of sugar ring and linker OCH_2), 3.25 (1H, br, H-2 of Gal), 2.88-2.66 (3H, brm, NHCH and $NHCH_2$), 2.01 (0.75H, brs, $COCH_3$), 1.59 (4H, brs, 2×linker CH_2), 1.36 (2H, brs, linker CH_2). Anal. Calcd for $(C_8H_{13}NO_5)_{0.20}(C_{23}H_{41}NO_{15})_{0.80}(H_2O)_{0.23}$: C, 47.81; H, 7.14; N, 2.79. Found: C, 47.59; H, 7.25; N, 2.78.

Synthesis of SLCC 1. Conjugate 5, CMP-NANA, and α2,3-SiaT, calf intestine alkaline phosphatase (CIAP) in a 50 mM sodium cacodylate buffer (pH 7.3, 1-5 mL) containing 2.5 mM $MnCl_2$ and 0.1% bovine serum albumin (BSA) was incubated at 37 ℃ for 1 day. The mixture was heated at 100 ℃ for a few minutes and then centrifuged. The supernatant was subjected to exhaustive dialysis against water using a cellulose ester membrane tube with MWCO of 300 kDa, which allowed differentiation of the conjugates (M_w>5000kDa) from the contamination of proteins and small molecule byproduct (the M_w of proteins used in the reaction including the enzyme were lower than 160 kDa). After filtration to remove the precipitate, lyophilization gave SLCC 1a-d as the white amorphous solids. The details of reactants, yield, and structural parameter of the products are listed in Table S1 (Supporting Information). ^1H NMR (D_2O, 27 ℃) of SLCC 1a: δ 4.47, 4.42, 4.39 (2.76H, 3×d, J=7.8Hz, respectively, H-1 of Gal and Glc), 4.05 (1H, d, J=8.0 Hz, H-3 of Gal in SiaLac branch), 3.94-3.47 (32.9H, m, protons of sugar ring and linker OCH_2), 3.25 (1.38H, br, H-2 of Gal in SiaLac branch), 3.20-2.75 (4.14H, brm, NHCH and $NHCH_2$), 2.70 (1H, brd, J=7.8 Hz, H-3$_{eq}$ of Neu5Ac), 2.06-1.97 (4.03H, brm, COCH$_3$), 1.74—1.61 (6.52H, brm, H-3$_{ax}$ of Neu5Ac and 2×linker CH$_2$), 1.40 (2.76H, brs, linker CH$_2$).SLCC 1b—d have the similar ^{1}H NMR data.The elemental analysis data of SLCC 1a—d are summarized in Supporting Information.

Preparation of chitosan fiber. Chitosan fiber was prepared according to a wet-spinning process as described previously.[38] Briefly, chitosan flake (10 g, same as that used for SLCC 1 preparation) was dissolved in an aqueous solution of AcOH (3 v/v %) to afford the dope (5 wt %), which was subsequently filtered through flannel and injected through a stainless-steel spinneret into a coagulation bath containing a mixture of 5% aqueous NaOH solution and ethanol (7:3v/v).The resulting filaments were stretched in water (15℃) at the ratio of 1.2—1.4 and exhaustively washed with hot water (60℃, in washing bath), deionized water, and methanol.The fibers were airdried and stored in a dry cabinet before use.Chitosan fiber was characterized by ^{1}H NMR spectroscopy, IR spectroscopy, and CHN elemental analysis.The obtained data is same as that of starting chitosan flakes.

Preparation of LCF 6. Modification of chitosan fiber with aldehyde 4 was carried out according to the N-alkylation procedure reported by Hirano et al.[38] Chitosan fiber was immersed in a solution of aldehyde 4 in methanol (5—10 mL).After removing of air bubbles on the filaments surface by stirring under reduced pressure for a few minutes, the mixture was incubated with shaking at room temperature overnight.A solution of NaCNBH$_3$ in methanol (0.5—3 mL) was added dropwise to the mixture with stirring.The reaction mixture was incubated with shaking at room temperature for an additional 1 day.The resultant filaments were collected by filtration, sufficiently washed with methanol, and air-dried to afford LCF 6a—c.The details of reactants, yield, and structural parameter of the products are listed in Table S2.^{1}H NMR (DCl/D$_2$O) data of LCF 6a—c are similar to that of conjugate 5.The ^{1}H NMR spectrum of 6a is typically illustrated in Figure 2.IR (KBr) of LCF 6a: ν = 3415 (OH and NH), 2930, 2879 (CH), 1639, 1406, 1362 (NH$_2$ and NH-COCH$_3$), 1091 cm^{-1} (COC). Compounds 6b and 6c have the similar IR data.The elemental analysis data of LCF 6a—c are summarized in Supporting Information.

Preparation of Ac—LCF 7. The N-acylation of LCF 6 was carried out according to the known procedures.[39-41] Briefly, LCF 6a—c were respectively immersed in methanol (5—10 mL).To each mixture, acetic anhydride was added with stirring.The mixtures were incubated with shaking at 40℃ overnight.Workups as described for preparation of LCF 6 afforded Ac—LCF 7a—c.The details of reactants, yield, and structural parameter of the products are listed in Table S3.IR (KBr) of Ac—LCF 7a: ν=3415 (OH and NH), 2939, 2878 (CH), 1639, 1567, 1406, 1362, 1318 (NH$_2$ and NHCOCH$_3$), 1091 cm^{-1} (COC). Compounds 7b and 7c have the similar IR data.The elemental analysis data of Ac—LCF 7a—c are summarized in Supporting Information.

Preparation of SLCF 2. Ac—LCF 7a—c were, respectively, immersed in a 50 mM sodium cacodylate buffer (pH 7.3, 5—10 mLmL) containing 2.5 mM MnCl$_2$, 0.1% BSA, and CMP-NANA.To each mixture, α2,3-SiaT and ClAP were added with stirring.The mixtures were incubated at 37℃ for 2 days and the filaments were collected by filtration, washed with deionized water and methanol, and air-dried to afford SLCF 2a—c.The details of reactants, yield, and structural parameter of the products are listed in Table S4.IR (KBr) of SLCF 2a: ν = 3415 (OH and NH),

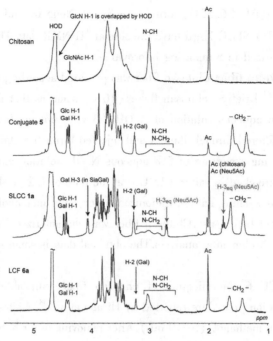

Fig. 2 ^{1}H NMR spectra (600 MHz, 27°C) of chitosan (DCl/D_2O), Lac–chitosan conjugate 5 (D_2O), SLCC 1a (D2O), and LCF 6a (DCl/D_2O).

2941 (CH), 1679 (COOH), 1632, 1591, 1428, 1386 (NH_2 and $NHCOCH_3$), 1060 cm^{-1} (COC). Compounds 2b and 2c have similar IR data. The elemental analysis data of SLCF 2a–c are summarized in Supporting Information.

Binding assay. The binding of SLCC 1 with the HA protein was analyzed by a BIAcore biosensorsystem based on the SPR technique. The HA protein was covalently immobilized by a standard aminecoupling method onto a sensor chip (CM-5, research grade) that was precoated with carboxymethyl dextran. The serial dilutions of SLCC 1a–d (~5 mg/mL) in running buffer [phosphate buffered saline (PBS), pH 7.2] were, respectively, injected over the immobilized HA at a flow rate of 20 μL/min for 5 min (association phase). During the dissociation phase, the sensor chip surface was exposed to running buffer at a flow rate of 20 μL/min. Conjugate 5 and monomeric SL were used as the negative controls. The K_d of the binding was calculated by a standard BIA evaluation software.

Hemagglutination inhibition (HAI) assay. HAI assays were performed as described by Suzuki et al.[11] Briefly, the 2-fold serial dilutions of SLCC 1a–d (in PBS, pH 7.2) were, respectively, added (25 μL) to 96-well microtiter plates followed by adding the influenza A/PR/8/34 (H1N1) virus suspensions (in PBS, 4 HAU, 25 μL) to each well. After 1 h of incubation (4°C), 50 μL of 0.5% (v/v) guinea pig erythrocytes suspension in PBS was added to each well. The hemagglutinations and MIC values were recorded after 2 h of incubations. Conjugates 5, monomeric SL, and fetuin were used for controls.

Virus capture assay. To examine the virus–capture capability of SLCF 2a–c, the simple filtration devices were built by packing the fiber (20 mg, cut into 0.5 cm length) onto the bottom of a 1 mL sterilized syringe (Terumo Co.) between 2 sheets of glass fiber. The filters were successively washed with 0.01% Tween 20 in PBS (10 mL) and 0.1% BSA in PBS (5 mL) for preventing the nonspecific adsorption of the virus. Influenza virus suspensions in PBS (512 HAU, 1 mL) containing 0.1% BSA were added to each syringe, and passed slowly through the filter. The virus-capture efficacy of the fibers was evaluated by hemagglutination assay[36] of the filtrates. Chitosan fiber and Ac-LCF 7a were used as the controls.

RESULTS AND DISCUSSION

Synthesis and characterizations of SLCC 1. The synthesis of SLCC 1 is shown in Scheme 1. With our glycoconjugation approach,[34,35] a lactose (Lac)-branched chitosan conjugate 5 was prepared by reductive N-alkylation of chitosan with an aldehyde 4, which was quantitatively obtained from an alkene 3 by ozonolysis. Complete N-substitution has been achieved in this reaction by the excess use of 4 (2 equiv), yielding the water-soluble conjugate 5 (degree of substitution of Lac branch, DS_{Lac} = 80%) that allowed the next enzymatic sialylation to proceed smoothly under the homogeneous aqueous condition. The transfer of sialic acid (Sia) residue from donor CMP-Neu5Ac to the lactose branch of 5 by an $\alpha 2,3$-sialyltransferase ($\alpha 2,3$-SiaT) afforded SLCC 1a-d in excellent yields. The degree of substitution of the SL branch (DS_{SL}) in SLCC 1 was adjusted by the molar ratio of CMP-Neu5Ac to the Lac branch of conjugate 5. The maximum DS_{SL} was obtained at 58% by using 1.6 equiv of CMP-Neu5Ac. More excessive use of the donor and enzyme had been tested without effect on further increasing the DS_{SL}, probably due to the increased steric hindrance of SLCC 1. Like the conjugate 5, SLCC 1a-d showed good solubility in the neutral water (solubility ≥ 1.0 mg/mL, PBS, pH 7.2) in contrast to the water-insoluble chitosan, owing to the incorporated hydrophilic sugar branches.

All the conjugate 1a-d and 5 were characterized by ^{1}H NMR spectroscopy, CHN elemental analysis, and GPC analysis. The structural compositions were verified by ^{1}HNMR assignments of several baseline-separated signals of the sugar branch, alkyl linker, and acetyl group of the chitosan backbone, as typically illustrated in Figure 2. The DS_{Lac} and DS_{SL} determined on the basis of an integral ratio of the branchderived peaks to chitosan acetyl peak were in good agreement with the CHN elemental analysis data. Moreover, the successive increase in the M_w of chitosan, conjugate 5, and SLCC 1d-a measured by GPC equipped with a LLS detector provided additional evidence of the incorporated sugar units (Supporting Information). However, an attempt for further characterization by ^{13}C NMR spectroscopy has failed because of the rigid chitosan backbone[42] and increased macromolecule size of the conjugates.

Bindings of SLCC 1 with influenza virus HA protein. To establish the binding capability of SL residues bound to chitosan scaffold, interactions between the conjugates and a viral HA protein (A/Equine/La Plata/93, H3N8) were analyzed by a BIAcore biosensor system based on the surface plasmon resonance (SPR) technique. The HA protein was covalently immobilized on a carboxymethylated dextran-coated sensor chip by the general amine-coupling method.[43] The solutions containing various concentrations of SLCC 1a-d were respectively injected over the sensor chip surface, and the binding affinity was determined. As illustrated in a typical SPR sensorgram (Figure 3, in resonance units, RU), significant bindings of SLCC 1a-d to the HA were observed, with the

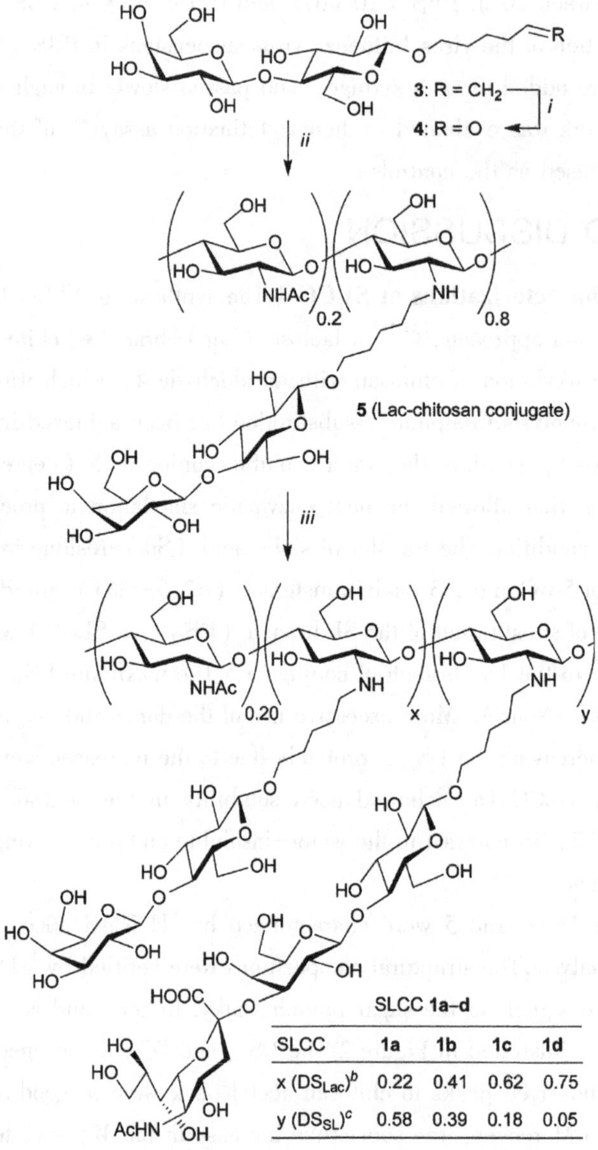

SLCC	1a	1b	1c	1d
x (DS_{Lac})[b]	0.22	0.41	0.62	0.75
y (DS_{SL})[c]	0.58	0.39	0.18	0.05

Scheme 1 Synthesis of SLCC 1[a]

[a]Reagents and conditions: (i) O_3, Me2S, −78℃, quantitatively; (ii) chitosan (80% GlcN, 20% GlcNAc, M_w = 5000kDa), $NaCNBH_3$, AcOH (aq.5 wt %), rt, 24 h, 73%; (iii) α−2,3−SiaT, CMP−Neu5Ac, sodium cacodylate buffer (pH 7.3), BSA, 37℃, 24 h, 91.98%.

[b]Degree of substitution (DS) of Lac branch.[c]DS of sialyllactose (SL) branch. Abbreviations: SiaT, sialyltransferase; CMP, cytidine 5′−monophosphate; BSA, bovine serum albumin.

decreased intensity (RU) in accordance with the DS_{SL} value due to the cluster glycoside effect (multivalent glyco−systems with the higher density of sugar moiety often achieve increased binding to the target proteins[44−47]).In contrast, no activity was detected for the conjugate 5, proving that the specific binding was derived from SL branches of SLCC 1.The interaction between monomeric SL and influenza HA has been reported to be typically low affinity with the dissociation constant (K_d) in the millimolar range (approximately 2–3 mM[48,49]). Here the SPR study also revealed a commercially available SL had no visible binding with the HA at the concentration of 1 mg/mL (1.57 mM) or less. The result of high avidity of SLCC 1a–d to HA (K_d = 0.59–1.93 μM) demonstrates multivalent presentation of SL ligands on chitosan backbone is highly effective to achieve strong binding.

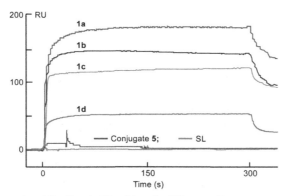

Analyte	Concentration		K_d (μM)[a]
	Sample (mg/mL)	SL branch (mM)	
SLCC **1a**	0.5	0.43	0.59
SLCC **1b**	0.7	0.44	0.66
SLCC **1c**	1.3	0.43	0.98
SLCC **1d**	4.4	0.43	1.93
Conjugate **5**	1.0	–	–
SL[b]	1.0	1.57	–

[a] The K_d was determined by BIAevaluation software. [b] Commercially available sialyllactose.

Fig. 3 Representative sensorgrams for the bindings of SLCC 1 with influenza virus HA protein analyzed by a BIAcore biosensor system

Note: The hemagglutinin (HA) was immobilized on a sensor chip (CM–5, research grade). The solution containing SLCC 1, Lac–chitosan; Conjugate 5, or sialyllactose (SL) was, respectively, injected over the sensor chip surface. Data was outputted in resonance unit (RU); Abbreviations: SPR, surface plasmon resonance; PBS, phosphate buffered saline; K_d, dissociation constant.

Inhibitions of influenza virus attachment to host cells by SLCC 1. The above binding study provides a baseline for the follow–up hemagglutination inhibition (HAI) assay, in which the inhibitory efficacy of SLCC 1 on the viral adhesion to host erythrocytes was verified (Table 1). SLCC 1a–d significantly inhibited the agglutination of erythrocytes induced by the viral attachment, in contrast to monomeric SL and conjugate 5 that had no activity at the concentration tested, suggesting that the high affinity bindings of SLCC 1a–d with virus surfaces HAs resulted in the effective inhibitions. The MIC value of SLCC 1 strongly relies on the DS_{SL}, and 1a, with the highest density of SL branch and affinity to the HA protein, was the most active inhibitor among the conjugates. Previous studies have demonstrated that a verity of multivalent Siacontaining molecules could act as active influenza virus HA inhibitors.[50-58] Here the fact that inhibitory efficiency of SLCC 1a (MIC = 54μM) is comparable with fetuin, a naturally occurring sialoglycoprotein which has been well used as an experimental control to define the potency of HA inhibitors,[50-54,59,60] highlights its potential as a new HA blocker in explorations of safe and effective anti-influenza agents.

Table 1 Inhibition of Influenza Virus-Induced Hemagglutination of Erythrocytes by SLCC 1[a]

sample	DS_{SL} (%)	MIC (μg/mL)[b]
SLCC 1a	58	63 (54 μM)
SLCC 1b	39	109 (70 μM)
SLCC 1c	18	219 (72 μM)
SLCC 1d	5	950 (93 μM)
SL[c]		NA[d]
conjugate 5		NA[d]
fetuin[e]		156 (31 μM)

[a] Influenza A/PR/8/34 (H1N1) virus and guinea pig erythrocytes were used. A/PR/8/34 is a typical human influenza virus that has been widely used in laboratory investigations. This virus binds specifically to structures of Siaα (2,3) Gal and Siaα (2,6) Gal. See ref[61] for more details. [b] Minimum inhibitory concentration: μg/mL of sample (μM of SL residue). The MIC value was calculated as the mean of a set of data obtained from at least 3 individual experiments, and the standard deviations were generally very low. [c] Commercially available sialyllactose. [d] No activity at the concentration of 1 mg/mL (for SL, equal to 1.6 mM). [e] On average, 1 molecule of fetuin ($M_w \approx 50$ kDa) contains approximately 10 Sia-terminated carbohydrate branches. See refs[59] and [60].

Preparation and characterizations of SLCF 2. The successful synthesis and bioactivity verifications of soluble SLCC 1 led us to embark on modification of chitosan fiber with the SL functionality. Chitosan fiber was prepared using a wetspinning procedure as described previously.[38] The chemical modifications of chitosan fiber using aldehydes (N-alkylation) or carboxylic anhydrides (N-acylation) have been well established by several groups.[38-41] According to these published methods and our own experience in the synthesis of SLCC 1, we first attached Lac residues onto the fiber via the N-alkylation with aldehyde 4 under a heterogeneous condition (Scheme 2). The adjusted DS_{Lac} of resultant fiber 6a-c (Lacmodified chitosan fiber, LCF) was 16, 7, and 2%, respectively, as estimated by the ^{1}H NMR spectra and elemental analyses (attempts to increase the DSLac over 16% had failed, probably due to the heterogeneous reaction condition). However, a preliminary test of sialylation (α2,3-SiaT, CMP-Neu5Ac) of LCF 6a had failed to yield product displaying satisfactory Sialoading (lower than 1%). We reasoned that electrostatic attractions between CMP-Neu5Ac and the alkalic primary amino groups abundantly present in LCF 6a had interfered with the enzymatic reaction and resulted in the low SL-loading. The heterogeneous N-acetylation of LCF 6a-c was therefore performed to convert the amines to amides to reduce the alkalinity. As a consequence, fiber products (Ac-LCF 7a-c) with the degree of acetylation over 72% were afforded by respective treatment of LCF 6a-c with acetic anhydride in menthol, and the desired SLCF 2a-c with the full SL-loadings (DS_{SL}=16, 7, and 2%, respectively) were obtained successfully via the subsequent enzymatic sialylation of Ac-LCF 7a-c. The SLCF 2a-c and Ac-LCF 7a-c were tested to be stable in neutral, alkalic, and acidic waters, but chitosan fiber and LCF 6a-c were soluble under acidic conditions due to the abundant free protonable amino groups.

All the fibers described here were characterized by CHN elemental analysis and IR spectroscopy. Chitosan fiber and LCF 6a-c were additionally characterized by ^{1}H NMR spectroscopy (the spectrum of 6a is shown in Figure 2). As typically illustrated by the IR spectra of 6a, 7a, and 2a in Figure 4, the increased intensity of bands at 2900 cm^{-1} due to the methylene absorption was

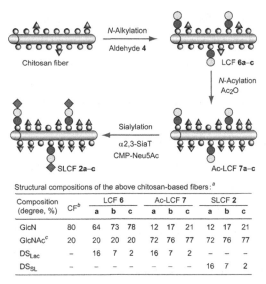

Structural compositions of the above chitosan-based fibers:[a]

Composition (degree, %)	CF[b]	LCF 6			Ac-LCF 7			SLCF 2		
		a	b	c	a	b	c	a	b	c
GlcN	80	64	73	78	12	17	21	12	17	21
GlcNAc[c]	20	20	20	20	72	76	77	72	76	77
DS_{Lac}	–	16	7	2	16	7	2	–	–	–
DS_{SL}	–	–	–	–	–	–	–	16	7	2

Scheme 2 Heterogeneous Modification of Chitosan Fiber

Note: [a]Estimated by CHN elemental analysis. For chitosan fiber and LCF 6a–c, also estimated by the NMR spectra. [b]Chitosan fiber. [c]Equal to degree of acetylation (DA). Abbreviations: SiaT, sialytransferase; CMP, cytidine 5′-monophosphate; DS, degree of substitution.

observed for all the modified fibers as compared to that of chitosan fiber, indicating the incorporated sugar moieties via the alkyl linker. The peaks at 1632, 1567, 1406, and 1362 cm^{-1} assigned to amide absorptions increased remarkably in the spectra of fibers 7 and 2, proving that the N-acetylation had indeed occurred with a high degree of substitution (no absorption band was detected for O-acetylation at ~700 cm^{-1}). Furthermore, the characteristic absorption of carboxyl group at 1679 cm^{-1} in the spectrum of SLCF 2 provided evidence for the incorporated-Sia residue.

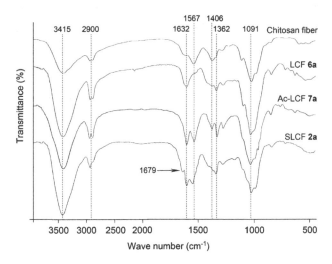

Fig. 4 IR spectra of chitosan fiber, LCF 6a, Ac–LCF 7a, and SLCF 2a

Additionally, scanning electron microscopy (SEM) analyses were performed to examine surface morphology of the fibers. As shown in Figure 5, the surface of the chitosan fiber is coarse and porous, which may provide access for chemical diffusion, indicating that the modification reactions can occur not only on the fiber surface, but within an individual filament. There is no remarkable difference in the surface pattern between chitosan fiber (Figure 5a) and SLCF 2a (Figure 5c), although a slightly distorted surface was observed for the latter. Fibers 6a-c, 7a-c, and 2b-c showed the similar surface pattern (data not shown), suggesting that the modification processes had little affect on the surface morphologies.

Capture of influenza virus by SLCF 2. To examine the virus-capture capability of SLCF 2, a simple filtration device, as illustrated in Figure 1b, was built by packing the fiber onto the bottom of a 1 mL sterilized syringe between two sheets of glass fiber. After successive treatment of the fiber with Tween 20 and BSA to prevent nonspecific adsorption,[62,63] influenza virus suspended in PBS buffer (512 HAU) was passed through the filter. The capture efficacy was evaluated by hemagglutination assay of the filtrate. As shown in Table 2, the virus concentration was dramatically decreased by the contact with SLCF 2a or 2b (HAU < 2 and = 8, respectively), highlighting their high capture efficiency. In common with SLCC 1, SLCF 2 also had the DS_{SL}-dependent binding activity, and 2c (DS_{SL} = 2%, HAU = 256), with the very low SL density, was insufficient to make effective capture. Chitosan fiber and Ac-LCF 7a also showed slightly adsorptive activities with the decreased 256 HAU. This is probably due to the nonspecific captures of the virus by the coarse surface of the fiber (Figure 5). The SEM observation revealed that the size of the porous pattern on the surface is roughly in the same range as that of the influenza virus particle (ca. 80-120 nm); therefore, the virus was also likely trapped by the porous surface structure. In addition, it should be noted that the current experiment is performed in an aqueous medium, although transmission of the virus is more likely to occur via the air. Nevertheless, this study clearly demonstrated that SLCF 2 was able to capture the virus via the specific binding. Further expanded bioevaluations, including the virus-capture test in a nonaqueous environment, are currently being pursued. Overall, although different kinds of sugarincorporated systems have been studied as probes for detections of pathogens and carbohydrate-protein interactions,[21-28] the above results reveal first that direct modification of a fibrous material with glyco-functionality can generate a unique system for virus decontamination.

Table 2 Capture of Influenza Virus by SLCF 2[a]

fiber	$DS_{SL}b$ (%)	HAU[c]	
		before filtration	after filtration
none		512	512
chitosan fiber		512	256
Ac-LCF 7a		512	256
SLCF 2a	16	512	<2
SLCF 2b	7	512	8
SLCF 2c	2	512	256

Note: [a] Influenza A/PR/8/34 (H1N1) virus was used, and the capture efficiency was evaluated by hemagglutination assay of the filtrate. [b] Degree of substitution (DS) of sialyllactose branch. [c] Hemagglutination unit (HAU): maximum dilution of the virus suspension capable of causing the hemagglutination of erythrocytes (guinea pig). All HAU numbers were confirmed by at least three individual experiments.

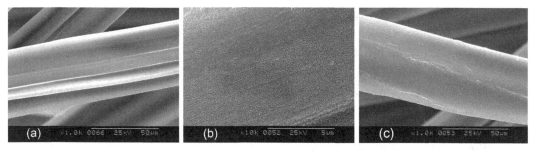

Fig. 5　SEM images of chitosan fiber (a and b) and SLCF 2a (c)

CONCLUSIONS

We report the potential of carbohydrate-functionalized chitosan-based materials, either as a water-soluble polymer or as a functional fiber, for viral adhesion inhibition and removal. This preliminary work sets the stage for further developments of the materials in real world applications such as drugs and filters for control of influenza and, more importantly, provides an appealing approach for presenting carbohydrate on chitosan backbone. In particular, chitosan offers a safe and effective molecular platform that can be employed as the carrier of proteins, nucleic acids, and drugs.[29-33] The unique characteristics of chitosan-based systems enable a wide range of applications especially in biomedical fields where, for example, the effectiveness has been demonstrated by biosensing[30] and delivering drugs such as insulin and progesterone.[33] We believe, in addition to the explorations of control measures against influenza, the present study is also of practical value in extensive medical development such as diagnosis and drug delivery. Currently, more extensive and in-depth bioevaluations of the SL-modified conjugate as well as the fiber is being pursued, and efforts are ongoing to develop these promising materials for real world applications.

ASSOCIATED CONTENT

Ⓢ Supporting information

Tables S1-S4 and elemental analysis data. This material is available free of charge via the Internet at http://pubs.acs.org.

ACKNOWLEDGMENTS

This work was supported financially by MOST (973 Program 2012CB518803), NSFC (31070726), and CAS (KSCX2-YWG-032).

References

[1] Fiore, A.E.; Shay, D.K.; Broder, K.; Iskander., J.K.; Uyeki, T.M.; Mootrey, G.; Bresee, J. S.; Cox, N.S. *MMWR Recomm. Rep.* 2008, 57 (No.RR-7): 1-60.

[2] Liu, J.; Xiao, H.; Lei, F.; Zhu, Q.; Qin, K.; Zhang, X.W.; Zhang, X.L.; Zhao, D.; Wang, G.; Feng, Y.; Ma, J.; Liu, W.; Wang, J.; Gao, G.F. *Science* 2005, 309: 1 206.

[3] Wang, T.T.; Palese, P. *Cell* 2009, 137: 983-985.

[4] Bright, R.A.; Medina, M.; Xu, X.; Perez-Oronoz, G.; Wallis, T.R.; Davis, X.M.; Povinelli, L.; Cox, N.J.; Klimov, A.I.*Lancet* 2005, 366: 1 175-1 181.

[5] Ferraris, O.; Lina, B.*J.Clin.Virol.*2008, 41: 13-19.

[6] Tamura, D.; Mitamura, K.; Yamazaki, M.; Fujino, M.; Nirasawa, M.; Kimura, K.; Kiso, M.; Shimizu, H.; Kawakami, C.; Hiroi, S.; Takahashi, K.; Hata, M.; Minagawa, H.; Kimura, Y.; Kaneda, S.; Sugita, S.; Horimoto, T.; Sugaya, N.; Kawaoka, Y.*J.Clin.Microbiol.*2009, 47: 1 424-1 427.

[7] Gerdil, C.*Vaccine* 2003, 21: 1 776-1 779.

[8] MacIntyre, C.R.; Cauchemez, S.; Dwyer, D.E.; Seale, H.; Cheung, P.; Browne, G.; Fasher, M.; Wood, J.; Gao, Z.; Booy, R.; Ferguson, N.*Emerg.Infect.Dis.*2009, 15: 233.241.

[9] Lingwood, C.A.*Curr.Opin.Chem.Biol.*1998, 2: 695-700.

[10] Olofsson, S.; Bergstro.m, T.*Ann.Med.*2005, 37: 154-172.

[11] Suzuki, Y.; Nagano, Y.; Kato, H.; Matsumoto, M.; Nerome, K.; Nakajima, K.; Nobusawa, E.*J.Biol.Chem.*1986, 261: 17 057-17 061.

[12] Stevens, J.; Blixt, O.; Tumpey, T.M.; Taubenberger, J.K.; Paulson, J.C.; Wilson, I.A.*Science* 2006, 312: 404-410.

[13] Weis, W.; Brown, J.H.; Cusack, S.; Paulson, J.C.; Skehel, J.J.; Wiley, D.C.*Nature* 1988, 333: 426-431.

[14] Boyce, T.G.; Swerdlow, D.L.; Griffin, P.M.*N.Engl.J.Med.*1995, 333: 364-368.

[15] Johannes, L.; Ro.mer, W.*Nat.Rev.Microbiol.*2010, 8: 105-116.

[16] Ling, H.; Boodhoo, A.; Hazes, B.; Cummings, M.D.; Armstrong, G.D.; Brunton, J.L.; Read, R.J.*Biochemistry* 1998, 37: 1 777-1 788.

[17] Astronomo, R.D.; Burton, D.R.*Nat.Rev.Drug Discovery* 2010, 9: 308-324.

[18] de la Fuente, J.M.; Penadés, S.*Biochim.Biophys.Acta* 2006, 1760: 636-651.

[19] Stevens, J.; Blixt, O.; Paulson, J.C.; Wilson, I.A.*Nat.Rev.Microbiol.*2006, 4: 857-864.

[20] Paulson, J.C.; Blixt, O.; Collins, B.E.*Nat.Chem.Biol.*2006, 2: 238-248.

[21] Kitov, P.I.; Sadowska, J.M.; Mulvey, G.; Armstrong, G.D.; Ling, H.; Pannu, N.S.; Read, R.J.; Bundle, D.R.Nature 2000, 403: 669-672.

[22] Van Kasteren, S.I.; Campbell, S.J.; Serres, S.; Anthony, D.C.; Sibson, N.R.; Davis, B.G. *Proc.Natl.Acad.Sci.U.S.A.*2009, 106: 18-23.

[23] Thomas, G.B.; Rader, L.H.; Park, J.; Abezgauz, L.; Danino, D.; DeShong, P.; English, D. S.*J.Am.Chem.Soc.*2009, 131: 5 471-5 477.

[24] Morton, V.; Jean, J.; Farber, J.; Mattison, K.Appl.*Environ.Microb.*2009, 75: 4 641-4 643.

[25] Kikkeri, R.; Lepenies, B.; Adibekian, A.; Laurino, P.; Seeberger, P.H.*J.Am.Chem.Soc.*2009, 131: 2 110-2 112.

[26] Huang, C.-Y.; Thayer, D.A.; Chang, A.Y.; Best, M.D.; Hoffmann, J.; Head, S.; Wong, C. H.*Proc.Natl.Acad.Sci.U.S.A.*2006, 103: 15-20.

[27] Kato, T.; Miyagawa, A.; Kasuya, M.; Hatanaka, K. *Open Chem. Biomed. Methods J.* 2009, 2: 13-17.

[28] Song, E.H.; Osanya, A.O.; Petersen, C.A.; Pohl., N.L.*J.Am.Chem.Soc.*2010, 132: 11 428-11 430.

[29] Kumar, M.N.; Muzzarelli, R.A.; Muzzarelli, C.; Sashiwa, H.; Domb, A.J. *Chem. Rev.* 2004, 104: 6 017-6 084.

[30] Yi, H.; Wu, L.-Q.; Bentley, W.E.; Ghodssi, R.; Rubloff, G.W.; Culver, J.N.; Payne, G.F.

Biomacromolecules 2005, 6: 2 881-2 894.
[31] Prabaharan, M.*J.Biomater.Appl.*2008, 23: 5-36.
[32] Suh, J.K.F.; Matthew, H.W.T.*Biomaterials* 2000, 21: 2 589-2 598.
[33] Agnihotri, S.A.; Mallikarjuna, N.N.*J.Controlled Release* 2004, 100: 5-28.
[34] Xu, Y.; Fan, H.; Lu, C.; Gao, G.F.; Li, X.*Biomacromolecules* 2010, 11: 1 701-1 704.
[35] Han, J.; Li, X.*Carbohydr.Polym.*2011, 83: 137.143.
[36] Suzuki, Y.; Suzuki, T.; Matsumoto, M.*J.Biochem.*1983, 93: 1 621-1 633.
[37] Franchi, M.; Checchi, L.*J.Clin.Periodontol.*1995, 22: 655-658.
[38] Hirano, S.; Nagamura, K.; Zhang, M.; Kim, S.K.; Chung, B.G.; Yoshikawa, M.; Takehiko, M.*Carbohydr.Polym.*1999, 38: 293-298.
[39] Choi, C.Y.; Kim, S.B.; Pak, P.K.; Yoo, D.I.; Chung, Y.S. *Carbohydr. Polym.* 2007, 68: 122-127.
[40] Hirano, S.; Zhang, M.; Nakagawa, M.; Miyata, T.*Biomaterials*2000, 21: 997-1 003.
[41] Yang, Y.M.; Hu, W.; Wang, X.D.; Gu, X.S.*J.Mater.Sci.: Mater.Med.*2007, 18: 2 117-2 121.
[42] Rinaudo, M.*Prog.Polym.Sci.*2006, 31: 603-632.
[43] Malmqvist, M.; Karlsson, R.*Curr.Opin.Chem.Biol.*1997, 1: 378-383.
[44] Lee, Y.C.; Lee, R.T.*Acc.Chem.Res.*1995, 28: 321-327.
[45] Lundquist, J.J.; Toone, E.J.*Chem.Rev.*2002, 102: 555-578.
[46] Geng, J.; Mantovani, G.; Tao, L.; Nicolas, J.; Chen, G.J.; Wallis, R.; Mitchell, D.A.; Johnson, B.R.; Evans, S.D.; Haddleton, D.M.*J.Am.Chem.Soc.*2007, 129: 15 156-15 163.
[47] Becer, C.R.; Gibson, M.I.; Geng, J.; Ilyas, R.; Wallis, R.; Mitchell, D.A.; Haddleton, D.M.*J.Am.Chem.Soc.*2010, 132: 15 130-15 132.
[48] Skehel, J.J.; Wiley, D.C.*Annu.Rev.Biochem.*2000, 69: 531-569.
[49] Collins, B.E.; Paulson, J.C.*Curr.Opin.Chem.Biol.*2004, 8: 617-625.
[50] Sigal, G.B.; Mammen, M.; Dahmann, G.; Whitesides, G.M.*J.Am.Chem.Soc.*1996, 118: 3 789-3 800.
[51] Tsuchida, A.; Kobayashi, K.; Matsubara, N.; Muramatsu, T.; Suzuki, T.; Suzuki, Y.*Glycoconjugate J.*1998, 15: 1 047-1 054.
[52] Umemura, M.; Itoh, M.; Makimura, Y.; Yamazaki, K.; Umekawa, M.; Masui, A.; Matahira, Y.; Shibata, M.; Ashida, H.; Yamamoto, K.*J.Med.Chem.*2008, 51: 4 496-4 503.
[53] Umemura, M.; Makimura, Y.; Itoh, M.; Yamamoto, T.; Mine, T.; Mitani, S.; Simizu, I.; Ashida, H.; Yamamoto, K.*Carbohydr.Polym.*2010, 81: 330-334.
[54] Ohta, T.; Miura, N.; Fujitani, N.; Nakajima, F.; Niikura, K.; Sadamoto, R.; Guo, C.T.; Suzuki, T.; Suzuki, Y.; Monde, K.; Nishimura, S.*Angew.Chem., Int.Ed.*2003, 42: 5 186-5 189.
[55] Oka, H.; Onaga, T.; Koyama, T.; Guo, C.T.; Suzuki, Y.; Esumi, Y.; Hatano, K.; Terunuma, D.; Matsuoka, K.*Bioorg.Med.Chem.*2009, 17: 5 465-5 475.
[56] Totani, K.; Kubota, T.; Kuroda, T.; Murata, T.; Hidari, K.I.; Suzuki, T.; Suzuki, Y.; Kobayashi, K.; Ashida, H.; Yamamoto, K.; Usui, T.*Glycobiology* 2003, 13: 315-326.
[57] Ogata, M.; Murata, T.; Murakami, K.; Suzuki, T.; Hidari, K.I.; Suzuki, Y.; Usui, T.*Bioorg.Med.Chem.*2007, 15: 1 383-1 393.
[58] Makimura, Y.; Watanabe, S.; Suzuki, T.; Suzuki, Y.; Ishida, H.; Kiso, M.; Katayama, T.; Kumagai, H.; Yamamoto, K.*Carbohydr.Res.*2006, 341: 1 803-1 808.
[59] Baenziger, J.U.; Fiete, D.*J.Biol.Chem.*1979, 254: 789-795.
[60] Green, E.; Adelt, G.; Baenziger, J.; Wilson, S.; van Halbeek, H. *J. Biol. Chem.* 1988, 263:

18 253-18 268.

[61] Gamblin, S.J.; Haire, L.F.; Russell, R.J.; Stevens, D.J.; Xiao, B.; Ha, Y.; Vasisht, N.; Steinhauer, D.A.; Daniels, R.S.; Elliot, A.; Wiley, D.C.; Skehel, J.J. *Science* 2004, 303: 1 838-1 842.

[62] Jeyachandran, Y.L.; Mielczarski, J.A.; Mielczarski, E.; Rai, B.J. *Colloid Interface Sci.* 2010, 341: 136-142.

[63] Hsu, B.B.; Yinn Wong, S.; Hammond, P.T.; Chen, J.; Klibanov, A.M. *Proc. Natl. Acad. Sci. U.S.A.* 2011, 108: 61-66.

（发表于《Biomacromolecules》, SCI, IF: 5.325）

Optimization of Polymerization Parameters for the Sorption of Oseltamivir onto Molecularly Imprinted Polymers[*]

YANG Ya-jun[1], LI Jian-yong[1], LIU Yu-rong[1], ZHANG Ji-yu[1], LI Bing[1], CAI Xue-peng[2]

(1. Key Lab of New Animal Drug Project, Gansu Province/Key Lab of Veterinary Pharmaceutics Discovery, Ministry of Agriculture/Lanzhou Institute of Animal Science and Veterinary Pharmaceutics, Chinese Academy of Agricultural Sciences, Lanzhou 730050, China; 2. Lanzhou Institute of Veterinary Science, Chinese Academy of Agricultural Sciences, Lanzhou 730000, China)

Abstract Molecularly imprinted polymers (MIPs) are tailor-made polymers with high selectivity for a given analyte, or group of structurally related compounds. The influence of the process parameters (the moles of functional monomer and cross-linker, the selection of functional monomer and solvent) on the preparation of oseltamivir (OS)-imprinted polymers was investigated. A mathematical method for uniform design to optimize these selected parameters and to increase the MIP selectivity for template molecules was applied. The optimal conditions to synthesize MIP were 0.69 mmol 30% acrylamide (AA) +70% 4-Vinylpyridine (4-VP) and 5.0 mmol ethylene glycol dimethacrylate (EGDMA) copolymerized in 5 mL toluene in the presence of 0.1 mmol OS. MIP showed high affinity and selectivity for separation of the template molecule from other compounds. In the present study, we have established an effective LC-MS/MS method to identify and quantify OS with good sensitivity, accuracy and precision.

Key words Molecularly imprinted polymer (MIP); Polymerization optimization; Selectivity; Uniform design; Oseltamivir; LC-MS/MS

INTRODUCTION

Molecularly imprinted polymers (MIPs) are tailor-made polymers with high selectivity for a given analyte, or group of structurally related compounds, which make them ideal materials for analytical separation science. As a selective sorbent, MIPs have been widely applied to enantiomer separation, solid-phase extraction, catalysis, finding novel compounds, or as chemical sensors, bio-

---

[*] Received: 1 February 2011 / Revised: 5 April 2011 / Accepted: 26 April 2011 / Published online: 10 May 2011. © Springer-Verlag 2011

sensors, etc.

The selectivity of MIPs is due to the following synthetic procedures: (1) the functional monomer and template are connected by a covalent linkage (for covalent imprinting) or they are positioned in proximity to each other through noncovalent interactions (for noncovalent imprinting). (2) The obtained conjugates or adducts are copolymerized with an excess of cross-linker in the presence of an initiator under thermal or photochemical conditions. (3) The template molecules within the polymer are removed and the resulting vacant sites, whose shape and functionality are determined by the template molecule, are thus "imprinted" on the polymer. Under appropriate conditions, these imprint sites retain the size, structure, and other physicochemical properties of the template molecule, and therefore efficiently and selectively bind this molecule or its analogs. Several factors influence the polymerization process, such as the choice of functional monomer[1-4], cross-linker[5-7], porogenic solvent[8,9], initiator[10], and the method of polymerization[11,12].

The uniform design of experiments with many parameters based on number theory was proposed by Fang and Wang[13]. Instead of performing experiments that cover all possible combinations of experimental parameters, the theory prefers relatively fewer experimental trials that are uniformly distributed within the parameter space. These experimental trials are determined using the number-theoretic method and are mathematically proven to be a good approximation of the complete combination of experimental parameters. Thus, the uniform design method presents the attractive advantage that, when compared with other conventional statistical experiment design methods such as orthogonal design, it reduces significantly the number of experiments needed to evaluate multiple parameters and their interactions. Uniform design generates a regression model based on the results and it is able to predict at what independent variables the dependent variable may gain the maximum. The uniform design method is suitable for multi-factor and multilevel experiments. Uniform design has been successfully applied in chemistry and chemical engineering for process-parameter optimization[14-16]. Most MIPs are synthesized using noncovalent imprinting methods and several factors contribute to the polymerization process. Therefore, we applied uniform design to the present investigation of a multiparameter polymerization process.

Pandemic influenza is generally considered as the most significant potential global public health emergency caused by a naturally occurring pathogen. Recent human cases of highly pathogenic strains of avian influenza (H5N1) have raised the concerns about the imminence of this threat. Furthermore, on June 11th, 2009, the World Health Organization signaled that a global pandemic of novel influenza A (H1N1) was underway by raising the worldwide pandemic alert level to Phase 6. As of September 20th, 2009, the World Health Organization has received reports of over 300,000 laboratory-confirmed cases of pandemic influenza H1N1 and of 3,917 deaths in 191 countries and territories[17]. Oseltamivir (OS; see Fig. 1) is considered the leading antivirus currently available to counter a serious epidemic or pandemic outbreak of influenza[18,19]. Oseltamivir phosphate (Tamiflu® is an ester prodrug which is rapidly and extensively hydrolyzed *in vivo* to its active metabolite oseltamivir carboxylate, a potent and selective inhibitor of influenza virus neuraminidase[20]. On the other hand, it is necessary for discovery of new drugs in addition to oseltamivir against such pandemic influenza as avian influenza H5N1 in the future. However, avian influenza virus is very infective for humans and traditional screening for antivirus drug is a tedious and higher dangerous work with strict requirements for laboratory conditions. So, a biological replacement method of screening antivirus drug is important.

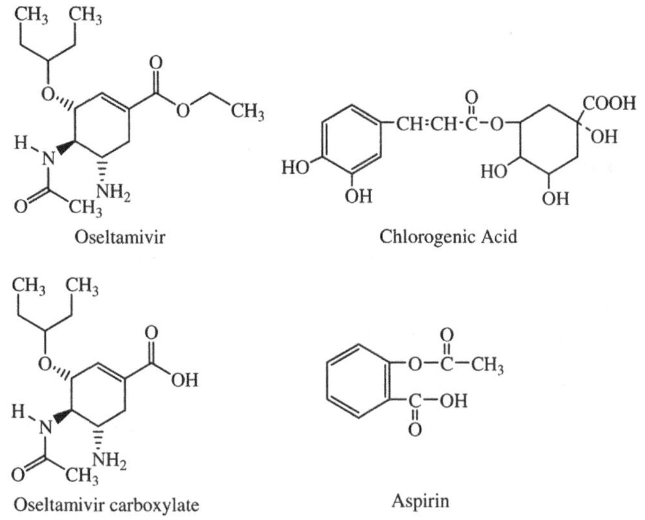

Fig. 1 Structure of OS and other chemicals

The novel application of immobilized enzymes, polyclonal antibodies (PcAb) and MIP for effective recognition of active compounds immersed in afforded us a new approach for the discovery of lead compounds or drug candidates, which will be very beneficial in drug development as non-biological screening method. Molecular recognition is defined as the ability of one molecule to attach to another molecule that has a complementary shape. The immobilized enzyme and PcAb methods in combination with FAC-MS are only used for screening. Although both methods have similarly advanced features in recognizing active compounds, the PcAb as a mimic of enzyme has less recognition ability than that of immobilized enzyme. When the immobilized enzyme is not available or not suitable for screening, PcAb can be used as an alternative method. Molecular imprinting is a rapidly evolving technique to prepare synthetic receptors. MIP can be used to mimic the enzyme or receptor for direct screening, separation and online identification of active compounds, and can speed up drug discovery based on traditional Chinese medicine. MIP has proven to be feasible mimics in terms of physical robustness, resistance to elevated temperatures and pressures, and inertness towards acids, bases, metal ions and organic solvents. This material can be well employed in various kinds of samples including some biological and pharmaceutical samples[21].

Since MIP properties are similar to drug target, MIP prepared with good active compounds as the template will be a good and practicable model of screening for drugs. Moreover, MIP is supposed to link LC-MS (liquid chromatography 秋 ass spectrometry) and can separate, screen and identify some special active compounds from compound library or Chinese traditional herbs directly. If OS is selected as the template, the MIP can screen for new compound with the similar spatial structure and antivirus activities as OS.

In this study, we use an OS template to prepare MIPs via noncovalent imprinting. The process parameters that influence polymerization (the moles of functional monomer and crosslinker, the selection of functional monomer and solvent), were optimized using the uniform design method to obtain an OSimprinted polymer that has a high sorption quantity for the template and could separate OS from matrix. This provides basis on further screening for more active analogs of OS from Chinese traditional herbs and other compound library.

EXPERIMENTAL

Reagents and solvents

Oseltamivir phosphate was obtained from Roche R&D Center (China) Ltd., neutralized and extracted using sodium hydroxide and ethyl acetate, and the extraction was vacuum dried at 50℃. Chlorogenic acid, phillyrin, and aspirin were purchased from National Institute for the Control of Pharmaceutical and Biological Products (Beijing, China). 4-Vinylpyridine (4-VP) and 5-M ammonium formate were purchased from Alfa Aesar (Lancashire, England) and Agilent Technologies (USA), respectively. Acrylamide (AA), ethylene glycol dimethacrylate (EGDMA) and azobisisobutyronitrile (AIBN) were purchased from Shanghai No.4 Reagent and H.V.Chemical Co., Ltd. (Shanghai, China). Before use, EGDMA was distilled under vacuum after being extracted with 10% sodium hydroxide brine and dried over anhydrous magnesium sulfate. AIBN was recrystallized from alcohol. 4-VP was extracted with 10% sodium hydroxide brine and dried over anhydrous magnesium sulfate. Toluene and tetrahydrofuran (THF) were dried by sodium and then redistilled. Dichloromethane (CH_2Cl_2) and glacial acetic acid were analytical - reagent grade. Methanol (MeOH) and acetonitrile (MeCN) were HPLC grade and obtained from Fisher Scientific (New Jersey, USA). Deionized water (18 MΩ) was prepared with a Direct-Q® 3 system (Millipore, USA).

The liquid phase consisted of MeOH.water-5 M ammonium formate (75 : 25 : 0.1).

Analysis equipment

We used an Agilent 1200 for LC. It consisted of a G1312B binary LC pump, a G1322A vacuum degasser, a temperature-controlled (G1330B) micro-well plate autosampler (G1367C) set at 10℃, and a G1316Bthermostatted column compartment set at 30℃ (Agilent Technologies, Germany). Data acquisition and quantification were performed using MassHunter software version B.01.04 (Agilent Technologies, USA). The compounds were analyzed using an Agilent Zorbax SBC_{18} (30 mm × 2.1 mm, 3.5 μm) column protected by a prefilter (4 mm, 5 μm, from GRACE, USA) at a flow rate of 400 μL · min^{-1}.

For LC-MS/MS analysis, we used an Agilent 6410A triple-quadrupole mass spectrometer (Agilent Technologies, USA) with an ESI (electrospray ionization) source interface operated in the positive-ion scan mode. The capillary potential of the MS was +4,000 V, the gas temperature was 350℃, the gas flow rate was 10 L · min^{-1}, the nebulizer pressure was 30 psi, and the dwell time was 200 ms.

Standard solutions

Primary OS stock solution (100 μg mL^{-1}) was prepared in MeOH, stored at 4℃, and was stable for 1 month (data not shown). Appropriate dilutions of OS were made in MeOH to produce working-stock solutions of 5, 10, 20, 40, 80, and 160 ng mL^{-1} on the day of analysis, and these stocks were used to generate the calibration curve.

Preparation of the polymers

We prepared several MIPs for OS under different conditions. The preparation procedure involved dissolving 0.1 mmol OS (template) and the appropriate functional monomer in 5 mL of porogen (toluene, THF, MeCN, or CH_2Cl_2) in a 100-mL flask. The flask was placed in an ultrasonic water bath for 30 min, and then shaken for 3 h. We next added the appropriate EGDMA (cross-linker) and 15 mg AIBN (initiator) to the solution and the mixture was purged with nitrogen for 5 min with the flask under an airtight seal. The polymerization proceeded in a water bath at 60℃ for 24 h. Next, the block polymer was ground using a laboratory mortar and pestle and particles between 45 and 60 μm were collected, whereas fine particles were removed by repeated sedimentation in acetone. The template trapped in the polymer matrix and unreacted reagents were removed by extraction with MeOH-glacial acetic acid (9 : 1, v/v). After extraction, the solution was analyzed using LC-MS/MS to ensure that the template molecules were removed from the MIP. The extracted particles were then washed with MeOH for 4 h to remove residual acetic acid. Finally, the particles were dried under vacuum at 50℃ and stored at ambient temperature until the evaluation of their sorption performance.

To adjust the polymerization parameters for optimized OS sorption performance of the MIPs, we used the U8 ($8^2\times4^2$) uniform design table to arrange the process parameters under investigation, such as the sort and quantity of functional monomers, cross-linkers, and porogenic solvents (see Tables 1 and 2). The OS - MIP was then synthesized using the optimized parameters. The nonimprinted polymer (NIP) was prepared in exactly the same way as o control, except that the template molecule was absent in the polymerization stage.

Morphological study of OS-MIP

For morphological studies, the samples of optimized, extracted OS - MIP and NIP were mounted onto metal stages, coated with gold by sputtering, and then observed using a scanning electron microscope (SEM, model JSM-5600LV and JSM-6701F, JEOL, Japan).

Table 1 Factors and levels of polymerization parameters

Levels	Factors			
	X_1 Moles of functional monomers (mmol)	X_2 Moles of EGDMA (mmol)	X_3 Kind of functional monomers	X_4 Kind of porogenic solvents
1	0.3	1.5	100% AA	MeCN
2	0.4	2.0	70% AA + 30% 4-VP	Tol.
3	0.5	2.5	30% AA + 70% 4-VP	THF
4	0.6	3.0	100% 4-VP	CH_2Cl_2
5	0.7	3.5		
6	0.8	4.0		
7	0.9	4.5		
8	1.0	5.0		

Sorption quantity

To investigate the sorption quantity of OS-MIPs for OS, we performed a static sorption experiment, in which 50 mg of OS-MIP, 50 mg of NIP, and 0.5 mg of OS were mixed with 5.0 mL standard solution in 10-mL conical flasks and shaken in the dark at 30℃ for 24 h. Each experiment was carried out in triplicate. These solutions were filtered and examined for the free concentrations using multiple-reaction monitoring (MRM) LC-MS/MS analysis. We calculated the sorption quantity by subtracting the free concentrations from the initial concentrations.

Specific affinity of OS-MIP-LC column

To investigate the specific affinity of OS-MIPs, we performed the following experiment. The polymer particles, which prepared by the optimized parameter and their diameter within 25-45 μm, were wet-packed into stainless steel columns (150 mm×4.6 mm i.d.) with MeOH under 2,000 psi pressure. The column was washed online with MeOH, MeOH-acetic acid 9∶1 (v/v), and MeOH until a stable baseline indicating removal of the template molecules. Then, specific affinity of OS-MIP column for OS and other compound were studied by LC-MS.

The OS-MIP column was protected by a prefilter (4 mm, 5 μm, from GRACE, USA) in this evaluation experiment. The autosampler was set at 10 °C, and temperature of column was 30℃. The liquid phase consisted of MeOH-MeCN-formate acid (75∶25∶0.01). The backpressure of this column was high, so the flow rate was maintained at 0.075 mL min^{-1}. The injection volume was 10 μL. LC-MS analysis used an Agilent 6410A triple-quadrupole mass spectrometer (Agilent Technologies, USA) with an ESI source interface operated in the positive-ion scan mode. The capillary potential of the MS was +4,000 V, the gas temperature was 350℃, the gas flow rate was 10 L · min^{-1}, the nebulizer pressure was 30 psi, and the dwell time was 200 ms.

Table 2 The pseudo-level uniform design with eight experiments for two factors in eight levels and two factors in four levels, U_8 ($8^2 \times 4^2$)

No.	X_1 (mmol)	X_2 (mmol)	X_3	X_4	Responses Adsorbent quantity (μg · (50 mg)$^{-1}$)
1	0.3	2.0	3 (30% AA + 70% 4-VP)	4 (CH$_2$Cl$_2$)	18.845
2	0.4	3.0	1 (100% AA)	3 (THF)	40.265
3	0.5	4.0	3 (30% AA + 70% 4-VP)	2 (Tol.)	176.744
4	0.6	5.0	1 (100% AA)	1 (MeCN)	72.539
5	0.7	1.5	4 (100% 4-VP)	4 (CH$_2$Cl$_2$)	20.306
6	0.8	2.5	2 (70% AA + 30% 4-VP)	3 (THF)	45.460
7	0.9	3.5	4 (100% 4-VP)	2 (Tol.)	149.033
8	1.0	4.5	2 (70% AA + 30% 4-VP)	1 (MeCN)	43.822

RESULTS AND DISCUSSION

Analysis of oseltamivir using LC-MS/MS

Various methods for quantifying OS have been reported, such as HPLC-UV[22,23], HPLC-fluorescence[24], micellar electrokinetic chromatography[25], LC-MS/MS[26,27], etc. In the present study, we identified and quantified the OS using LC-MS/MS. First, total ionic chromatographic (TIC), MS, and MS/MS spectra were recorded after direct infusion of 1 μL of OS primary stock solution into the ESI source. MS spectra, MS/MS spectra, and parameters of OS are presented in Fig. 2. The ions monitored were OS m/z = 312.3, $[M+H]^+$ 313.3, and $[M+Na]^+$ 335.3, and the ion transitions monitored included $[M+H]^+ \rightarrow$ 166.2, 208.2, among others. Thus, in the following MRM analysis product ion 166.2 was used as the quantifying ion whereas ion 208.2 was used as the qualitative ion.

The calibration curve was obtained by injection of the standard working-stock solution of 5 ~ 160 ng mL^{-1} of OS. The calibration curve was generated by determining the best fit to the peak areas versus concentrations, and analysis using the MassHunter software yielded the relation y = 463.4301x+1285.2845 (R^2 = 0.9995). We used the calibrationcurve to quantify the OS in the eluting solution and static sorption solution.

Parameter selection

The formation of template-functional monomer complexes is a key step in the synthesis of MIPs. The chosen functional monomer should match the functionality of the template. To favor the formation of template-functional monomer complexes, and thus the imprinting effect, the functional monomer is normally used in excessive molar quantities relative to that of the template[28]. In addition, for an imprinted polymer the cross-linker should not be overlooked because a high degree of cross linking (70%-98%) is necessary to achieve specificity. The functions of the cross-linker are to stabilize the imprinted binding site, control the morphology, and influence the mechanical stability of the polymer matrix[29]. The solvent serves to bring all the components in the polymerization into proximity in a single phase, and it is also for creating the pores in the polymer. In addition to its influence on the polymer morphology, the solvent govern the strength of noncovalent interactions for noncovalent imprinting polymerization. To accentuate the binding strength, the best imprinting porogens are solvents with very low dielectric constants, such as toluene. The use of more polar solvents weakens the interaction forces, resulting in poorer recognition. Thus the solvent must be chosen carefully to maximize the formation process of template-functional monomer complexes[9].

In covalent imprinting, templates are bound to vinyl moieties by covalent linkages. For this purpose, esters and amides of acrylic acid or methacrylic acid (MAA) are most often used, and the synthesis proceeds easily in most cases. For noncovalent imprinting, vinyl monomers bearing appropriate functional groups are designed and synthesized[30]. The aim of the current research is to prepare a noncovalent OS-MIP to find OS and/or its analogs from frond or microorganism. Therefore, we used OS as a template, AA and/or 4-VP as the functional monomer, EGDMA as the cross-linker, and AIBN to initiate the reaction, and the procedure was initiated at 60℃. According to the

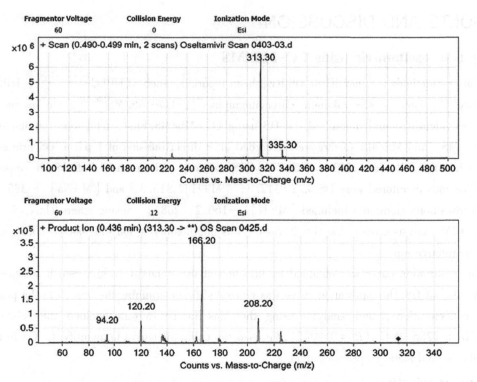

Fig. 2　MS and MS/MS of OS. An ESI interface was used

synthesis principle of the polymers and the results of the preliminary experiments, we selected the following independent variables: four polymerization parameters (see Table 1), the moles of functional monomer (X_1, 0.3–1.0 mmol), the moles of EGDMA (X_2, 1.5–5.0 mmol), the type of functional monomer (X_3, 1 represents 100% AA, 2 represents 70% AA + 30% 4-VP, 3 represents 30% AA + 70% 4-VP, and 4 represents 100% 4-VP), and porogenic solvent (X4, 1 represents MeCN, 2 represents toluene, 3 represents THF, and 4 represent CH_2Cl_2).

Uniform design

The mole ratio of cross linking agent to functional monomer is important. If the ratio is too small, the guestbinding sites are located so closely to each other that they cannot work independently. In extreme cases, guest binding on one site completely inhibits guest binding on neighboring sites. At extremely large mole ratios, however, the imprinting efficiency is damaged, especially when the cross linking agent interacts noncovalently with the functional monomer and/or templates[30]. Therefore, we applied the mixed-level table U_8 ($8^2 \times 4^2$) to arrange the uniform design of experiments (see Table 2)[31]. In this study, the factors X_1 and X_2 were graded into eight levels, and the factors X_3 and X_4 were graded into four levels. The parameters investigated were graded into either or four levels and incorporated into eight polymerization trials, and each experiment was performed in triplicate.

The objective response of eight polymers is taken to be the sorption quantity of OS−MIPs for OS. In a 10−mL conical flask we mixed 50 mg of imprinted particles and 5.0 mL of standard MeOH solution with 0.5 mg of OS and oscillated the mixture in the dark at 30℃ for 24 h. These solutions were filtered and examined for free concentrations using LC−MS/MS. We calculated the sorption quantity Q by subtracting the free concentration from the initial concentration (see Table 2).

The sorption quantity of OS−MIPs polymerized under the various conditions ranges from 18.8450 μg·(50 mg)$^{-1}$ to 176.7436 μg·(50 mg)$^{-1}$. The best trial was No.3, where the sorption quantity of OS is 176.7436 μg·(50 mg)$^{-1}$. The data of Table 2 also indicate that the sorption quantity of OS is higher for polymers synthesized in toluene than for polymers synthesized in other porogenic solvents.

Regression analysis

With the mean sorption quantity of OS for OS−MIPs being the objective response, we used SAS 8.2 for Windows for the regression analysis of the experimental data. The binary parameters Z_3 and Z_4 represent X_3 and X_4, respectively (X_3 and X_4 are ordinal data in this study—see Table 3).

The equation for regression analysis is

$$Y = -130.41 + 272.34X_1 + 38.01X_2 + 8.56Z_{31} + 60.05Z_{41} - 80.36Z_{42} - 197.14X_1^2 \quad (1)$$

With $R = 0.9999$, $R^2 = 0.9999$, and adjusted $R^2 = 0.9992$.

The multiple correlation coefficient $R = 0.9999$ indicated a highly close agreement between experimental and predicted values. The predictive model might reproduce the observed values reasonably well, which was confirmed by the coefficient of adjusted determination $R^2 = 0.9992$ which indicated a good fit, and meant that the model should explain 99.92% of the variability in the data. This can ensure a satisfactory adjustment of the model to the experimental data.

The analysis of the variance for the regression model was given in Table 4. The statistical significance of the model equation was also confirmed by an F test. The accuracy of the regression equation provided by the residual mean square was acceptable. Table 5 compared the recovery results of the uniform design of experiments with values calculated from the regression model.

Optimized parameters

The mathematical optimization of the resulting regression equation (Eq. 1) was carried out using a quadratic multinomial regression built in SAS 8.2 for Windows. The regression equation was a quadratic multinomial. The equation would have maximal output when X_1 was 0.69, X_2 was 5.0, Z_{31} was 1, Z_{41} was 1 and Z_{42} was 0. However, there was no binary system parameter Z_{32} in Eq.1, maybe it has little contribution for regression analysis. The trial had maximal sorption quantity when X_4 was 2 and X_3 was 3 in Table 1. So Eq.1 and the trials results in Table 2 generated the following optimal process parameters: 0.69 mmol functional monomer, 5.0 mmol EGDMA, 30% AA + 70% 4-VP for the functional monomer, and toluene for porogenic solvent. The model predicts that the maximal sorption quantity of OS is 222.30 μg·(50 mg)$^{-1}$.

Table 3 Factors X_3 and X_4 are described as Z_3 and Z_4 in the binary system

Levels	X_3		X_4	
	Z_{31}	Z_{32}	Z_{41}	Z_{42}
1	0	1	0	1
2	1	0	1	0

Levels	X_3		X_4	
	Z_{31}	Z_{32}	Z_{41}	Z_{42}
3	1	1	1	1
4	0	0	0	0

Table 4 ANOVA for the model

Source	df	Sum of squares	Mean squares	Overall F	P
Regression (SR)	6	24,896	4,149.32	1,477.01	0.0199
Residual (SE)	1	2.8092	2.8092		
Total (ST)	7				

MAA is by far the most widely and successfully used monomer because it has carboxylic acid and can interact via hydrogen bonds with the amide, carbamate, and carboxyl groups on amino acid derivatives. The results of evaluating the functional monomer indicate that AA can form much stronger hydrogen bonds with the template molecule than MAA[32]. The basic functional monomer 4-VP was also introduced for noncovalent molecular imprinting[33,34]. In the OS imprinting processes, AA and 4-VP may interact noncovalently with the amine and carboxyl group of OS through hydrogen bonds and ionic bonds, and this cooperative interaction can produce a stable complex between the template molecule and AA + 4-VP. Therefore, binding sites recognizing the template molecule more efficiently would form the MIP with AA + 4-VP as combined functional monomers. Thus, the presence of the template - monomer complex produces the specific binding sites in OS-MIPs.

Relative to the mole quantity of template molecule, the functional monomers are normally used in excessive amounts to maximize complex formation and thus the imprinting effect. In the present study, Eq.1 indicates that the sorption quantity is quadratic in moles of functional monomer, which likely results from it having a reactive methacrylate ester with a short spacer that allows infinite conformation possibilities as well as a degree of rigidity in the resultant polymer. In this experiment, a high cross-link ratio should not only decrease the swelling and increase the sorption quantity of the polymer, but also lead to materials with adequate mechanical stability. Polar and nonprotic solvents are preferred in the noncovalent imprinting process because they cannot compete in binding to sites between the template and the functional monomer, thus in favoring the formation of the template-functional monomer complex. In other work, it was demonstrated that selectivity increased as the porogen decreased in hydrogen bonding capacity[12], which explains why toluene is more suitable than other solvents for the current experiment. The results also showed that when toluene was used as the porogen solvent, the quantity of OS adsorbed by OS-MIP was greater than for other solvents.

To verify the use of uniform design for optimizing the polymerization parameters, we compared the OS-MIP with the NIP, both prepared under optimal conditions. Explicitly, the following procedures were carried out to synthesize OSMIP: 0.1 mmol OS and 0.69 mmol 30% AA + 70% 4-VP (functional monomer) were dissolved in 5 mL of toluene in a 100-mL flask, and mixed by placing in an ultrasonic water bath for 30 min, then shaken for 3 h. Next, 5.0 mmolEGDMA (cross-linker) and 15 mg AIBN were added to the solution. The subsequent steps were the same as described in the section "Preparation of the polymers". The corresponding NIP was prepared under the same conditions, but without the template molecule. OS-MIP and NIP were analyzed as described in the following sections.

Table 5 Comparison of the recovery results of the uniform design of experiments with the values calculated from the regression model

No.	Experimental result ($\mu g \cdot 50$ mg)$^{-1}$	Value from the equation ($\mu g \cdot (50$ mg)$^{-1}$)	Relative error (%)
1	18.845	18.129	3.80
2	40.265	40.704	-1.09
3	176.744	177.125	-0.22
4	72.539	71.714	1.14
5	20.306	20.644	-1.66
6	45.460	44.567	1.96
7	149.033	148.098	0.63
8	43.822	44.035	-0.49

Morphology observation

To further elucidate the nature of the imprinting effect, the physical characteristics of the polymers were studied. Figure 3a1 and b1 showed that both OS-MIP and NIP were amorphic particles with diameters between 45 and 60 μm. The SEM image in Fig. 3a2, b2, a3 and b3 clearly showed that pores were embedded in the OS-MIP and NIP networks, and that there were substantial differences in morphology between OS-MIP and NIP. NIP had a smoother structure with fewer and smaller cavities (Fig. 3b2 and b3) than were found on OS-MIP (Fig. 3a2 and a3), possibly revealing the effect of imprinting on the surface area of OS-MIP. These results were also consistent with previous results reported by Brüggemann[35], which indicated that MIPs had larger relative surface areas than the corresponding NIPs.

Sorption quantity

Using the OS-MIP and NIP prepared as described in the section "Optimized parameters", we carried out static sorption experiments of OS-MIP and NIP in triplicate. The average sorption quantity of OS-MIP (NIP) was 220.95 $\mu g \cdot (50$ mg$)^{-1}$ 12.08 $\mu g \cdot (50$ mg$)^{-1}$, and the calculated sorption quantities were in agreement with the experimental values. These results demonstrated that optimization of the polymerization parameters by uniform design was effective and that OS-MIP synthesized under optimal conditions exhibited a significantly higher sorption quantity of OS compared with the NIP.

When static sorption quantity of NIP particles was compared with that of the various imprinted polymer particles in Table 5, statistical analysis showed that the value of NIP was not significantly different from the values of Nos.1 and 5 ($P>0.05$), while the value of NIP was significantly different from other values in Table 5 ($P<0.05$). It indicated that NIP particles prepared as optimized method can sometimes have stronger static sorption capacity than some of MIP particles prepared as usual. This also indicated it was necessary to optimize polymerization.

OS-MIP and NIP prepared as described in the section "Optimized parameters", the average sorption quantity of OSMIP for template was significantly different from the value of NIP. It indicated that spatial structure of MIP was different from NIP, although these two polymers had the same material composition. The reason of this difference was that MIP contained many cavities. There were some fixed array functional groups on these cavities inner surface. Shape, size, and functional groups of cavities were complementation with OS. Sorption quantity of OS-MIP for template was mainly from noncovalent binding of OS and functional groups, which fixed array on these cavities inner surface.

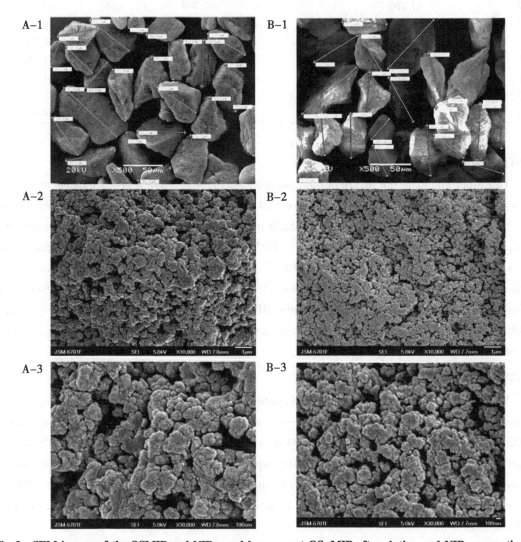

Fig. 3 SEM images of the OSMIP and NIP. a and b represent OS-MIP after elution and NIP, respectively

Specific affinity of OS-MIP-LC column

MIP will be combined with LC and MS in following study for a biological replacement method of screening for antivirus drug and it is necessary for the specific affinity of LC column packed with OS-MIP to be verified. In the present study, the main method was for direct separation of OS from matrix solution containing OS and other chemicals such as chlorogenic acid, phillyrin, and aspirin. The packed OS-MIP polymers were amorphous particles, so the background pressure of this column was very high. If the flow rate is the same as commercial HPLC column, the pressure will be higher than wet-packed pressure. This pressure would destroy the inner structure of column, and may change the structure of polymer and affect the separation. So MeOH-MeCN-formate acid (75 : 25 : 0.01) was used as the mobile phase at the flow rate of 0.075 mL·min^{-1}.

The MIP-LC-MS separation result was shown in Fig. 4. This illustrated that there were two peaks on the TIC. OS was completely separated from matrix components. The first peak contained aspirin ([M+Na]$^+$ ion at 203.00, [M-H+Na+Na]$^+$ ion at 225.00), chlorogenic acid ([M+Na]$^+$ ion at 377.20), phillyrin ([M+Na]$^+$ ion at 557.20) and diisooctyl phthalate (plasticizer) ([M+Na]$^+$ ion at 413.30). Diisooctyl phthalate was generated from plastic-tips used for matrix solution preparation. The ion m/z 629.10 was a triple polymer of sodium aspirin. Chlorogenic acid, phillyrin, aspirin, and diisooctyl phthalate were simultaneously detected by ESI-MS. It indicated that the OS-MIP-LC column did not have selectivity for those three components, and did not separate them each other. The second peak contained OS ([M+H]$^+$ ion at 313.20) predominantly. Ion m/z 335.20 is [M+Na]$^+$ of OS. Ions m/z 225.10 and m/z 301.10 as ESI+common background ions are dicyclohexyl urea (m/z 225.10, [M+H]$^+$) and dibutylphthalate (plasticizer; m/z 301.10, [M+Na]$^+$). The retention time of OS was longer than those of chlorogenic acid, phillyrin, aspirin and diisooctyl phthalate in this OS-MIP-LC column. The reasons for this are that the structures of chlorogenic acid, phillyrin, aspirin, and diisooctyl phthalate are different from OS. The cavities of polymer were not appropriate for those four components, so this OS-MIP-LC column did not have affinity and selectivity for them but for OS. This column can be used in the following study to search for more active OS analogs from Chinese traditional herbs.

CONCLUSION

In the present study, we demonstrated that the statistical methodology of uniform design is effective and reliable for selecting the factors with significant contribution and finding the optimal conditions for synthesizing OSimprinted polymers.

We used MIPs prepared under optimal conditions as a sorbent and developed a high sorption quantity for the template molecule (OS). MIPs can be used to effectively enrich and separate OS from the complex matrix. Thus, we concluded that uniform design is a useful technique to optimize the polymerization parameters and to obtain MIPs with high efficiency for extracting given molecules from complex matrices. On the other hand, the specific affinity test of MIPs for the template molecule itself and other different compounds such as chlorogenic acid, phillyrin, aspirin and diisooctyl phthalate showed that MIPs had a very high affinity and selectivity to separate the template molecule and its analogs from other compounds. This column with MIPs can, therefore, be used to search for more active OS analogs from Chinese traditional herbs.

Although there are many methods for OS detect, it is very difficult to quantify OS effectively. In the present study, we have established an effective LC-MS/MS method to identify and quantify OS using LC-MS/MS with good sensitivity, accuracy, and precision.

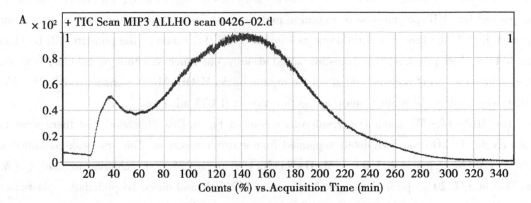
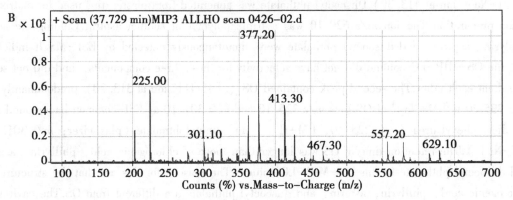
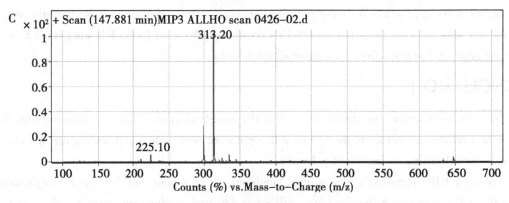

Fig. 4 Separation of OS-MIP column for OS from mixture

ACKNOWLEDGMENTS This work was supported by the National Natural Science Foundation of China (No. 30771590) and the Postdoctoral Science Foundation of China (No. 20070420443).

References

[1] Hantash J, Bartlett A, Oldfield P, Denes G, O'Rielly R, Roudiere D, Menduni S (2006) J Chromatogr A 1125: 104–111.

[2] Xu ZF, Liu L, Deng QY (2006) J Pharm Biomed Anal 41: 701–706.

[3] Karlsson JG, Karlsson B, Andersson LI, Nicholls IA (2004) Analyst 129: 456–462.

[4] Wang JF, Zhou LM, Liu XL, Wang QH, Zhu DQ (2000) Chin J Chem 18: 621-625.
[5] Syu MJ, Nian YM, Chang YS, Lin XZ, Shiesh SC, Chou TC (2006) J Chromatogr A 1122: 54-62.
[6] Sibrian-Vazquez M, Spivak DA (2004) J Polym Sci A Polym Chem 42: 3 668-3 675.
[7] Cummins W, Duggan P, McLoughlin P (2006) Biosens Bioelectron 22: 372-380.
[8] Dong WG, Yan M, Liu Z, Wu GS, Li YM (2007) Sep Purif Technol 53: 183-188.
[9] Sellergren B, Shea KJ (1993) J Chromatogr A 635: 31-49.
[10] Mijangos I, Navarro-Villoslada F, Guerreiro A, Piletska E, Chianella I, Karim K, Turner A, Piletsky S (2006) Biosens Bioelectron 22: 381-387.
[11] Vaughan AD, Sizemore SP, Byrne ME (2007) Polymer 48: 74.81.
[12] Lu Y, Li CX, Wang XD, Sun PC, Xing XH (2004) J Chromatogr B 804: 53-59.
[13] Fang KT (1980) Acta Math Appl Sin 3: 363-372.
[14] Wr.blewska A (2006) Appl Catal A Gen 309: 192-200.
[15] Liang YZ, Fang KT, Xu QS (2001) Chemom Intell Lab Syst 58: 43-57.
[16] Yang YZh, Jiang M, Xu J, Ma YH, Tong J (2011) Appl Compos Mater. doi: 10.1007/s10443-011-9188-9.
[17] Weekly update, Pandemic (H1N1) 2009—update 67, Global Alert and Response (2009) World Health Organization.http://www.who.int/csr/don/2009_09_25/en/index.html.
[18] Mayor S (2006) BMJ 332: 196.
[19] Jefferson T, Demicheli V, Rivetti D, Jones M, Pietrantonj CD, Rivetti A (2006) Lancet 367: 303-313.
[20] Hill G, Cihlar T, Oo C, Ho ES, Prior K, Wiltshire H, Barrett J, Liu BL, Ward P (2002) Drug Metab Dispos 30: 13-19.
[21] Xu X, Zhang L, Chen K (2006) Drug Discov Today 3: 247-253.
[22] Lindegardh N, Hien TT, Farrar J, Singhasivanon P, White NJ, Day NPJ (2006) J Pharm Biomed Anal 42: 430-433.
[23] Bahrami G, Mohammadi B, Kiani A (2008) J Chromatogr B 864: 38-42.
[24] Li XN, Zhu JR, Wang H, Zhou Z, Ge QH (2002) Chin J Clin Pharm 11: 259-262.
[25] Jabbaribar F, Mortazavi A, Jalali-milani R, Jouyban A (2008) Chem Pharm Bull 56: 1 639-1 644.
[26] Lindegardh N, Hanpithakpong W, Wattanagoon Y, Singhasivanon P, White NJ, Day P (2007) J Chromatogr B 859: 74-83.
[27] Wiltshire H, Wiltshire B, Citron A, Clarke T, Serpe C, Gray D, Herron W (2000) J Chromatogr B 745: 373-388.
[28] O'Mahony J, Molinelli A, Nolan K, Smyth MR, Mizaikoff B (2005) Biosens Bioelectron 20: 1 884-1 893.
[29] Lehmann M, Dettling M, Brunner H, Tovar GEM (2004) J Chromatogr B 808: 43-50.
[30] Komiyama M, Takeuchi T, Mukawa T, Asanuma H (2003) Molecular imprinting: from fundamentals to applications.Wiley.
[31] Fang KT (1994) Uniform design and uniform designs table.Science, Beijing.
[32] Yu C, Mosbach K (1997) J Org Chem 62: 4 057-4 064.
[33] Kempe M, Mosbach K, Fischer L (1993) J Mol Recognit 6: 25-29.
[34] Kempe M, Mosbach K (1994) J Chromatogr A 664: 276-279.
[35] Bruggemann O (2001) Biomol Eng 18: 1-7.

(发表于《Anal Bioanal Chem》, SCI, IF: 3.841)

Enzyme-Assisted Extraction of Naphthodianthrones from *Hypericum perforatum* L. by $^{12}C^{6+}$-ion Beam-Improved Cellulases*

LI Zhao-zhou[1], WANG Xue-hong[1], SHI Guang-liang[3], BO Yong-heng[3], LU Xi-hong[2], LI Xue-hu[2], SHANG Ruo-feng[1], TAO Lei[3], LIANG Jian-ping[1]**

(1. Key Laboratory of New Animal Drug Project, Gansu Province/Key Laboratory of Veterinary Pharmaceutical Development, Ministry of Agriculture/Lanzhou Institute of Husbandry and Pharmaceutical Sciences of CAAS, Lanzhou 730050, China; 2. Institute of Modern Physics, Chinese Academy of Sciences, Lanzhou 730050, China; 3. College of Veterinary Medicine, Gansu Agricultural University, Lanzhou 730050, China)

Abstract: The production of cellulase by *Aspergillus niger* and *Trichoderma viride* strains was significantly enhanced via irradiation with a $^{12}C^{6+}$-ion beam. Successive mutants showed a clearance zone on cellulose Congo red medium and increased cellulase activities. By using these cellulases, enzyme-assisted extractions of the naphthodianthrones hypericin and pseudohypericin from *Hypericum perforatum* L. were examined for the first time. The strong disruption of plant tissue structure through cellulase maceration was observed with scanning electron microscopy. Optimum conditions for improving the extraction yield of naphthodianthrones with the purified cellulases were as follows: the dried plant powder was treated with a cellulase solution of 150 U/g substrate and incubated at pH 5 for 6 h at 40 C. The extraction yields of purified hypericin and pseudohypericin were up to 166.5 lg/g and 260.6 lg/g, respectively, in the cellulase-treated sample, representing higher yields than sonication and Soxhlet extractions. Cellulase-assisted extraction was proven to be highly efficient in the degradation of plant cell-wall polysaccharides and could be used in natural product extraction on a large scale.

Key words: Enzyme - assisted extraction; Naphthodianthrones; $^{12}C^{6+}$ - ion beam; Mutant screening; Cellulase

---

* Received 7 February 2011, Received in revised form 26 October 2011, Accepted 10 November 2011, Available online 18 November 2011

** *Abbreviations*: PDA, potato dextrose agar; MCC, microcrystalline cellulose; WB, wheat bran; CMC, carboxymethyl cellulose; HIRFL, Heavy Ion Research Facility in Lanzhou; FPA, filter paper activity; H/C, ratio of hydrolysis halo diameter to colony diameter; HPLC, high - performance liquid chromatography; HIC, hydrophobic interaction chromatography.

*** Corresponding author.liangjp100@ sina.com© 2011 Elsevier B.V.All rights reserved.

1 INTRODUCTION

Hypericum perforatum L. is an herbaceous perennial plant of the Hypericeae family, whose members have been widely used for their therapeutic efficacy and biological activity. Today, methods to extract active components from *H. perforatum* L. that give a high yield of total extract and high content of the desired pharmacologically active compounds is the focus of many research activities. Although the active ingredients of *H. perforatum* L. have not been clearly defined, hypericin and pseudohypericin, which are the main naphthodianthrones present in the *Hypericum* species, are generally regarded as the marker substances for the standardization of the herbal product; they have also been reported to be the main contributors to the pharmacological effects of *H. perforatum* L. As powerful naturally occurring photosensitizers, hypericin and pseudohypericin were first isolated from the *Hypericum* species [1,2]. They possess high antiviral activity *in vitro* and *in vivo*, and may therefore be interesting compounds in the search for new tools against *Herpes simplex* virus type 1, human immunodeficiency virus type 1, and some other viruses[3]; they have also shown antidepressant activity[4]. In recent years, increased interest in hypericin and pseudohypericin as potential clinical anticancer agents has arisen since studies established its powerful *in vivo* and *in vitro* antineoplastic activity upon irradiation[5,6].

Several methods have been described for the extraction of hypericin and pseudohypericin from *H. perforatum* L., including conventional maceration, sonication, Soxhlet extraction, accelerated solvent extraction, supercritical fluid extraction, and pressurized-liquid extraction. However, there are still disadvantages associated with these methods, such as a long extraction time, large organic solvent consumption, high cost, and low efficiency; in addition, some methods require special instruments and have limited digestion of cell wall polysaccharides. In view of this, a cost-effective, highly specific, and easy-to-use extraction process is urgently needed to improve the efficiency, yield, separation, and purification of these compounds.

Degradation of cell-wall polysaccharides is a fundamental step to improve the release of active compounds from different medicinal raw materials under the same extraction conditions. Conventional solvent extraction only extracts the most accessible and feebly attached compounds from plant material. However, cellwall degrading enzymes can break down the structural integrity of the cell wall so as to increase solvent accessibility and release of active compounds from intracellular compartments; it can also catalyze hydrolytic degradation of the cell-wall polysaccharides that are presumed to retain active compounds in the polysaccharides-lignin network by hydrogen or hydrophobic bonding[7]. Another mechanism may be the direct enzyme-catalyzed breakage of the chemical bonds between the active compounds and the plant cell wall polymers. Additionally, enzyme-assisted extraction possesses the advantage of being environmentally friendly, highly efficient, and easily operated, owing to the relatively mild reaction conditions; thus, it has been represented as an alternative method of natural product extraction.

Among the cellulolytic fungi, *Aspergillus niger* and *Trichoderma viride* have been extensively studied due to their ability to secrete extra-cellular cellulases, which act synergistically during the conversion of cellulose, hemi-cellulose, and lignin to glucose[8]. However, the high cost of cellulase production still hinders the use of enzymes at the industrial scale for the enzyme-assisted extraction of active compounds from herbal medicines[9]. Using a heavy ion beam as a novel and efficient mutagen provides a new approach to obtaining improved strains that are capable of producing high levels of cellulases. The beam has attracted increased attention since some studies reported complex biological effects of heavy-ion implantation, such as lower germination and survival rates and genetic variations in plants and microorganisms[10,11]. It has a broad mutation spectrum and high mutation frequency compared with mutation from other inducing methods[12].

The purpose of this work was to examine the options for maximizing the extraction of selected types of active compounds from *H. perforatum* L. by defining the optimal conditions of enzyme-assisted extraction with heavy-ion irradiation-improved cellulases from *A. niger* and *T. viride*[7]. For this purpose, we screened and selected mutants of *A. niger* and *T. viride* that produced cell walldegrading hydrolytic enzymes with high catalytic activities, and various parameters of enzyme-assisted extraction, such as pH, temperature, incubation duration, and enzyme concentration, were optimized to obtain high yields of the above natural products in an economical and environmentally friendly way. To our knowledge, this is the first report of enzyme-assisted extraction of active compounds from *H. perforatum* L. by highly complementary and improved cellulases from two different fungal strains, and it is also the first study of a mutation breeding of *A. niger* and *T. viride* using a $^{12}C^{6+}$ heavy-ion beam.

2 EXPERIMENTAL MATERIALS AND METHODS

2.1 Materials and apparatus

2.1.1 Fungal strains and media

A. niger and *T. viride* were purchased from Gansu Culture Collection of Industrial Microorganisms, PR China, and then have been improved cellulase production with the irradiation of an electron beam. The resulted strains were cultured on potato dextrose agar (PDA) slants (potato leachate, 1000 mL; glucose, 20.0 g; agar, 20.0 g and pH 5.5–5.7) at 30℃ for 3–5 days, sealed with sterile liquid paraffin, stored at 4℃ when spores were formed, and used for further irradiation with a heavy ion beam.

Modified Mandels' medium was chosen as the basal medium, and it contained (g/L): KH_2PO_4, 2; $(NH_4)_2SO_4$, 1.4; urea, 0.3; $MgSO_4 \cdot 7H_2O$, 0.3; $CaCl_2$, 0.3; $FeSO_4 \cdot 7H_2O$, 0.005; $MnSO_4 \cdot H_2O$, 0.00156; $ZnSO_4$, 0.0014 and $CoCl_2$, 0.0002. It was mixed with 0.5% PDA liquid medium. Moreover, the medium used for screening consisted of (g/L): microcrystalline cellulose (MCC), 5; agar, 15; $(NH_4)_2SO_4$, 5; $MgSO_4 \cdot 7H_2O$, 0.5; K_2HPO_4, 1; NaCl, 0.1; $CaCl_2$, 0.1; yeast extract, 0.2; Congo red, 0.002 and pH 6.4–6.6. Additionally, the fermentation medium used for cell wall degradation contained (g/L): wheat bran (WB), 5; peptone 0.3 and basal medium, 10 mL. After inoculation, all of the culture materials were maintained at a temperature of 30℃ for 3–5 days.

2.1.2 Plant material

H. perforatum L. was collected during the flowering stage in the region of Longnan city, Gansu province, PR China, and it had not been sprayed with herbicide, burnt, mowed, grazed by livestock, or mechanically cut for several years prior to the experiment. The herb was authenticated by associate researcher Luo Yongjiang from the Lanzhou Institute of Animal Science and Veterinary Pharmaceutical Science, Chinese Academy of Agricultural Sciences. Voucher specimens were deposited in the herbarium of this institute.

2.1.3 Chemicals and reagents

Hypericin and pseudohypericin were purchased from the Alexis Corporation (Lausen, Switzerland). Stock solutions of 1.0 and 2.0 mg/mL in methanol were prepared under sonication for pseudohypericin and hypericin, respectively. Chromatographic grade acetonitrile and methanol were purchased from Merck KGaA (Darmstadt, Germany) and double-distilled water was used in all experiments. All other reagents were of analytical grade.

2.1.4 Apparatus and chromatographic conditions

A $^{12}C^{6+}$-ion beam of 100 MeV/u was supplied by the Heavy Ion Research Facility in Lanzhou (HIRFL) at the Institute of Modern Physics, Chinese Academy of Sciences (Lanzhou, PR China).

The high-performance liquid chromatography (HPLC) analysis was carried out on a Varian liquid chromatographic system (Varian, USA) equipped with a Prostar 210 pump, LC workstation version 6.41 system software, and a Prostar 325 UV - Vis detector. Chromatographic separation was performed on a C18 reversedphase column (5 lm, 4.6 mm 250 mm HC, Agilent Corporation, USA). The mobile phase consisted of 10% ammonium acetate - acetic acid buffer (0.5 M, pH 3.7), 40% methanol, and 50% acetonitrile (by volume). Effluents were monitored at a wavelength of 590 nm. The flow rate was 1 mL/min, the injection volume was 20 ll, and the column temperature was maintained at room temperature (22℃)[13].

2.2 Irradiation with heavy ion beam

Spores of the fungal strains were harvested, diluted to 10^8 spores/ml, and irradiated in 35-mm disposable petri dishes. We then exposed the spores to a $^{12}C^{6+}$-ion beam at 5–500 Gy. The dose rate was adjusted to approximately 4 Gy/min. After irradiation, all of the cultures were sealed, diluted fivefold with 30% glycerolphysiological saline, and then stored at 4℃ for further studies. The controls were treated similarly but were not irradiated.

2.3 Mutant screening

A very simple and highly efficient method was used for a preliminary screen for hyper-cellulase-producing mutants. The MCC screening medium was inoculated with the mutated spore suspensions and cultured at 30℃. Three to five days later, we could estimate the cellulase activities by the ratio of the hydrolysis halo diameter to the colony diameter (H/C). Subsequently, the strains with higher ratios (H/C > 1.5) were selected and cultured in a 96-well microplate containing the PDA medium for the further assay of enzyme activities and the secondary screening of mutants. Although the standard filter paper assay is widely used to determine total cellulase activity, it is not suitable for the high-throughput determination of enzyme activity. By referring to the related literature, we developed a microplate-based method for assaying large sample volumes to screen the mutants. The reaction volume was reduced 25 times from the 1.5 mL used in the IUPAC method, and the absorbance was recorded with a Multiskan MK3 microplate reader (Thermo Labsystems) using a test wavelength of 540 nm[14].

2.4 Cellulase activity assays

Cellulase activities were determined as previously reported. Filter paper activity (FPA) was measured with the standard procedure recommended by the Commission on Biotechnology, IUPAC[15]. Monitoring of the total reducing sugars was determined colorimetrically by the DNS method (3,5-dinitrophenol) using glucose as a standard[16]. Exoglucanase and β-glucosidase activities were estimated using regenerated amorphous cellulose and p-nitrophenyl-β-D-glucopyranoside as substrates, respectively[17,18].

2.5 Purifying the main components of the cellulases complex

All of the purification steps were performed at 4℃ unless otherwise specified. Crude cellulase solutions from the fermentation medium inoculated with the two fungal strains stated above were obtained by centrifuging at 5 000 rpm for 10 min. The crude extracts were then concentrated 10-fold by ultrafiltration using a diether sulfone membrane (Millipore) with a 10 KDa cutoff and precipitated with $(NH_4)_2SO_4$ at 80% saturation.

Enzyme purification was followed by anion exchange chromatography, hydrophobic interaction chromatography (HIC), cation exchange chromatography, and gel filtration. First, the crude enzyme preparation was applied to a Sephadex DEAE A-50 column previously equilibrated with 25 mM Tris - HCl buffer (pH 6.8) containing 1 mM DL-dithiothreitol. Elution was achieved with a linear gradient of 0.15-0.5 M NaCl in the equilibration buffer at a flow rate of 0.5 mL/min.

Eluted fractions from the column were analyzed for endoglucanase, exoglucanase, and β-glucosidase activities. Subsequently, the fraction showing mainly β-glucosidase activity was processed on an HIC column (eluted using a linear $(NH_4)_2SO_4$ gradient (pH 5, 0-1 M) at a flow rate of 0.5 mL/min) and a Sephadex G-75 filtration step (eluted with three bed volume of 50 mM CH_3COONa (pH 5) at a flow rate of 0.3 mL/min); the proteins with endoglucanase activity were loaded on an anion exchange column (eluted with a linear NaCl gradient (0-0.5 M) at a flow rate of 1 mL/min) and concentrated using a 10-KDa ultrafiltration membrane, and the purified product with exoglucanase activity was further separated by a cation exchange column (eluted with a linear NaCl gradient (0-0.5 M) at a flow rate of 1 mL/min) and an HIC column (eluted using a linear $(NH_4)_2AC$ gradient (pH 5, 0-1 M) at a flow rate of 0.5 mL/min)[19]. In addition, the protein concentrations of each fraction were measured using the Bio-Rad protein estimation kit with bovine serum albumin as a standard.

2.6 Enzyme-assisted extraction and pretreatment

The *H. perforatum* L. plant was air dried at room temperature indoors and away from sunlight. After being finely ground, the dried powder was treated by a mixture of purified cellulases including endoglucanase I, endoglucanase II, and exoglucanase from *A. niger* as well as β-glucosidase from *T. viride*. Briefly, 50 g of dried powder was suspended in enzyme media to 200 mg/mL containing the each cellulase at the dosage of 150 U/g substrate. After adjusting to pH 6.5, the solution was incubated at 30℃ for 6 h with continuous stirring. Then, we filtered the solution by gravity filtration through a funnel containing a fluted filter paper and freeze-dried the resulting sediment on the same day.

A methanolic extract was prepared by extraction of the biomass (dry weight/ml) with an ethanol - water solution (80 : 20). A homogeneous suspension was obtained by shaking on a flat-bed orbital shaker for 6 h. After filtration under vacuum, the filter cake was washed with three portions of methanol (3 × 15 mL). The filtrate was defatted at room temperature with n-hexane (30 mL, three times) and it was sequentially fractionated by liquid-liquid extraction with ethyl ether and ethyl acetate (50 mL of each solvent, three times). All solvents were previously saturated with H_2O, and the resulting upper phases were concentrated on a rotary evaporator until completely dry, and then dissolved in methanol and fractionated over Sephadex LH-20 (column: 60 cm × 3 cm) using dichloromethane/methanol/water (4 : 5 : 1, v/v/v) as the eluent. The column retained pseudohypericin and hypericin, which were eluted at the end. Subsequently, we separated hypericin and pseudohypericin completely by Sephadex LH-60 (column: 60 cm × 3 cm) using ethyl ether/methanol (70 : 30, v/v) as the elution system[20]. The resulting fractions were dissolved in methanol and submitted for HPLC analysis.

2.7 Scanning electron microscopy

To achieve a complete understanding of the effect of the enzyme on the structural changes of the plant cells, scanning electron micrographs were taken of the samples before and after enzyme treatment. The sample particles were fixed on a specimen holder with aluminum tape, then sputtered with gold in a JEOL JEC-1200 sputter-coater, and examined with a JEOL JSM-5600 LV scanning electron microscope under high vacuum conditions at an accelerating voltage of 5 kV (10 lm, 1500× magnification).

2.8 Statistical analysis

All treatments in the experiments were replicated a minimum of three times, and all experiments were repeated twice with similar results. Mean values of all data were obtained from triplicate experiments. Significance of differences was evaluated at the $p < 0.05$ level.

3 RESULTS AND DISCUSSION

3.1 Cellulase production and enzyme purification

We selected mutated strain colonies with the high H/C ratios (H/C > 1.5) and significantly high enzyme activities after irradiation with a heavy ion beam. The data from the crude enzyme preparations are summarized in Table 1. The results indicate that even after nine generations, cellulase production remained stable, and growth and morphological changes of mutants were not observed (coefficients of variation no more than 5%). Thus, the mutants were steady and inheritable.

We were interested in developing the methodology for purifying the cellulase components from these two fugal strains and therefore were prompted to apply these methods. After continuous culture for 5 days to 6 days, the components of the cellulolytic system from *A. niger* or *T. viride* were prepared according to the method above and then applied to the resins. The purified cellulase proteins exhibited several times higher activity than the crude enzyme solution. It was found that exoglucanase and endoglucanase were the main components of the cellulases from A.niger, and the main component of T.viride cellulases was β-glucosidase. A summary of the purification is presented in Table 1.

Table 1 Enzyme activities and purification steps of cellulases isolated from
Aspergillus niger and Trichoderma viride

Fraction	Fungal straina	Protein concentration (mg/mL)	FPA (IU/mL)	Exoglucanase (IU/mL)	Endoglucanase (IU/mL)	β-Glucosidase (IU/mL)	Yield (%)
Crude enzyme	AnP	11.3±0.2	132.4±3.9	124.6±5.3	110.8±3.5	16.6±0.5	100
	AnM	22.8±0.4	421.3±10.5	252.6±4.5	397.6±11.5	24.7±0.6	100
	TvP	15.2±0.5	72.6±2.3	9.5±0.3	33.8±1.2	216.7±9.1	100
	TvM	31.7±1.4	485.6±15.1	16.6±0.7	68.9±3.3	435.4±17.2	100
HIC and Sephadex G-75 columns	AnM	1.4±0.1	25.8±1.2	0	0	1.5±0.07	43.5±2.1
	TvM	116.3±2.1	1781.6±43.6	0	0	1597.3±40.5	48.2±2.5
Anion exchange column and ultrafiltration membrane	AnM	108.6±2.4	2006.7±36.7	0	1893.8±53.2	0	36.6±1.2
	TvM	2.2±0.1	33.7±1.3	0	4.8±0.2	0	34.3±1.7
Cation exchange and HIC columns	AnM	76.5±2.1	1413.5±18.8	847.5±28.2	0	0	23.4±1.4
	TvM	1.8±0.1	27.5±1.1	0.94±0.04	0	0	22.2±1.1

Note: AnP and TvP represent parental strains from Aspergillus niger and Trichoderma viride, respectively; AnM and TvM stand for mutant strains from Aspergillus niger and Trichoderma viride, respectively.

The fungal strains used in this study showed differential cellulase production and growth according to the medium used. Fig. 1a and b show the cellulase activity values of culture supernatants of A. niger and T.viride fermented in basal media containing different carbon sources. It was found that the maximum enzyme activity and the best growth conditions were obtained using WB as carbon source, and that cellulase activity increased greatly after irradiation of the fungi. The most significant induction of cellulase activity was observed in the presence of WB, whereas the other carbon sources (carboxymethyl cellulose, CMC; MCC) gave moderate extracellular enzyme production. This could be ascribed to the heterogeneous nature and structural complexity of WB; in addition, some disaccharides and oligosaccharides might be liberated during the hydrolysis of the substrate and induce the cellulase activity [21]. In addition, we found that some mutants exhibited cellulase activities regardless of the substrates used because the enzyme-induced mechanism might be blocked after the irradiation.

3.2 Microscopic morphology of enzyme-treated H.perforatum L.

We used electron microscopy to evaluate the degradative effects of cellulase treatment on the extraction of active constituents from H. perforatum L. A comparison between the change in microstructure of the herbal powder with and without enzymatic hydrolysis was carried out. There was no rupture or significant destruction to the microstructure without cellulase treatment in Fig. 2a. However, the surface of the sample was mostly destroyed and the microstructure was disorganized after enzymolysis (see Fig. 2b). This observation suggests that cellulase treatment can weaken or break down the cell wall, rendering the intracellular material more accessible for extraction. During the rupture process, rapid exudation of the chemical substance within the cell into the surrounding solvents took place.

3.3 Enzymatic extraction of active constituents from *H.perforatum* L.

3.3.1 Effect of pH value of enzyme solutions

Enzyme activities are influenced by pH, and most enzymes are active over specific ranges of pH. An increase or decrease in the pH affects the charges on the amino acids within the active site such that the enzyme will not be able to form an enzyme-substrate complex. Additionally, pH changes can stop enzyme activity by denaturing (altering) the three-dimensional shape of the enzyme by breaking ionic and hydrogen bonds. We varied the pH values of the enzyme solution in a range from 4 to 6 to investigate the effect of incubation acidity on the extraction yields in order to select the optimum pH value for cellulase activity. Fig. 3a shows that the yields of two active compounds varied irregularly with different pH values and achieved a maximum at pH 5. Thus, pH 5 was chosen as the optimum pH for obtaining optimal extraction yields of the two active compounds from *H.perforatum* L.

3.3.2 Effect of incubation temperature

The effect of temperature on extraction yield was studied, and the results are shown in Fig. 3b. The reaction rates of the enzyme solutions gradually accelerated with the increase of temperature up to an optimum, above which some enzymes are denatured. As seen in Fig. 3b, the extraction yields varied with the changing temperature, and the optimum yields occurred at a temperature of 40℃. Therefore, 40℃ was considered to be the best choice for the enzyme incubation. However, the fact that this temperature does not corroborate the data of crude cellulase characteristics (optimum temperatures for cellulase production from *A.niger* and *T.viride* in the fermentation medium were 30 and 35℃, respectively) implies that some other kinds of enzymes or factors may also be exerting effects in the incubation, either directly or indirectly.

3.3.3 Effect of incubation duration

Fig. 3c shows the results of the effect of the enzyme incubation time on the extraction yields of the two active compounds from *H.perforatum* L. Extending the incubation time resulted in a dramatic increase in the extraction yield up to 6 h. The yields began to decrease with longer incubation, but no statistical differences were noted. Nevertheless, 6 h was considered to be long enough for the hydrolysis of the cell wall in *H.perforatum* L. because shorter incubation led to only a partial release of the active components from the herbal medicine, and longer incubation failed to improve the extraction efficiency.

3.3.4 Effect of enzyme concentration

The effect of using different concentrations of the enzyme preparation on the extraction of the compounds is shown in Fig. 3d (6 h incubation at 40℃ in the enzyme solution at pH 5). With the increasing enzyme concentration, the yields of these two compounds reached their peaks at an enzyme concentration of 150 U/g substrate (166.5 lg/g hypericin and 260.6 lg/g pseudohypericin). The yields started to decrease with higher concentrations, because higher enzyme loads resulted in faster total hydrolysis, and end-product inhibition probably occurred. Consequently, 150 U/g was selected as a suitable concentration for the extraction of the active compounds.

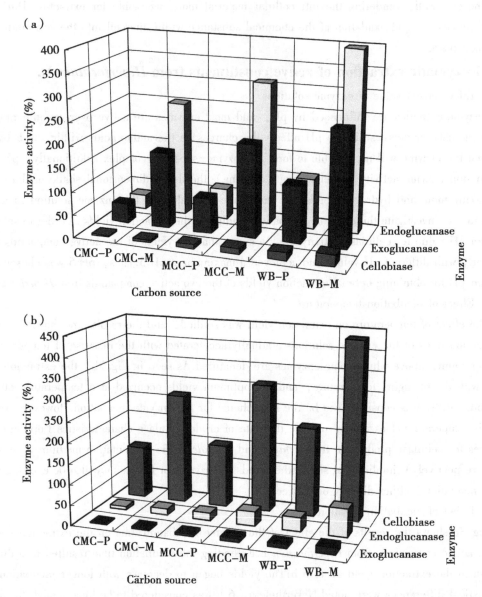

Fig. 1 Cellulase activities of the medium supernatants of *Aspergillus niger* and *Trichoderma viride* strains fermented in basal media containing different carbon sources

Note: CMC-P, WB-P, and MCC-P represent, respectively, parental strains with CMC, WB, and MCC as carbon sources; CMC-M, WB-M, and MCC-M stand for mutant strains with CMC, WB, and MCC as carbon sources, respectively.

3.4 Comparison with other extraction methods

Several methods have been used to extract active compounds from *H. perforatum* L., such as Soxhlet and sonication extractions. A comparison between these two extraction procedures and enzyme-assisted extraction was conducted. The Soxhlet extraction was carried out with 50 g of dried

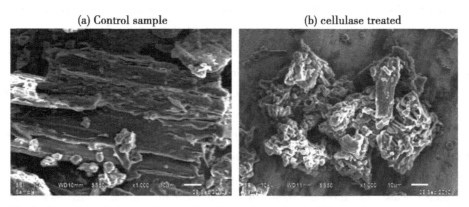

Fig. 2 Scanning electron micrographs of herbal powder in different treatment processes

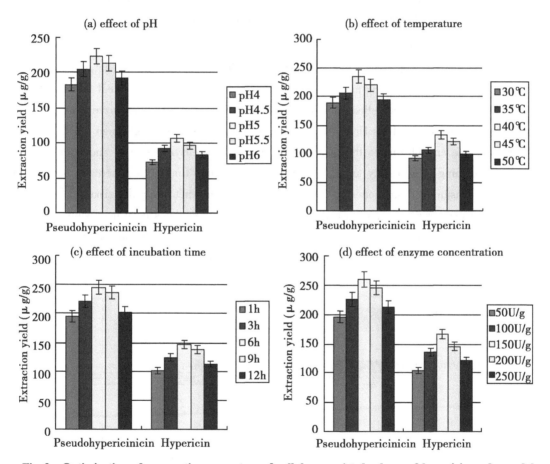

Fig. 3 Optimization of enzymatic parameters of cellulase-assisted release of hypericin and pseudohypericin from *H.perforatum* L

plant power and 2 000 mL of methanol for a period of 24 h. For extraction by ultrasonication, the dried plant material (50 g) and methanol (2 000 mL) were sonicated for 30 min using a supersonic bath (Kunshan Hechuang Ultrasonic Instruments, PR China) at a frequency of 35 kHz. After the extraction, the samples were purified, processed as described above, and then analyzed by

HPLC to calculate the extraction yields of the two active compounds from *H.perforatum* L. As indicated in Fig. 4, our optimized procedure of enzyme-assisted release using the enzyme preparation proved to be better than the other two methods, and it gave the highest extract yields. Quantities of 260..6 lg/g pseudohypericin and 166.5 lg/g hypericin were extracted and separated from the enzyme-treated plant material. Sonication extraction produced 180.3 lg/g pseudohypericin and 113.5 lg/g hypericin. The Soxhlet extraction showed the lowest extraction yields (156.2 lg/g pseudohypericin and 93.1 μg/g hypericin), partly because of the higher temperatures used, which accelerate the decomposition of some active compounds from *H.perforatum* L. Comparison of the data indicates that cell-wall degrading enzymes can weaken or break down the cell wall more effectively than sonication because the cell-wall breaks to a greater extent, increasing the ease and efficiency of extraction. The decreased solvent consumption provides a greener option than traditional non-enzymatic extraction. However, the overall extraction steps are not fully environmentally and/or operator-friendly since the organic solvents were still needed in the extraction procedure.

3.5 Mutant screening method

The screening strategy is a critical step for finding desired mutants from a large mutant library. Therefore, an efficient screening method is a prerequisite. We performed a typical facilitated screening method on solid agar that relied on product solubilization followed by an enzymatic reaction that gave rise to a zone of hydrolysis. However, a plate-screening method using Congo red is not quantitative or sensitive enough due to the poor correlation between enzyme activity and halo size. Thus, secondary screening is necessary to validate the primary screen findings. Furthermore, this method can be time-consuming, and zones of hydrolysis are sometimes not easily discernable[22]. In view of these facts, some improved high-throughput mutant screening methods have been studied and published[23-25], and the next step for the research team will be to improve the current method or design a new one to screen or select cellulase mutants more efficiently and effectively.

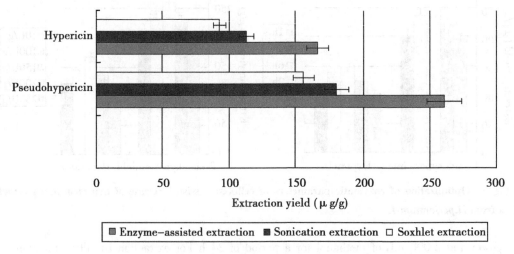

Fig. 4 Comparison of extraction yields (%) of hypericin and pseuohypericin using different extraction methods

3.6 Degradation effect by different enzymes

A plant cell wall consists of a rigid skeleton of cellulose embedded in a gel-like matrix composed of pectic compounds, hemi-cellulose, and glycoprotein. The fact that certain enzymes could ease the release of active compounds from *H.perforatum* L. could be beneficial for extraction. As indicated above, cellulases can be used to facilitate extraction in extract processing. But certain cellulases with limited enzyme activity may not be sufficiently efficient at cell-wall degradation, so multi-enzyme preparation and its application in enzyme-assisted extraction will require further studies. However, it has been reported that yields of extracted compounds when complex enzyme samples were used were similar or even much lower than those in individual enzymes[26,27]. Thus, it is clear that more research is needed to investigate the synergistic mechanisms of action of different enzymes in enzyme-assisted extraction. Besides focusing on enzyme-facilitated cell wall degradation, targeting the enzymatic modification of the target compounds could be an alternative approach. The key point for this approach is to find an enzyme that could modify the properties, such as solubility, of target compounds for more efficient extraction without impacting bioactivity[28]. Additionally, a higher extraction performance might be achieved by using multifunctional enzymes or the enzymatic transformation of primary metabolites into target compounds[29].

3.7 Scale-up of enzyme-assisted extraction

Enzyme-assisted extraction has been used successfully at laboratory scale and large scale in a limited number. The main limitations for the application of enzyme in industrial processes have been placed on the cost, enzymatic hydrolysis conditions and extraction yield obtained.

One of the bottlenecks in the enzyme-assisted extraction remains the expensive enzyme consumption in the hydrolysis step. A viable approach to improving the practicality and economic feasibility is the use of enzymes with enhanced activities and productions. Furthermore, the enzymatic hydrolysis of cellulose would be more economical if the enzymes could be recovered from the reaction mixture in active forms and reused many times. Several strategies have been developed to improve the process, including ultrafiltration, re-adsorption onto fresh substrates, and enzyme immobilization. From our perspective, enzyme recycling using the strategies alone or in combination could potentially decrease enzyme costs in the bioconversion process.

Secondly, the enzymes are sensitive to environmental conditions. Their activities are influenced by the temperature, hydrolysis time, substrate species, acidity, nutrient availability and enzyme concentration. Therefore, the hydrolysis should be under tightly controlled conditions to minimize the variance of extraction yields.

In addition, currently available enzymes can not hydrolyze plant cell walls completely, limiting the yield increase. There will be a great potential for enzyme-assisted extraction when combining with ultrasonication[30,31], microwave[32], or superfinegrinding[33].

If the above limitations can be overcome, the extraction process assisted by enzymes could be widely utilized for extraction of natural products in industrial scale. The results of the present study may provide valuable data for process design and industrial scale-up application.

4 CONCLUSIONS

Stable mutants with enhanced production of cellulases were generated through irradiation of *A. niger* and *T. viride* strains with a $^{12}C^{6+}$ heavy-ion beam. The cellulases derived from the mutants showed higher activities than the parent strains. We optimized the incubation duration, pH, temperature, and enzyme concentration for cellulase-assisted extraction of hypericin and pseudohypericin at 6 h, 5, 40℃, and 150 U/g substrate using the purified cellulases. At the optimum conditions, the extraction yields of hypericin and pseudohypericin were 166.5 lg/g and 260.6 lg/g, respectively, which are higher than sonication and Soxhlet extractions. The study can serve as a reference for the enzyme-assisted extraction of other compounds from herbal plants.

ACKNOWLEDGEMENTS

This study was supported by research Grants from the Hundred Talent Program of the Chinese Academy of Sciences (No.0861010ZY0) and the National Key Technologies Research and Development Program of China during the 11th Five-Year Plan Period (No.2006BAD31B05).

References

[1] H.Brockmann, M.N.Haschad, K.Maier, F.Pohl, Hypericin, the photodynamically active pigment from *Hypericum perforatum*, Naturwissenschaften 27 (1939) 550–555.

[2] H.Brockmann, W.Sanne, Pseudohypericin, a new red *Hypericum* pigment, Naturwissenschaften 40 (1953) 461.

[3] D.Meruelo, G.Lavie, D.Lavie, Therapeutic agents with dramatic antiretroviral activity and little toxicity at effective doses: aromatic polycyclic diones hypericin and pseudohypericin, Proc.Natl Acad.Sci.USA 85 (1988) 5 230–5 234.

[4] V.Butterweck, T.Bökers, B.Korte, W.Wittkowski, H.Winterhoff, Long-term effects of St.John's wort and hypericin on monoamine levels in rat hypothalamus and hippocampus, Brain Res. 930 (2002) 21–29.

[5] P.Agostinis, A.Vantieghem, W.Merlevede, P.A.M.De Witte, Hypericin in cancer treatment: more light on the way, Int.J.Biochem.Cell Biol.34 (2002) 221–241.

[6] A.Vantieghem, Y.Xu, W.Declercq, P.Vandenabeele, G.Denecker, J.R.Vandenheede, W.Merlevede, P.A.De Witte, P.Agostinis, Different pathways mediate cytochrome c release after photodynamic therapy with hypericin, Photochem.Photobiol.74 (2001) 133–142.

[7] M.Pinelo, B.Zornoza, A.S.Meyer, Selective release of phenols from apple skin: mass transfer kinetics during solvent and enzyme-assisted extraction, Sep.Purif.Technol.63 (2008) 620–627.

[8] D.E.Eveleigh, Cellulase: a perspective, Philos.Trans.R.Soc.London, Ser.A 321 (1987) 435–447.

[9] H.Brunner, Radiation induced mutations for plant selection, Appl.Radiat.Isot.46 (1995) 589–594.

[10] D.Li, F.Nie, L.Wei, B.Wei, Z.Chen, Screening of high-yielding biocontrol bacterium Bs-916 mutant by ion implantation, Appl.Microbiol.Biotechnol.75 (2007) 1 401–1 408.

[11] L.Wu, Z.Yu, Radiobiological effects of a low-energy ion beam on wheat, Radiat.Environ.Biophys.40 (2001) 53–57.

[12] L.Zhou, W.Li, L.Yu, P.Li, Q.Li, S.Ma, X.Dong, G.Zhou, C.Leloup, Linear energy transfer dependence of the effects of carbon ion beams on adventitious shoot regeneration from in vitro leaf explants

of *Saintpaulia ionahta*, Int.J.Radiat.Biol.82（2006）473-481.

[13] U.Ruckert, K.Eggenreich, W.Likussar, R.Wintersteiger, A.Michelitsch, A highperformance liquid chromatography with electrochemical detection for the determination of total hypericin in extracts of St. John's wort,Phytochem.Anal.17（2006）162-167.

[14] Z.Xiao, R.Storms, A.Tsang, Microplate-based filter paper assay to measure total cellulase activity, Biotechnol.Bioeng.88（2004）832-837.

[15] T.K.Ghose, Measurement of cellulase activities, Pure Appl.Chem.59（1987）257-268.

[16] S.Takashima, H.Iikura, A.Nakamura, M.Hidaka, H.Masaki, T.Uozumi, Overproduction of recombinant *Trichoderma reesei* cellulases by *Aspergillus oryzae* and their enzymatic properties, J.Biotechnol.65（1998）163-171.

[17] Y.H.Zhang, J.Hong, X.Ye, Cellulase Assays, in: J.R.Mielenz（Ed.）, Biofuels: Methods and Protocols, Humana Press Inc., Totowa, New Jersey, USA, 2010, pp.213-231.

[18] K.Kovács, G.Szakacs, G.Zacchi, Comparative enzymatic hydrolysis of pretreated spruce by supernatants, whole fermentation broths and washed mycelia of *Trichoderma reesei and Trichoderma atroviride*, Bioresour.Technol.100（2009）1 350-1 357.

[19] J.Zhou, Y.H.Wang, J.Chu, Y.P.Zhuang, S.L.Zhang, P.Yin, Identification and purification of the main components of cellulases from a mutant strain of *Trichoderma viride* T 100-14, Bioresour.Technol. 99（2008）6 826-6 833.

[20] A.Karioti, F.F.Vincieri, A.R.Bilia, Rapid and efficient purification of nap hthodianthrones from St. John's wort extract by using liquid-liquid extraction and SEC,J.Sep.Sci.32（2009）1 374-1 382.

[21] J.Thongekkaew, H.Ikeda, K.Masaki, H.Iefuji, An acidic and thermostable carboxymethyl cellulase from the yeast Cryptococcus sp.S-2: Purification, characterization and improvement of its recombinant enzyme production by high cell-density fermentation of *Pichia pastoris*, Protein Expr.Purif.60（2008）140-146.

[22] M.Maki, K.T.Leung, W.Qin, The prospects of cellulase-producing bacteria for the bioconversion of lignocellulosic biomass, Int.J.Biol.Sci.5（2009）500-516.

[23] G. Fia, G. Giovani, I. Rosi, Study of beta-glucosidase production by wine-related yeasts during alcoholic fermentation. A new rapid fluorimetric method to determine enzymatic activity, J. Appl. Microbiol.99（2005）509-517.

[24] S.R.Decker, W.S.Adney, E.Jennings, T.B.Vinzant, M.E.Himmel, Automated filter paper assay for determination of cellulase activity, Appl.Biochem.Biotechnol.107（2003）6 890-703.

[25] R.Kasana, R.Salwan, H.Dhar, S.Dutt, A.Gulati, A rapid and easy method for the detection of microbial cellulases on agar plates using Gram's iodine,Current Microbiology 57（2008）503-507.

[26] Y.Zu, Y.Wang, Y.Fu, S.Li, R.Sun, W.Liu, H.Luo, Enzyme-assisted extraction of paclitaxel and related taxanes from needles of *Taxus chinensis*, Sep.Purif.Technol.68（2009）238-243.

[27] H.B.Sowbhagya, K.T.Purnima, S.P.Florence, A.G.Appu Rao, P.Srinivas, Evaluation of enzyme-assisted extraction on quality of garlic volatile oil, Food Chem.113（2009）1 234-1 238.

[28] S.Chen, X.H.Xing, J.J.Huang, M.S.Xu, Enzyme-assisted extraction of flavonoids from *Ginkgo biloba* leaves: Improvement effect of flavonol transglycosylation catalyzed by *Penicillium decumbens* cellulase, Enzyme Microb.Technol.48（2011）100-105.

[29] M.Xu, Q.Sun, J.Su, J.Wang, C.Xu, T.Zhang, Q.Sun, Microbial transformation of geniposide in *Gardenia jasminoides* Ellis into genipin by *Penicillium nigricans*, Enzyme Microb.Technol.42（2008）440-444.

[30] P.Bermejo, J.L.Capelo, A.Mota, Y.Madrid, C.Cámara, Enzymatic digestion and ultrasonication: a

powerful combination in analytical chemistry, TrAC, Trends Anal.Chem.23 (2004) 654-663.

[31] G.Vale, R.Rial-Otero, A.Mota, L.Fonseca, J.L.Capelo, Ultrasonic-assisted enzymatic digestion (USAED) for total elemental determination and elemental speciation: a tutorial, Talanta 75 (2008) 872-884.

[32] Y.C.Yang, J.Li, Y.G.Zu, Y.J.Fu, M.Luo, N.Wu, X.L.Liu, Optimisation of microwave-assisted enzymatic extraction of corilagin and geraniin from *Geranium sibiricum* Linne and evaluation of antioxidant activity, Food Chem.122 (2010) 373-380.

[33] S.Y.Jin, H.Z.Chen, Superfine grinding of steam-exploded rice straw and its enzymatic hydrolysis, Biochem.Eng.J.30 (2006) 225-230.

(发表于《Separation and Purificaton Technology》, SCI, IF: 2.773)

Lonicera japonica Thunb.: Ethnopharmacology, Phytochemistry and Pharmacology of an Important Traditional Chinese Medicine*

SHANG Xiao-fei[1,2], PAN Hu[1], LI Mao-xing[2,3,*],
MIAO Xiao-lou[1], DING Hong[4,**]

(1. Engineering & Technology Center of Traditional Chinese Veterinary Medicine of Gansu, Key Laboratory of Veterinary Pharmaceutics Discovery, Ministry of Agriculture, Lanzhou Institute of Husbandry and Pharmaceutical Sciences, Chinese Academy of Agricultural Sciences, Lanzhou 730050, China; 2. Department of Pharmacy, Lanzhou General Hospital of PLA, Key Laboratory of the Prevention and Treatment for Injury in Plateau of PLA, Lanzhou 730050, China; 3. Key Laboratory of Chemistry and Quality for Traditional Chinese Medicines of the College of Gansu Province, Gansu College of Traditional Chinese Medicine, Lanzhou 730000, China; 4. Northwest Normal University, Lanzhou 730050, China)

Abstract: *Ethnopharmacological relevance*: *Lonicera japonica* Thunb. (*Caprifoliaceae*), a widely used traditional Chinese medicine, was known as *Jin Yin Hua* (Chinese: 金银花),

---

* *Abbreviations*: ACV, acyclovir; AIV, avian influenza virus; ALT, alanine transarninase; AST, aspartate amino transferase; CAT, catalase; Cd, cadmium; CGN, carrageenan; COX, cycloxygenase; ConA, concanavalin A; CPE, cytopathologic effect; DAD, diode-array detection; DPPH, 1,1-diphenyl-2-picrylhydrazyl; ELSD, evaporative light scattering detectors; EtOAc, ethyl acetate; ERK, extracellular signal-regulated kinase; GC.MS, gas chromatography.mass spectrometry; GSH, glutathione; HDL-C, high density lipoprotein cholesterin; HIV-1, human immunodeficiency virus-1; HPLC, high performance liquid chromatography; HSV, Herpes simplex virus; HUVEC, Human Umbilical Vein Endothelial Cells; IC_{50}, 50% inhibition concentration; JNK, Jun nuclear kinase; LED, Least Effective Dose; *Lonicera japonica*, *Lonicera japonica* Thunb.; LPS, lipopolysaccharide; MAPK, mitogen activated protein kinase; MDA, malondialdehyde; MEC, minimum effective concentration; MeOH, methyl alcohol; MIC, minimum inhibitory concentration; MPO, myeloperoxidase; MTT, methyl thiazolyl tetrazolium; MUFA, monounsaturated fatty acid; NDV, newcastle disease virus; NO, nitric oxide; PAPR2, proteinase-activated receptor 2; PDT, photodynamic therapy; PMNs, polymorphonuclear leukocytes; PUFA, polyunsaturated fatty acid; ROS, reactive oxygen species; RSV, respiratory syncytial virus; SARS coronavirus, severe acute respiratory syndromes coronavirus; SFA, saturated fatty acid; SI, selectivity index; SOD, superoxide dismutase; TCM, traditional Chinese medicine; TEAC, Trolox equivalent antioxidant capacity; TI, therapeutic index; TLC, thin layer chromatography; TNF-α, tumor necrosis factor-α; TOF-MS, time-of-flight mass spectrometry.

** Corresponding author. © 2011 Elsevier Ireland Ltd. All rights reserved.

Ren Dong and *Japanese honeysuckle*. It was taken to treat the exopathogenic wind-heat, epidemic febrile diseases, sores, carbuncles and some infectious diseases. At the same time, *Lonicera japonica* could be used as healthy food, cosmetics, ornamental groundcover, and so on.

Aim of the review: The present paper reviewed the ethnopharmacology, the biological activities, toxicology and phytochemistry of *Lonicera japonica*.

Materials and methods: Information on *Lonicera japonica* was gathered via the Internet (using Google Scholar, Baidu Scholar, Elsevier, ACS, Medline Plus, CNKI and Web of Science) and libraries. Additionally, information also was obtained from some local books and brilliant scholars on ethnopharmacology.

Results: More than 140 chemical compounds have been isolated, and the main compositions are essential oils, organic acids and flavones, etc. *Lonicera japonica* and its active principles possess wide pharmacological actions, such as anti-inflammatory, antibacterial, antiviral, antioxidative and hepatoprotective activities.

Conclusions: As an important traditional Chinese medicine, further studies on *Lonicera japonica* can lead to the development of new drugs and therapeutics for various diseases, and how to utilize it better should be paid more attentions.

Key words: *Lonicera japonica*; Ethnopharmacology; Chlorogenic acid; Anti-inflammatory; Antiviral activity

1 INTRODUCTION

Lonicera japonica Thunb. (*Caprifoliaceae*), also known as *Japanese honeysuckle*, *Jin Yin Hua* or *Ren Dong*, is native in the East Asian (He et al., 2010). Now as an ornamental groundcover, *Lonicera japonica* commonly planted in many areas for sprawling habit, numerous sweetly fragrant white flowers, attractive evergreen foliage, and become naturalized in Argentina, Brazil, Mexico, Australia, New Zealand and United States. Due to *Lonicera japonica* has escaped from cultivation in several places, becoming a major nuisance, and is restricted in parts of North America and New Zealand (Starr et al., 2003). But in China, 1500 years ago, *Lonicera japonica* has been planted largely in Fengqiu county of Henan province, and the flowers of *Lonicera japonica* have been used as the local and traditional medicine in clinical practice for the treatment of exopathogenic wind-heat, epidemic febrile diseases, sores, carbuncles, furuncles and some infection diseases. Scine 1995, *Lonicera japonica* has been listed in the Pharmacopoeia of the People's Republic of China and more than 500 prescriptions containing *Lonicera japonica* have been used to treat various diseases in China (http://www.zysj.com.cn). The modern pharmacological studies showed that *Lonicera japonica* and its active principles possessed wide pharmacological actions, such as antibacterial, anti-inflammatory, antiviral, antiendotoxin, blood fat reducing, antipyretic and other activities (Wang, 2008c). Most of these actions matched to those traditional uses seriously. At the same time, it was also used as food, healthy beverage in the world (Wang, 2010). Along with *Lonicera japonica* being used and cultivated in more and more countries, the chemical compounds have been extensively studied. Essential oils, organic acids, flavones, saponins, iridoids and inorganic elements as the main compositions were isolated and identified. Among of them, essential oils and chlorogenic acid have been proved with some good pharmacological effects, and were though as the active compounds of *Lonicera japonica*. In current Chinese Pharmacopoeia (Committee for the Pharmacopoeia of PR China, 2010), chlorogenic acid (1) has been officially used as the indicator compound to characterize the quality of this herb.

In this review, the advances in ethnopharmacology, phytochemistry, biological and pharmacological activities, and toxicology of *Lonicera japonica* are displayed, and the increasing data supports the utilization and the exploitation for new drug.

2 BOTANY AND ETHNOPHARMACOLOGY

2.1 Botany

According to the description of Wagner et al. (1999), *Lonicera japonica* is sprawling and twining lianas; young stems pubescent; leaves ovate, elliptic, oblong or broadly lanceolate, blades 3-8 cm long, 1-3.5 cm wide, pubescent, becoming glabrate above, entire or young lower leaves sometimes lobed; flowers 2 in axillary cymes, bracts 1-2 cm long, bracteoles suborbicular, ca. 1 mm long; corolla white, turning yellowish or tinged pink, 2-lipped, 2 - 3 cm long; berries bluish black, globose, 6-7 mm in diameter. The flowering duration of individual plant is usually 5-8 days, but the flowering period is from May to September in the field, can be divided into six stages, i.e. the juvenile bud stage, the third green stage, the second white stage, the complete white stage, the silver flowering stage and the gold flowering stage. *Lonicera japonica* often grows in hillside scrub, rocks pile and roadside, and the highest altitude is 1500 m. Due to its beautiful flowers and strong roots, *Lonicera japonica* was cultivated for people to watch, conserve water and soil in world. In traditional Chinese medicine, due to the outline form of sprawling and twining lianas, and the different flower colors, the dried flowers or flower buds of *Lonicera japonica* was named as *Jin Yin Hua* and *Ren Dong* in TCM. Both the chemical contents and compositions of *Lonicera japonica* flowers vary in a flowering-dependent characteristic with the collection time (Fig. 1) (Wang et al., 2009).

Fig. 1 The flower and habitat of *Lonicera japonica*

2.2 Ethnopharmacology

With a wide spectrum of biological and pharmacological properties, *Lonicera japonica* played a very important role in TCM. 3 000years ago, our ancestors have adopted it to cure some illnesses. Due to the effects of curing fever and swelling of body, 'Ming Yi Bie Lu' and 'Shen Nong Ben Cao Jin' has listed it as 'top grade' (Tao, 1986; Gu et al., 2007).Then, 'Ben Cao Gang Mu', the famous classical book of Chinese materia medica, has recorded that it could be applied to clear away the heat-evil, treat the swellings and dysentery, protect body and prolong life (Li, 1979; Wang, 2010).In addition, more than ten classical medicine books in China also have recorded this plant, and it has been used as the main composition in some famous prescription to treat various diseases (Table 1).

Lonicera japonica has been planted and used as the local medicine in many places, especially in East Asian.In China, it is widely distributed in drainage areas of the Yellow River and Yangzi River, and largely cultivated in Longhui, Fengqiu, Pingyi and Fei counties of Hunan, Henan and Shandong provinces.According to the quality analyses, *Lonicera japonica* planted in Fengqiu county has the highest contents of chlorogenic acid with 4%-6% (Wang, 2010).Since 1995, *Lonicera japonica* has been listed in the Pharmacopoeiaof the People's Republic of China (Committee for the pharmacopoeia of PR China, 1995), and made to some preparations to treat chronic enteritis, pneumonia, acute tonsillitis, nephritis, acute mastitis, leptospirosis in clinic. Among of them, '*Jin Yin Hua Jiu* (Wine)' has been used to clear away the heat-evil and expel superficial evils; '*Jin Yin Hua Tang*' has been applied to clear heat and detoxicating, and so on (Table 1).Recently, *Lonicera japonica* also has been employed extensively to prevent and treat some serious viral diseases of human and veterinary, such as SARS coronavirus, H1N1 (Swine) flu virus, and being called the 'bouvardin' (Jiao, 2009).

Lonicera japonica was also employed as healthy beverage to improve body and prevent ills in China.In Qing dynasty, according to 'Yan Shou Dan Fang', it was used to moisturize the skin and rejuvenation (Chen, 2008).Modern pharmacological researches thought that these effects may be related to the active compositions volatile oils, chlorogenic acid and flavones.Myung et al. (2004) suggested that *Lonicera japonica* prepared by extraction of 70% methanol or 70% acetone followed by gamma irradiation treatment have a bright color, good tyrosinase inhibition, xanthine oxidase inhibition, and nitrite scavenging activities.It could use for the food or cosmetic industry as natural source of bioactive compound. And with other compositions, *Lonicera japonica* has been made healthy beverage through various technology, such as '*Jin Yin Hua tea*', '*Jin Yin Hua nutritive beverage*', '*Jin Yin Hua acidophilous milk*', '*Jin Yin Hua Wine*', '*Jin Yin Hua oral liquid*' (He et al., 2010) (Table 1).

At the same time, *Lonicera japonica* has been made to the article of everyday uses and cosmetics, such as '*Jin Yin Hua floral water*', '*Jin Yin Hua facial mask*', especially it could be made to toothpaste which have the effects of preventing and treating the oral cavity's diseases (Jiao, 2009).Zhang (2008) investigated the antibacterial and antisepticize activities in cosmetics of the flower extracts of *Lonicera japonica*, the results showed that it had the marked antibacterial and antisepticize activities, and could be applied in cosmetics extensively.Shu et al. (2008) suggested that essential oils isolated from *Lonicera japonica* by supercritical extraction method would cover the smell from cigarettes and improve the quality.So *Lonicera japonica* would bring the social and economic values well.

Table 1 The traditional and clinical uses of *Lonicera japonica* in China

Preparation name	Main compositions	Traditional and clinical uses	References
Bao An Yan Shou Fang	Flos Lonicerae, Radix Glycyrrhizae	Curing some infection diseases	Yi Fang Yi Jian*
Bei Mu San	Flos Lonicerae, Bulbus Fritillariae Thunbergii	Curing mammary abscess	Pu Ji Fang, Vol. 325*
Fu Fang Jin Yin Hua Jiao Nang	Flos Lonicerae, Fructus Forsythiae, Radix Scutellariae	Clearing heat and detoxicating. Curing headache, fever, cough and toothache	Chinese Medicine Dictionary*
Gan Ju Tang	Flos Lonicerae, Radix Glycyrrhizae, Flos Chrysanthemi	Curing all furunculosis	Chuan Mo You De Book*
Hua Gan Xiao Du Tang	Flos Lonicerae, Fructus Gardeniae, Radix Glycyrrhizae, Radix Angelicae, Radix Angelicae Sinensis	Curing the pain of coastal regions	Bian Zhen Lu, Vol. 13*
Jia Wei Sheng Hua Tang	Flos Lonicerae, Fructus Forsythiae, Radix Glycyrrhizae, Olibanum, Myrrha and Radix Ginseng	Curing deficiency of both QI and blood after childbirth	Jin Jian, Vol. 48*
Jie Du Xian Cao	Fresh branches and leaves of *Lonicera japonica*	Curing late syphilis	Shang Yi Da Quan, Vol.34*
Jin Pu Tang	Flos Lonicerae, Herba Taraxaci, Semen Benincasae, Radix Aucklandiae, Radix et Rhizoma Rhei	Clearing away heat evil, promoting diuresis and Qi to activate blood	Zhu Ri Dong Fang*
Jin Qi San	Flos Lonicerae, Radix Astragali, Radix Glycyrrhizae, Radix Rehmanniae, Radix Paeoniae Alba, Radix Angelicae Sinensis	Curing women' acute pain and trug of hypogastrium	Pu Ji Fang, Vol.335*
Jin Qiang Gao	Flos Lonicerae, Radix et Rhizoma Rhei, Herba Violae, Radix Arnebia, Radix Angelicae Sinensis, Eupolyphaga seu Steleophage, Cortex Phellodendri, Radix Glycyrrhizae, Radix Saposhnikoviae	Curing wound infection	Zhong Yi Shang Ke Xue Jiang Yi*
Jin Yin Hua San	Flos Lonicerae, Herb Schizonepetae, Fructus Cnidii, Radix seu Rhizoma Nardostachyos, Radix Angelicae, Semen Arecae, Natril Sulfas	Treating chancre sore	Pu Ji Fang, Vol. 301*
Jin Yin Hua Gao	Flos Lonicerae, Radix Glycyrrhizae, Herb Leonuri	Treating pregnancy carbuncle	Chen Su An Fu Ke Bu Jie, Vol.3*
Jin Yin Hua Jiu (Wine)	Fresh leaf of *Lonicera japonica*	Curing superficial infection and furunculosis	Gu Fang Hui Jin*

* Cited from the Website: http://www.zysj.com.cn.

** Cited from 'Chinese Pharmacopoeia'.

*** Cited from the Website: http://www.sda.gov.cn.

(continued)

Preparation name	Main compositions	Traditional and clinical uses	References
Jin Yin Hua Tang Jiang	Vine of *Lonicera japonica*	Clearing heat and detoxicating. Curing fever, sore throat and so on	Chinese Medicine Dictionary*
Jin Yin Wine	Flos Lonicerae, Herba Taraxaci	Curing breast's bump	Xian Nian Ji, Vol.3*
Ju Hua Jin Yin Hua Tang	Flos Lonicerae, Flos Chrysanthemi, Radix Platycodi, Radix Ophiopgonis, Radix Glycyrrhizae	Treating pharyngo-laryngitis chronica	Dan Yan Fang*
San Xing Tang	Flos Lonicerae, Herba Taraxaci, Radix Glycyrrhizae	Curing carbuncle in mouth	Bian Zheng Lu, Vol. 13*
Sheng Hua Tang	Flos Lonicerae, Radix Ginseng	Curing ulcer, the deficiency of Qi and blood	Dong Tian Ao Zhi, Vol. 14*
Shu Feng Qing Gan Tang	Flos Lonicerae, Radix Paeoniae Lactiflora, Herba Schizonepetae, Radix Saposhnikoviae, Rhizoma Chuanxiong, Herba Menthae Haplocalycis, Flos Chrysanthemi, Fructus Gardeniae, Radix Bupleuri, Fructus Forsythiae, Radix Glycyrrhizae, Radix Angelicae Sinensis	Clearing away heat evil, promoting diuresis and removing heat to brighten vision	Yi Zong Jin Jian, Vol. 65*
Wan Shan Wan	Flos Lonicerae, Radix Glycyrrhizae, Fructus Forsythiae, Spica Prunellae	Curing hemorrhoid	Yan Fang Xin Bian*
Xiao Du San	Flos Lonicerae, Fructus Forsythiae, Herba Schizonepetae, Radix Angelicae, Fructus Arctium, Radix Saposhnikoviae, Cortex Dictamni, Radix Paeoniae Lactiflora, Radix Glycyrrhizae, Fructus tribuli	Expelling wind and dispelling dampness, clearing away the heat-evil and expelling superficial evils	Shang Yi Da Quan*
Xiao Hua Tang	Flos Lonicerae, Radix Trichosanthis, Radix Angelicae Sinensis, Radix Glycyrrhizae, Gynura Bicolor, Medulla Tetrapanacis	Clearing away the heat-evil and expelling superficial evils, curing mammitis	Dong Tian Ao Zhi, Vol. 7*
Yin Hua Tang	Flos Lonicerae, Rhizoma Menispermi, Radix Trichosanthis, Bulbus Fritillariae Thunbergii, Radix Angelicae, Radix Saposhnikoviae, Radix Paeoniae Lactiflora, Olibanum, Myrrha, Radix Glycyrrhizae	Clearing away the heat-evil and expelling superficial evils, curing the phyma and body pain	Gan Zu Wang Fang*
Yin Hua Tea	Fresh Flos Lonicerae	Curing children' parotitis and furunculosis sudariferus	Chang Yong Zhong Cao Yao Shou Ce*

(continued)

Preparation name	Main compositions	Traditional and clinical uses	References
Chinese Pharmacopoeia Kang Gan Ke Li	Flos Lonicerae, Radix Paeoniae Lactiflora, Rhizoma Dryopteris Crassirhizomae	Curing headache, fever, cough and pharyngalgia	Chinese Pharmacopoeia**
Li Yan Jie Du Ke Li	Flos Lonicerae, Radix Isatis, Fructus Forsythiae, Herba Menthae Haplocalycis, Fructus Arctium, Fructus Crataegi, Radix Platycodi, Folium Isatidis, Bombyx Batryticatus, Radix Scrophulariae, Radix Scutellariae, Radix Rehmanniae, Radix Trichosanthis, Radix et Rhizoma Rhei, Bulbus Fritillariae Thunbergii, Radix Ophiopgonis	Curing anemopyretic tonsillitis, acute tonsillitis and anemopyretic laryngalgia	Chinese Pharmacopoeia**
Qin Guo Wan	Flos Lonicerae, Fructus Canarli, Radix Scutellariae, Rhizoma Menispermi, Radix Ophiopgonis, Radix Scrophulariae, Radix Paeoniae Alba, Radix Platycodi	Curing the swell of throat, celostomia, dry mouth and xeropulmonary cough	Chinese Pharmacopoeia**
Qin Re Jie Du Kou Fu Ye	Flos Lonicerae, Gypsum Fibrosum, Radix Scrophulariae, Radix Rehmanniae, Fructus Forsythiae, Fructus Gardeniae, Radix Scutellariae, Radix Gentianae, Radix Isatis, Rhizoma Anemarrhenae, Radix Ophiopgonis	Clearing away the heat-evil and expelling superficial evils	Chinese Pharmacopoeia**
Xiao Yin Pian	Flos Lonicerae, Radix Rehmanniae, Mudanpi, Radix Paeoniae Lactiflora, Radix Angelicae Sinensis, Radix Sophorae Flavescentis, Radix Scrophulariae, Fructus Arctium, Perioatracum Cicadae, Covtex Diatamni, Radix Saposhnikoviae, Folium Isatidis, Flos Carthami	Removing heat to cool blood, dispelling wind and arresting itching, curing pruritus	Chinese Pharmacopoeia**
Shuang Huang Lian Shuan	Flos Lonicerae, Radix Scutellariae, Fructus Forsythiae	Curing upper respiratory tract infection and pneumonia	Chinese Pharmacopoeia**
Shuang Huang Lian Ke Li	Flos Lonicerae, Radix Scutellariae, Fructus Forsythiae	Dispelling the evil in the superficies with drugs of pungent taste and cool nature, and curing fever, cough and pharyngalgia	Chinese Pharmacopoeia**
Xiao Er Re Su Qin Kou Fu Ye	Flos Lonicerae, Radix Scutellariae, Radix Isatis, Radix Puerariae, Fructus Forsythiae, Radix Bupleuri, Radix et Rhizoma Rhei	Curing children headache, fever, nasal obstruction, cough and pharyngalgia	Chinese Pharmacopoeia**

(continued)

Preparation name	Main compositions	Traditional and clinical uses	References
Ying Huang Kou Fu Ye	Flos Lonicerae, Radix Scutellariae	Curing upper respiratory tract infection, acute tonsillitis and pharyngitis	Chinese Pharmacopoeia**
Zhi Zi Jin Hua Wan	Flos Lonicerae, Fructus Gardeniae, Rhizoma Coptidis, Radix Scutellariae, Cortex Phellodendri, Radix et Rhizoma Rhei, Rhizoma Anemarrhenae, Radix Pinelliae	Curing the swell of throat, constipation, conjunctival congestion, etc.	Chinese Pharmacopoeia**
Yin Qiao Jie Du Pian	Flos Lonicerae, Fructus Forsythiae, Herba Menthae Haplocalycis, Herba Schizonepetae, Fructus Arctium, Radix Platycodi, Folium lophatheri, Radix Glycyrrhizae	Curing pharwind–heat type common cold and headache, fever, cough	Chinese Pharmacopoeia**
Yin Qiao Tang	Flos Lonicerae, Fructus Forsythiae, Radix Scutellariae, Radix Bupleuri Chinensis, Herba Artemisiae, Fructus amomi, Almond, Semen Coicis, Radix Adenophorae, Rhizoma Phragmitis	Treating SARS in clinic	http://www.satcm.gov.cn
Healthy food	Flos Lonicerae, Propolius, Herba Menthae Haplocalycis	Moistening and cleaning throat	http://www.sda.gov.cn***
Feng Jiao Jin Yin Hua Qin Liang Tang Jin Yin Hua Qin Liang Tang	Flos Lonicerae, Fructus Canarli, Fructus momordicae, Semen Sterculiae Lychnophorae, Herba Menthae Haplocalycis	Moistening and cleaning throat	http://www.sda.gov.cn***
Jin Yin Hua Li Yan Pian	Flos Lonicerae, Fructus momordicae, Rhizoma Imperatae, Herba Menthae Haplocalycis	Moistening and cleaning throat	http://www.sda.gov.cn***
Jin Yin Hua Jiao Nang	Flos Lonicerae	Moistening and cleaning throat	http://www.sda.gov.cn***
Jin Yin Hua Zhen Zhu Jiao Nang	Flos Lonicerae, Radix Salviae Miltiorrhizae, Herba Taraxaci, Mudanpi, Fructus Gardeniae, Radix et Rhizoma Rhei, Margarita	Curing acnes	http://www.sda.gov.cn***
Jin Yin Hua Luo Han Guo Han Pian	Flos Lonicerae, Fructus momordicae, Flos Chrysanthemi, Semen Sterculiae Lychnophorae, Herba Menthae Haplocalycis	Moistening and cleaning throat	http://www.sda.gov.cn***
Jin Yin Hua Pan Da Hai Chong Ji	Flos Lonicerae, Herba Taraxaci, Flos Chrysanthemi, Herb Houttuyniae, Fructus momordicae, Herba Menthae Haplocalycis, Radix Platycodi, Semen Sterculiae Lychnophorae	Moistening and cleaning throat	http://www.sda.gov.cn***

3 PHYTOCHEMISTRY

More than 140 compounds have been isolated and identified from *Lonicera japonica* so far. As one of the important chemical composition, essential oils were analyzed through GC - MS method, and linalool, hexadecanoic acid, octadecadienoic acid, ethyl palmitate and dihydrocarveol were the main compounds. At the same time, *Lonicera japonica* was abounds with flavones, organic acids, triterpenoid saponins, and iridoids (Table 2 and Fig. 2). Some of them displayed many bioactivities *in vivo* or *in vitro* (Table 3). And the different chemical compositions of *Lonicera japonica* will provide the foundation well of the different pharmacology activities.

3.1 Essential oils

As one of the important compositions, essential oils exist in the aerial parts of *Lonicera japonica*, flower (fresh and dry), leaves and vines. Due to the difference habitat, the harvest time, medicinal parts, extraction methods and the processing, the contents and components of essential oils are different.

3.1.1 The different habitat

From the dry flower of *Lonicera japonica* in Henan province, 27 compounds were identified, mainly including aromadendrene, linalool and geraniol, and most of them belong to monoterpenes and sesquiterpenes. But from the dry flower of *Lonicera japonica* in Shandong province, more than 65 compounds were identified, the main compound was hexadecanoic acid, and the others were aldehydes, ketones, acids, esters and so on (Wang, 2010). In 2009, Lin (2009) found that the main compounds of essential oils in Fujian province were hexadecanoic acid, methyl linolenate, methylpalmita, etc. At the same time, Ikeda et al. has studied the volatile components of the concrete from flowers of *Lonicera japonica* with GC - MS in Japan. 150 compounds, made up of 36 hydrocarbons, 28 alcohols, 21 aldehydes, 12 ketones, 38 esters, 15 miscellaneous were identified. The important components that characterize the volatiles of *honeysuckle* flowers were identified as linalool, (Z)-jasmone, (Z)-jasmin lactone, methyl jasmonate, and methyl epi-jasmonate. In addition, changes of the volatile components emitted from the living flowers throughout the whole day were investigated by dynamic headspace analysis using GC and GC-MS, and the strongest odor was found to be emitted in the middle of the night (Ikeda et al., 1994). In 2010, Feng (2010) identified the content of chlorogenic acid from *Lonicera japonica* in different habitats by TLC. Results showed the content were 3.43%, 3.14%, 1.62% and 2.11% in Shanxi, Shandong, Hubei and Henan, respectively.

So the difference habitat would change the proportion of the chemical compounds. And the chemical compositions have strong relationship with the habitat of traditional Chinese medicine.

3.1.2 The different harvest time

Yang and Zhao (2007) have investigated the compositions of essential oils from *Lonicera japonica* between June and August. The results displayed that 32, 54 and 74 compounds were identified by GC - MS from flowers at June, July and August, respectively. The mainly increased compositions were low molecular number and low-boiling point compounds. But the compounds with high content were similar at each month, which were linalool (13.47%, 13.47%, 7.92%), dibu-

Table 2 The compounds isolated from *Lonicera japonica* (the structure of main compounds illustrated in Fig. 2)

No.	Compounds	Resource	References
Organic acids			
1	Chlorogenic acid	Whole plant	Yip et al. (2006)
2	Isochlorogenic acid	Whole plant	Yip et al. (2006)
3	Caffeic acid	Flowers	Choi et al. (2007)
4	Hexadecanoic acid	Whole plant	Huang et al. (1996)
5	Myristic acid	Whole plant	Huang et al. (1996)
6	3,5-O-dicaffeoylquinic acid	Whole plant	Iwanhashi and Negoroy (1986)
7	4,5-O-dicaffeoylquinic acid	Whole plant	Iwanhashi and Negoroy (1986)
8	3,4-O-dicaffeoylquinic acid	Whole plant	Iwanhashi and Negoroy (1986)
9	1,3-O-dicaffeoylquinic acid	Whole plant	Iwanhashi and Negoroy (1986)
10	3-Ferulicoylquinic	Whole plant	Iwanhashi and Negoroy (1986)
11	4-Ferulicoylquinic	Whole plant	Iwanhashi and Negoroy (1986)
12	5-O-caffeoylquinic acid	Whole plant	Qi et al. (2009)
13	4-O-caffeoylquinic acid	Whole plant	Qi et al. (2009)
14	Caffeoyl-CH_2-O-quinic acid	Whole plant	Qi et al. (2009)
15	1,5-O-dicaffeoylquinic acid	Whole plant	Qi et al. (2009)
16	1,4-O-dicaffeoylquinic acid	Whole plant	Qi et al. (2009)
17	Methylated dicaffeoylquinic acid	Whole plant	Qi et al. (2009)
18	Oleanolic acid 28-α-O-l-rhamnopyranosyl-(1→2)-[β-D-xylopyranosyl(1→6)]-β-d-glu-copyranosyol ester	Flowers	Choi et al. (2007)
19	3,5-O-dicaffeoylquinic acid methyl ester	Flower buds	Lee et al. (2010)
20	Methyl chlorogenate	Flower buds	Peng et al. (2000)
21	3-O-caffeoylquinic acid butyl ester	Flower buds	Lee et al. (2010)
22	3-O-caffeoylquinic acid	Flower buds	Ren et al. (2008)
23	3-caffeoylquinic acid methyl ester	Flower buds	Peng et al. (2000)

(continued)

No.	Compounds	Resource	References
24	3,5-dicaffeoylquinic acid buthyl ester	Flower buds	Peng et al. (2000)
25	Vanillic acid 4-O-β-D- (6-O-benzoylglucopyranoside)	Flower buds	Lee et al. (2010)
26	Protocatechuic acid	Flowers	Choi et al. (2007)
			Yip et al. (2006)
27	Chlorogenic acid butyl ester	Flower buds	Wang (2008c)
28	Chlorogenin tetraacetate	Flower buds	Lou et al. (1996)
29	5-Feruloylquinic acids	Aerial Parts	Shanghai Institute of Pharmaceutical Industry (1975)
30	Methyl 3,5-di-O-caffeoylquinic acid	Whole plant	Chang et al. (1995)
31	Methyl 3,4-di-O-caffeoylquinic acid	Whole plant	Chang et al. (1995)
32	Caffeic acid methyl ester	Whole plant	Ma et al. (2005)
Flavones			
33	Chrysoeriol	Flowers	Choi et al. (2007)
34	Chrysoeirol-7-O-neohesperidoside	Aerial parts	Choi et al. (2007)
35	Luteolin	Flowers	Choi et al. (2007)
		Leaves	Yip et al. (2006)
			Kumar et al. (2005)
36	Chrysoeriol 7-O-β-D-glucopyranoside	Flowers	Choi et al. (2007)
37	Isorhamnetin 3-O-β-D-glucopyranoside	Flowers	Choi et al. (2007)
38	Isorhamnetin 3-O-β-D-rutinoside	Flower buds	Wang (2008c)
39	Kaempferol 3-O-β-D-glucopyranoside	Flowers	Choi et al. (2007)
40	Kaempferol 3-O-β-D-rutinoside	Flower buds	Wang (2008c)
41	Quercetin 3-O-β-D-glucopyranoside	Flowers	Choi et al. (2007)
			Gao and Mu (1995)
42	Luteolin 7-O-α-D-glucoside	Flowers	Choi et al. (2007)
			Gao and Mu (1995)
43	Luteolin-7-O-β-D-galactoside	Flowers	Choi et al. (2007)
			Gao and Mu (1995)

(continued)

No.	Compounds	Resource	References
44	Hyperoside	Aerial parts	Zhang et al. (2006)
45	Lonicerin	Whole plant	Lee et al. (1995)
46	Hydnocarpin	Aerial parts	Son et al. (1992)
			Gao and Mu (1995)
47	Quercetin	Aerial parts	Son et al. (1992)
48	Astragalin	Aerial parts	Son et al. (1992)
49	Isoquercitrin	Aerial parts	Son et al. (1992)
50	Rhoifolin	Aerial parts	Son et al. (1992)
51	Flavoyadorinin-B	Flower buds	Lee et al. (2010)
52	Rutin	Flower buds	Ren et al. (2008)
53	Tricin-7-O-β-D-glucoside	Flower buds	Ren et al. (2008)
54	Chrysin	Leaves	Kumar et al. (2005)
55	Eriodictyol	Aerial parts	Zhang et al. (2006)
56	Apigenin	Aerial parts	Zhang et al. (2006)
57	Corymbosin	Aerial parts	Huang et al. (1996)
58	5-Hydroxy-3, 4, 7-trimethoxylflavone	Aerial parts	Huang et al. (1996)
59	Ochnaflavone	Whole plant	Son et al. (2006)
		Aerial parts	Son et al. (1992)
60	Ochnaflavone 4′-O-methylether	Aerial parts	Son et al. (1992)
61	3-O-methyl loniflavone 5,5″, 7,7″-tetrahydroxy 3-methoxy 4′,4″-biflavonyl ether	Leaves	Kumar et al. 2005
62	Loniflavone 5,5″, 7,7″, 3′-pentahydroxy 4′, 4″-biflavonyl ether	Leaves	Kumar et al. (2005)
Iridoids			
63	Loganin	Whole plant	Lee et al. (1995a)
64	Sweroside	Flower buds	Song et al. (2006)
			Machida et al. (1995)
65	7-O-ethyl sweroside	Flower buds	Song et al. (2006)

(continued)

No.	Compounds	Resource	References
66	7-Epi vogeloside	Flower buds	Song et al. (2006)
67	Secoxyloganin	Flower buds	Song et al. (2006)
68	Secoxyloganin 7-butyl ester	Flower buds	Song et al. (2006)
69	7-Dimethyl-secologanoside	Flower buds	Song et al. (2006)
70	Centauroside	Flower buds	Song et al. (2006)
71	Secologanic acid	Flower buds	Qi et al. (2009)
72	Secologanin	Flower buds	Machida et al. (1995)
73	Secologanin dimethyl acetal	Flower buds	Machida et al. (1995)
74	Kingiside	Flower buds	Son et al. (1994)
75	Vogeloside	Flower buds	Kakuda et al. (2000)
76	Epi-vogeloside	Flower buds	Kakuda et al. (2000)
77	Dehydrornornorronisi	Flower buds	Li et al. (2003)
78	Ketologanin	Flower buds	Song (2008)
79	7α-Morroniside	Flower buds	Song (2008)
80	7β-Morroniside	Flower buds	Song (2008)
81	Secologanoside	Flower buds	Song (2008)
82	Lonijaposide A	Flower buds	Song et al. (2008)
83	Lonijaposide B	Flower buds	Song et al. (2008)
84	Lonijaposide C	Flower buds	Song et al. (2008)
85	Lonijaposide D	Flower buds	Song (2008)
86	Lonijaposide E	Flower buds	Song (2008)
87	Lonijaposide F	Flower buds	Song (2008)
88	Lonijaposide G	Flower buds	Song (2008)
89	Lonijaposide H	Flower buds	Song (2008)
90	Lonijaposide I	Flower buds	Song (2008)
91	Lonijaposide J	Flower buds	Song (2008)
92	Lonijaposide K	Flower buds	Song (2008)

(continued)

No.	Compounds	Resource	References
93	Lonijaposide L	Flower buds	Song (2008)
94	l-Phenylalaninosecologanin	Stems, leaves	Machida et al. (2002)
95	7-O-(4-β-D-glucopyranosyloxy-3-methoxy-benzoyl) secologanolic acid	Stems, leaves	Machida et al. (2002)
96	6′-O-(7α-hydroxyswerosyloxy) loganin	Stems, leaves	Machida et al. (2002)
97	(Z)-aldosecologanin	Stems, leaves	Machida et al. (2002)
98	(E)-aldosecologanin	Stems, leaves	Machida et al. (2002)
99	Loniceracetalide A	Flower buds	Kakuda et al. (2000)
100	Loniceracetalide B	Flower buds	Kakuda et al. (2000)
Saponins			
101	3-O-α-L-arabinopyranosyl-28-O-[β-D-glucopyranosyl(1→6)-β-D-glucopyranosyl] oleanolic acid	Aerial parts	Kawai et al. (1988)
102	3-O-[α-L-rahmnopyranosyl(1→2)-β-L-arabinopyranosyl]-28-O-β-D-glucopyran-osyl hederagenin	Aerial parts	Kawai et al. (1988)
103	3-O-[α-L-rahmnopyranosyl(1→2)-α-L-arabinopyranosyl]-28-O-[β-D-glucopyranosyl(1→6)-β-D-glucopyranosyl] oleanolic acid	Aerial parts	Kawai et al. (1988)
104	3-O-[α-L-rahmnopyranosyl(1→2)-α-L-arabinopyranosyl]-28-O-[6-acetyl-β-D-glucopyranosyl(1→6)-β-D-glucopyranosyl] hederagenin	Aerial parts	Kawai et al. (1988)
105	3-O-α-L-rhamnopyranosyl-(1→2)-α-L-arabinopyranosy hederagenin 28-O-β-D-xylpyranosyl(1→6)-β-D-glucopyranosyl ester	Flower buds	Lou et al. (1996)
106	3-O-α-L-arabinopyranosy hederagenin 28-O-α-D-rhamnopyranosyl(1→2)[β-D-xyl pyranosyl(1→6)-β-D-glucopyranosyl ester	Flower buds	Lou et al. (1996)
107	3-O-α-L-rhamnopyranosyl-(1→2)-α-L-arabinopyranosy hederagenin 28-O-α-D-rhamnopyranosyl(1→2)[β-D-xyl pyranosyl(1→6)-β-D-glucopyranosyl ester	Flower buds	Lou et al. (1996)

(continued)

No.	Compounds	Resource	References
108	3-O-β-D-glucopyranosyl-(1→4)-α-L-glucopyranosyl-(1→3)-α-L-rhamnopyranosyl-(1→2)-α-L-arabinopyranosyl hederagenin28-O-β-D-glucopyranosyl-(1→6)-β-D-glucopyranosyl ester	Flower buds	Chen et al. (2000)
109	Hederagenin-3-O-l-rhamnopyranosyl-(1→2)-α-L-arabinopyranoside	Flower buds	Chen et al. (2000)
110	3-O-β-L-rhamnopyranosyl-(1→2)-α-L-arabinopyranosy hederagenin 28-O-β-D-glucopyranosyl-(1→6)-β-D-glucopyranosyl ester	Flower buds	Chen et al. (2000)
111	3-O-β-D-glucopyranosyl-(1→3)-α-L-rhamnopyranosyl-(1→2)-α-L-arabinopyranosyl hederagenin 28-O-β-D-glucopyranosyl-(1→6)-β-d-glucopyranosyl ester	Flower buds	Chen et al. (2000)
112	Loniceroside A	Whole plant	Son et al. (1994) Lee et al. (1995a)
113	Loniceroside B	Whole plant	Son et al. (1994) Lee et al. (1995a)
114	Loniceroside C	Aerial parts	Kwak et al. (2003)
115	Loniceroside D	Flower buds	Lin et al. (2008)
116	Loniceroside E	Flower buds	Lin et al. (2008)
117	Macranthoidin A	Flower buds	Ren et al. (2008)
118	Macranthoidin B	Flower buds	Ren et al. (2008)
119	Dipsacoside B	Flower buds	Ren et al. (2008)
120	Hederagenin-28-O-[β-D-glucopyranosyl-(1→6)-β-D-glucopyranosyl] ester	Flower buds	Ren et al. (2008)
121	Macranthoside B	Flower buds	Ren et al. (2008)
122	Macranthoside A	Flower buds	Ren et al. (2008)
123	3-O-[α-L-rhamnopyranosyl-(1→2)-α-L-arabinopyranosyl] hederagenin	Flower buds	Ren et al. (2008)
124	Saponin 1	Flower buds	Qi et al. (2009)
125	Saponin 4	Flower buds	Qi et al. (2009)

(continued)

No.	Compounds	Resource	References
126	Hederagenin 3-O-α-L-arabinopyranoside	Flowers	Choi et al. (2007)
127	Hederagenin	Whole plant	Lee et al. (1995a)
128	Oleanolic acid	Flower buds	Wang (2008c)
Others			
129	Lonijaposide A1	Flowers	Kumar et al. (2006)
130	Lonijaposide A2	Flowers	Kumar et al. (2006)
131	Lonijaposide A3	Flowers	Kumar et al. (2006)
132	Lonijaposide A4	Flowers	Kumar et al. (2006)
133	Lonijaposide B1	Flowers	Kumar et al. (2006)
134	Lonijaposide B2	Flowers	Kumar et al. (2006)
135	5-Hydroxymethyl-2-furfural	Flowers	Choi et al. (2007)
136	1-O-methyl-myo-inositol	Flower buds	Wang (2008c)
137	Nonacontane	Flower buds	Wang (2008c)
138	β-Sitosterol	Flower buds	Wang (2008c)
139	Sucrose	Flower buds	Wang (2008c)
140	Glucose	Flower buds	Wang (2008c)
141	Shuangkangsu	Flowers	Li (2008a)
142	(+)-N-(3-methylbutyryl-β-D-glucopyranoyl)-nicotinate	Flower buds	Song (2008)
143	(+)-N-(3-methylbut-2-enoyl-β-D-glucopyranoyl)-nicotinate	Flower buds	Song (2008)
144	5′-O-methyladenosine	Flower buds	Song (2008)
145	Guanosine	Flower buds	Song (2008)
146	Adenosine	Flower buds	Song (2008)
147	Syringin	Flower buds	Song (2008)

Table 3 The activities of some compounds from *Lonicera japonica*.

Compounds	Effects	In vivo	In vitro	Reference
Caffeic acid	Antioxidative activity		Showed marked antioxidant and scavenging activities with IC_{50} values of 5.72 μM for DPPH radicals, and 3.18 μM for $ONOO^-$	Choi et al. (2007)
Chlorogenic acid	Anti-tumor activity		With IC_{50} values of 55 μmol/L and corresponding cell ($HepG_2$ cell) viabilities was 62%±4%. And the cytotoxicities of chlorogenic acid were partially eliminated by the antioxidant effect of N-acetyl-l-cysteine (NAC)	Yip et al. (2006)
	Antibacterial activity		Compared to the Gram-positive bacteria, the chlorogenic acid to Gram-negative bacteria's bacteriostasis activeness was stronger; the minimum inhibitory concentration of chlorogenic acid to shigella and salmonella was 0.125 mg/mL, almost the same to 0.1 mg/mL kanamycin	Xu (2008)
			MIC was 0.025 g/mL, 0.025 g/mL, 0.1 g/mL and 0.8 g/mL against Escherichia coli, Sarcina luteus, Bacillus subtillis and Staphylococcusaureus	Wu (2005)
	Antioxidative activity		The DPPH scavenging activity was 74% at the dose of 0.1 g/mL	Wu (2005)
	Antiviral activity		At the doses of 0.05 mg/mL, 0.1 mg/mL, 0.4 mg/mL, 0.8 mg/mL and 0.8 mg/mL, it respectively inhibited respiratory syncytia virus, coxsackie B3 virus, adeno-associated 7 viruses, adeno-associated 3 viruses and Coxsackie B5 virus	Hu et al. (2001)
			The 0% toxic dose, minimum effective concentration and therapeutic index against to human cytomegalovirus were 100 g/mL, 1 g/mL and 100, respectively	Chen et al. (2009)
	Anti-inflammatory activity		It (5, 10, 15 mmol/L) decreased the expression of NF-kB P65 induced by LPS at 4 h ($P<0.05$), and the concentration of NO at 6 h. At the same time, it would increase the decrease of activity of GSH-Px induced by LPS at 6 h ($P<0.05$)	Huo et al. (2003)

63

(continued)

Compounds	Effects	In vivo	In vitro	Reference
	Hypoglycemic activity		1 mM inhibited about 40% of glucose-6-phosphatase activity ($P < 0.05$) in the microsomal fraction of hepatocytes. It promoted a significant reduction ($P < 0.05$) in the plasma glucose peak at 10 and 15 min during the oral glucose tolerance test, probably by attenuating intestinal glucose absorption. This suggested a possible role for it as a glycaemic index lowering agent and highlighting it as a compound of interest for reducing the risk of developing type 2 diabetes	Bassoli et al. (2008)
Dicaffeoylquinic acids	Antiviral activity		Results showed that 3, 5 – dicaffeoylquinic acid and two analogues were potent and selective inhibitors of HIV–1 IN in vitro. All of the dicaffeoylquinic acids were found to inhibit HIV–1 replication at concentrations ranging from 1 to 6μM in T cell lines, whereas their toxic concentrations in the same cell lines were > 120μM. In addition, it inhibited HIV-1 IN in vitro at submicromolar concentrations. So the dicaffeoylquinic acids as a class are potent and selective inhibitors of HIV-1 IN and form important lead compounds for HIV drug discovery	Robinson et al. (1996)
Hederagenin	Anti-inflammatory activity	100 mg/kg showed anti-inflammatory activity in the same model with 42% and 23% inhibition rates ($P < 0.001$)		Lee et al. (1995a)
Hyperoside	Antibacterial activity		It showed a excellent antibacterial effect on SA strains with a low MIC of 0.5–1 mg/mL, and the MIC of 2 mg/mL for strains of Escherichia coli, Pseudomonas aeruginosa, Klebsiella pneumoniae were observed	Tang (2008)

(continued)

Compounds	Effects	In vivo	In vitro	Reference
	The antibacterial activity with synergistic effect		The FIC index indicated that the addition effects were found in 55%, 30%, 25% and 15% of MRSA strains ($n = 20$) when hyperoside combined with oxacillin, benzylpenicillin, gatifloxacin and levofloxacin, respectively. Results suggested that hyperoside could enhance the anti-MRSA efficiency of these β-lactams or quinolones. It combined with chlorogenic acid showed an obvious cumulate bactericidal action on *Pseudomonas aeruginosa* ATCC27853 with a FIC index of 0.75	Tang (2008)
	Hepatoprotective	80 g/mL exhibited the best protective effects for hepatocytes injured by CCl_4, characterized as the levels of ALT, AST and MDA decreasing, elevation of GSH level and survival hepatocytes increasing with little damage in cell structure. The significant hepatoprotective effects for CCl_4-attacked rats were found in three dosages of hyperoside (10 mg/kg, 20 mg/kg and 30 mg/kg) with the obvious improve biochemical indexes and liver histopathology examination. The results of animals received hyperoside of 30 mg/kg were almost similar to that of normal controls		Tang (2008)
	Anti-tumor activity		5 g/mL, 10 g/mL, 20 g/mL, 25 g/mL have inhibitory effect on Hep-2 cells when it was used as photosensitizer in PDT or as radiosensitizer in radiotherapy. And this indicate that hypericin may be a hopeful agent to treat laryngeal carcinoma	Sun (2002)

Compounds	Effects	In vivo	In vitro	Reference
Isorhamnetin 3-O-β-D-glucopyranoside	Antioxidative activity		Showed marked antioxidant and scavenging activities with IC$_{50}$ values of 11.76 μM for DPPH radicals, and 3.34 μM for ONOO$^-$	Choi et al. (2007)
Loganin				
Loniceroside	Anti-inflammatory activity	100 mg/kg presented anti-inflammatory activity against mouse ear edema induced by croton-oil and arachidonic acid with 20% and 19% inhibition rates		Lee et al. (1995a)
Loniceroside A	Anti-inflammatory activity	100 mg/kg showed anti-inflammatory activity against mouse ear edema induced by croton-oil and arachidonic acid with 34% and 23% inhibition rates. Although far less potent than prednisolone (52% and 36%), it was comparable to aspirin at the dose of 100 mg/kg. 100 mg/kg/day also could reduce adjuvant-induced arthritis in rats ($P < 0.05$). The reference compound showed potent activity at a dose of 20 mg/kg/day ($P < 0.01$). At the same time, it (100 mg/kg, p.o.) against mouse ear edema provoked by croton oil with 30.2% inhibition rate		Lee et al. (1995a) Kwak et al. (2003)

(continued)

(continued)

Compounds	Effects	In vivo	In vitro	Reference
Loniceroside C	Anti-inflammatory activity	At the doses of 50, 100, 200 mg/kg (p.o.), it showed anti-inflammatory activity against mouse ear edema provoked by croton oil with 15%, 31% and 28.7% inhibition rates, respectively		Kwak et al. (2003)
Lonicerin	Anti-inflammatory activity	It presented anti-inflammatory activity against mouse ear edema induced by croton-oil with 39% ($P < 0.001$)		Lee et al. (1995a)
Luteolin	Antioxidative activity Anti-inflammatory activity		It effectively inhibited the lipopolysaccharide (LPS)-induced tumor necrosis factor-, interleukin-6 and inducible nitric oxide production in vitro, protect against LPS-induced lethal toxicity by inhibiting pro-inflammatory molecule expression in vivo and reducing leukocyte infiltration in tissues	Park et al. (2005), Xagorari et al. (2001), Kotanidou et al. (2002)
	Anti-tumor activity		The MTT assay showed that was 20% viability when HepG2 hepatocellular carcinoma cells were incubated at 100 mol/L. IC_{50} values of 40 μmol/L and corresponding cell viabilities of 53%±5%. The cytotoxicities were partially eliminated by the antioxidant effect of N-acetyl-l-cysteine	Yip et al. (2006)
	Anti-5-lipoxygenase activity		It presented the 5-lipoxygenase inhibitory activities with 97% inhibition at 20 μM. Nordihydroguaiaretic acid was used as a reference compound with 100% inhibition at 20 μM	Lee et al. (2010)
Luteolin 7-O-β-D-glucopyranoside	Antioxidative activity		Showed marked antioxidant and scavenging activities with IC_{50} values of 9.97 μM for DPPH radicals, and 3.18 μM for $ONOO^-$	Choi et al. (2007)

(continued)

Compounds	Effects	In vivo	In vitro	Reference
Ochnaflavon	Anti-inflammatory activity		It inhibited cyclooxygenase-2 (COX-2) dependent phases of prostaglandin D2 (PGD2) generation in bone marrow – derived mastcells with IC_{50} values of 0.6 μM. Western blotting showed that the decrease in quantity of the PGD2 product was accompanied by a decrease in the COX-2 protein level. And this compound could consistently inhibit the production of leukotriene C4, with an IC_{50} value of 6.56 μM. So ochnaflavone has a dual cyclooxygenase – 2/5 – lipoxygenase inhibitory activity. It also strongly inhibited degranulation reaction, with an IC_{50} value of 3.01 μM	Son et al. (2006)
			At 10 μM, ochnaflavone showed the suppressive activity against lymphocyte proliferation induced by Con A or LPS	Lee et al. (1995b)
Protocatechuic acid	Antioxidative activity		Showed marked antioxidant and scavenging activities with IC_{50} values of 7.21 μM for DPPH radicals, and 1.47 μM for $ONOO^-$	Choi et al. (2007)
	Anti-tumor activity		It was capable of stimulating the c–Jun N-terminal kinase (JNK) and p38 subgroups of the mitogen – activated protein kinase (MAPK) family. It induced cell death was rescued by specific inhibitors for JNK and p38, with IC_{50} values of 60 mol/L	Yip et al. (2006)
Quercetin 3-O-β-D-glucopyranoside	Antioxidative activity		Showed marked antioxidant and scavenging activities with IC_{50} values of 4.60 μM for DPPH radicals, and 1.76 μM for $ONOO^-$	Choi et al. (2007)

(continued)

Compounds	Effects	In vivo	In vitro	Reference
Rutin	Anti-apoptotic activity	Improved I/R – induced contractile myocardial function and reduced infarct size (32.0% ± 6.0%). Rutin administration also inhibited apoptosis in myocardial tissues in I/R rats by increasing Bcl-2/bax ratio and decreasing active caspase-3 expression. These results suggest that rutin reduced oxidative stress – mediated myocardial damage in vitro and in vivo model, which might be useful in treatment of myocardial infarction	Rutin decreased expression of both cleaved from of caspase-3 ($P <$ 0.01, at 20 μM) and increased Bcl-2/Bax ratio in H9c2 cells. The protective effect of rutin was inhibited by PI3K inhibitor or ERK inhibitor. It increased phosphorylation of ERK and Akt in H9c2 cells. These anti-apoptotic effects of rutin were confirmed both by annexin-V and TUNEL assay	Jeong et al. (2009)
Shuangkangsu	Antiviral activity		It inhibited markedly influenza B virus and influenza A3 virus ($P <$ 0.5). $IC_{50} < 0.31$ mg/embroy, therapeutic index (TI) > 32. And also could inhibit respiratory syncytial virus (RSV) ($P < 0.005$), $IC_{50} = 0.9$ mg/mL, TI = 6.2	Li (2008a)

tylphthalate (10.26%, 7.54%, 7.67%) and carvacrol (7.92%, 10.09% and 6.67%). Then, the highest levels of chlorogenic acid were found at the second white and complete white flowering stages. The results indicated that the best time to harvest *Lonicera japonica* flowers for essential oils was the silver flowering stage, and for chlorogenic acid was the second white or complete white flowering stage (Wang et al., 2009).

3.1.3 The different medicinal parts

Wu et al. (2009) has analyzed the components of essential oils in different parts of *Lonicera japonica*. 85 compounds were identified with GC - MS method, and only seven of which were mutual in the buds, leaves and stems, the main compounds were benzaidehyde, hexadecanoic acid, diethyl phthalate and hexadecanoyl, respectively. At the same time, the marked difference was showed in the main compositions of the buds, leaves and stems. The content of alkanes in stems was the highest, and in leaves was the lowest; the content of aldehydes in leaves was the highest, and in buds was the lowest; the content of the acids in buds was the highest, and in stems was not found. This difference of the chemical compositions implied the leaves and stems could not be used as the succedaneum of the buds in TCM.

In 2008, essential oils from the buds, silver flower and golden flower of *Lonicera japonica* have been studied with GC - MS method. 39, 48 and 39 compounds were identified respectively and only 10 of which were mutual. At the process of buds changing to golden flowers, the contents of alkanes, alcohols and ketones increased gradually, but the contents of acids and esters reduced gradually (Wang et al., 2008a, b). At the same time, essential oils from the flowers and stems of *Lonicera japonica* have been studied. 36 constituents are isolated and identified in all, of which 28 from flowers and 26 from stems. 18 compounds are found simultaneously in both crude drugs and they account for 85.23%, 83.42%, respectively. And palmatic acid and linoleic acid are the highest principles (Li et al., 2002a). So, different medical parts should be selected for different diseases.

3.1.4 The different extraction methods

With the wide action and utilization of the essential oil from *Lonicera japonica*, different extraction methods have been employed to extract it. In 2009, Du et al. (2009) indicated that with the fresh flowers homogenate method, the main compounds were propylbenzene (12.40%), ethyl benzene (8.58%), benzoic aldehyde (8.04%), translinalool (4.72%) and isophytol (2.94%), but with the steam distillation method, the main components were cyclohexano (8.06%), cyclohexylisooxalic ester (3.45%), methylcyclohexane (12.35%) and n-hexadecanoic acid (12.56%). These suggested that different extraction methods would result in different contents and compositions of essential oils.

3.1.5 The different processing

In the fresh flowers, the content of linalool was more than 14%, and other oils compositions were low-boiling unsaturated terpenes. But in the dry flowers, the content of hexadecanoic acid was more than 26%, and linalool less than 0.4%. Obviously, the fragrant composites have been lost by heating and lighting in the drying process (Ji et al., 1990).

In a word, different habitat, harvest time, medicinal parts, extraction methods, and drying process would result in different chemical compositions and the different quality of *Lonicera japonica* flowers.From above studies, it can be suggested that the middle of China was the best habitat; the complete white and silver flower period were the preferable harvest time; the best medicinal part was flower, and leaver and stem could be used as supplement for some particular object; low temperature and no-lighting was in favor of the essential oil in the dry and extract processes.

3.2 Organic acids

Organic acids is another important compositions of *Lonicera japonica*, and it mainly contains chlorogenic acid (1), isochlorogenic acid (2), caffeic acid (3), hexadecanoic acid (4), etc. (Zhang et al., 2000).In 1996, hexadecanoic acid (4) and myristic acid (5) have been isolated from the chloroform extracts of *Lonicera japonica* (Huang et al., 1996), and tetraacetyl-phthalein chlorogenic acid was obtained from the aqueous extracts of *Lonicera japonica* (Lou et al., 1996).At the same time, six isomers of isochlorogenic acids have been identified, including 3,5-O-dicaffeoylquinic acid (6), 4,5-O-dicaffeoylquinic acid (7), 3,4-O-dicaffeoylquinic acid (8), 1,3-O-dicaffeoylquinic acid (9), 3-ferulicoylquinic (10) and 4-ferulicoylquinic (11) (Iwanhashi and Negoroy, 1986).

As a major bioactive component of the flowers, chlorogenic acid (1) has been received much attention.Studies showed the chlorogenic acid have stronger bacteriostasis activity to Gramnegative bacteria than Gram-positive bacteria.The minimum inhibitory concentration (MIC) of chlorogenic acid (1) to shigella and salmonella was 0.125 mg/mL, almost the same to 0.1 mg/mL kanamycin (Xu, 2008).And the MIC values against *Escherichia coli*, *Sarcina luteus*, *Bacillus subtillis* and *Staphylococcus aureus* were 0.025 g/mL, 0.025 g/mL, 0.1 g/mL and 0.8 g/mL (Wu, 2005). Meanwhile, at the doses of 0.05 mg/mL, 0.1 mg/mL, 0.4 mg/mL, 0.8 mg/mL and 0.8 mg/mL, chlorogenic acid (1) presented the significant antiviral activity to respiratory syncytia virus, coxsackie B3 virus, adeno-associated 7 viruses, adeno-associated 3 viruses and Coxsackie B5 virus respectively (Hu et al., 2001).The 0% toxic dose, minimum effective concentration and therapeutic index against to human cytomegalovirus were 100 g/mL, 1 g/mL and 100, respectively (Chen et al., 2009).

3.3 Flavones

Up to now, about 30 flavones have been isolated from *Lonicera japonica*.Gao and Mu (1995) isolated quercetin-3-O-β-D-glucoside (41), luteolin-7-O-α-D-glucoside (42), luteolin-7-O-β-D-galactoside (43), and hyperoside (44) from *n*-butanol extracts.Corymbosin (57) and 5-hydroxy-3',4',7'-trimethoxylflavone (58) were isolated from the chloroform extracts (Huang et al., 1996).In 2005, Kumar et al.isolated two new biflavones, 3'-O-methyl loniflavone [5,5″,7,7″-tetrahydroxy 3'-methoxy 4',4″-biflavonyl ether] (57) and loniflavone [5,5″,7, 7″,3'-pentahydroxy 4,4″-biflavonyl ether] (58), from the leaves of *Lonicera japonica* in India (Kumar et al., 2005).At the same time, colorimetric method was used to compare the contents of flavones from the different habitat.The results showed that the contents were more than 4.31% from *Lonicera japonica* in Rizhao (Shandong). And the contents were 1.83%, 2.02%, 2.47% and 0.65% in Pinyi (Shandong), Fenqiu (Henan), Xinmi (Henan) and Nanjin (Jiangsu), respectively.Xing et al. (2002) used the ultraviolet spectrophotometry to determine the contents of flavones in different medical parts.The results indicated that the contents in leaves were the highest, and another is flower, stems, and this distribution feature of flavones also was similar to hexadecanoic acid (4).So the leaves and flowers of *Lonicera japonica* also would be used in TCM to treat various diseases.

Due to the difficulty in the separation and purification of flavones, the pharmacology activities of these compounds had not been studied systematically, except hyperoside and the crude extract. Tang (2008) found that hyperoside (44) combined with β-lactams or quinolones could enhance the antibacterial effects on some common pathogenic bacteria.At the sub-MIC (<0.5 mg/mL), hyperoside (44) could enhance the antibacterial effects of hydrophilic quinolones on bacteria SA26592 (pUTnorA).It also could relieve the cell injury induced by CCl_4 in hepatocyte L-02 with the decrease of ALT, AST and MDA, and increase of GSH and cell survival rate.It also showed a significant hepatoprotective effect in CCl_4-attatcked rats.The biochemical indexes and liver histopathology examination of rats treated with hyperoside of 30 mg/kg were almost similar to that of normal animals.

3.4 Iridoids

In the last decades, more than 30 iridoids have been found from *Lonicera japonica* and HPLC with evaporative light-scattering detector or multi-spectrum detection could be used to analyze theses compounds.In 2008, 9 iridoids, loganin (63), sweroside (64), secoxyloganin (67), secologanin (72), kingiside (74), ketologanin (78), 7α-morroniside (79), 7β-morroniside (80) and secologanoside (81), with 12 pyridinium inner salt alkaloids lonijaposides A-L (82-93) were isolated from the flower buds.And in an *in vitro* assay, lonijaposide C (84) showed inhibitory activity against the release of glucuronidase in rat polymorphonuclear leukocytes (PMNs) induced by the platelet-activating factor (PAF) with an inhibition rate of 69.5% at 10 μM, while compounds lonijaposides A (82) and B (83) give inhibitory activities with 11.0% and 35.8% inhibition rates at the same concentration, respectively.This suggested that the ethanol-2-yl unit at N-1 and the acid form of C-11 may increase the activity (Song, 2008; Song et al., 2008). Then, four new iridoid glycosides, named L-phenylalaninosecologanin (94), 7-O-(4-β-D-glucopyranosyloxy-3-methoxy-benzoyl) secologanolic acid (95), 6'-O-(7α-hydroxyswerosyloxy) loganin (96) and (Z) -aldsecologanin (97), also were isolated, together with a known one, newly named (E)-aldsecologanin (98), from the stems and leaves (Machida et al., 2002).From the dry flower buds of *Lonicera japonica*, 10 known iridoids and loniceracetalides A, B (99, 100) (secoiridoid glycosides) have been identified (Kakuda et al., 2000).

Using pharmacophore-assisted docking, Ehrman et al. (2010) screened for compounds which may be active against four targets involving in inflammation.The results showed that iridoids, as the active composition, presented the marked anti-inflammation against COX, p38 and JNK.At the same time, the pharmacology researches suggested that iridoids also have good anti-tumor, antiinflammation, antioxidant activity and hepatoprotective effects (Shang, 2010).

3.5 Saponins

Most of saponins from *Lonicera japonica* belong to the oleanane type and hederagenin type.In

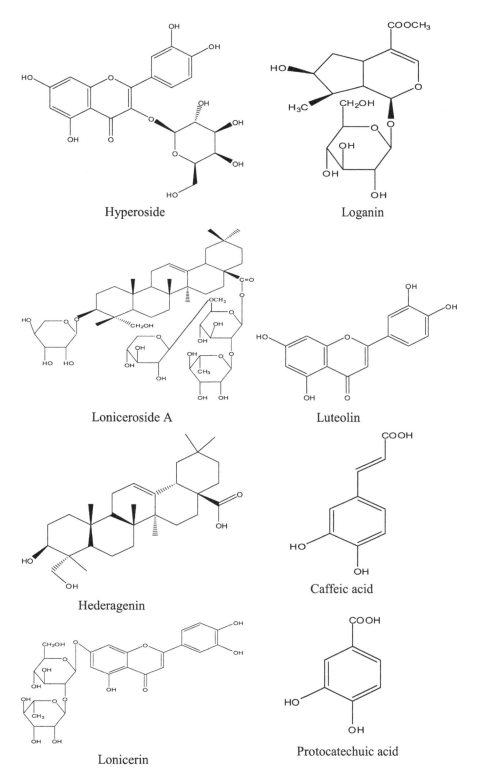

Fig. 2 The chemical structure of main compounds from *Lonicera japonica*

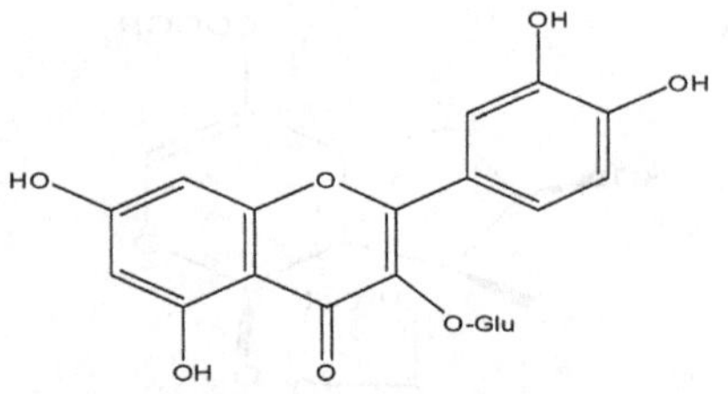

Quercetin 3-O-β-D-glucopyranoside

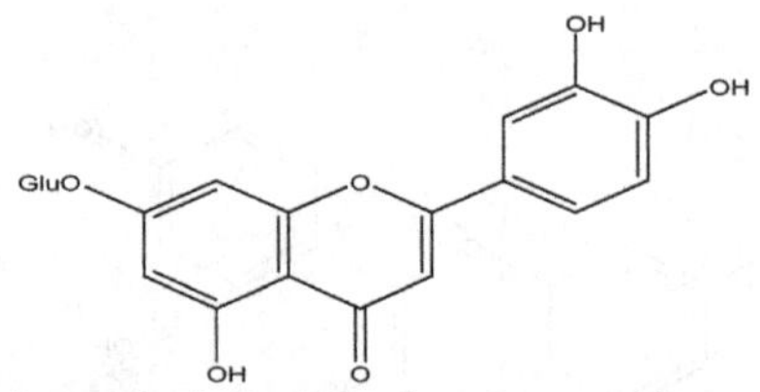

Luteolin 7-O-β-D-glucopyranoside

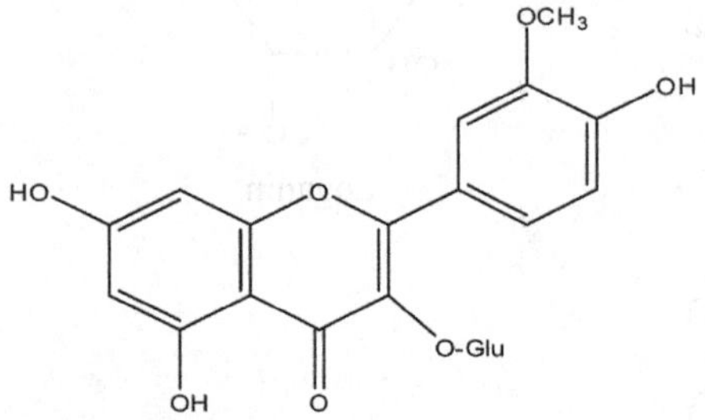

Isorhamnetin 3-O-β-D-glucopyranoside

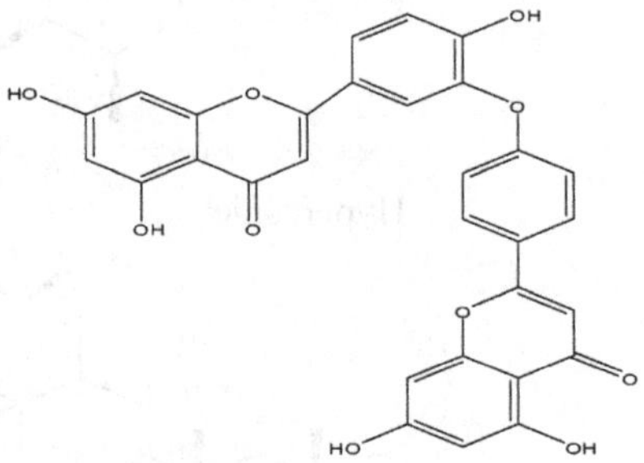

Ochnaflavon

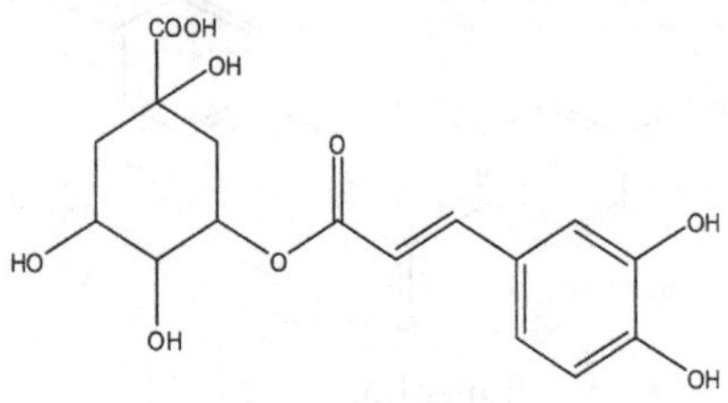

Chlorogenic acid

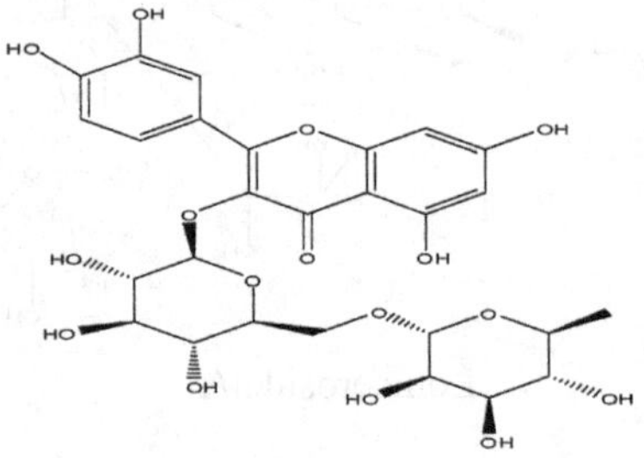

Rutin

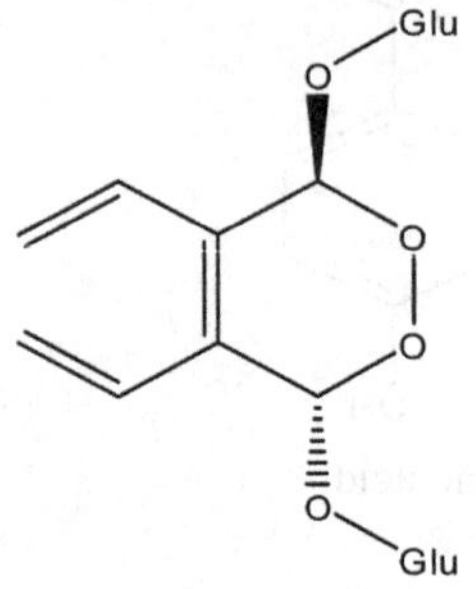

Shuangkangsu

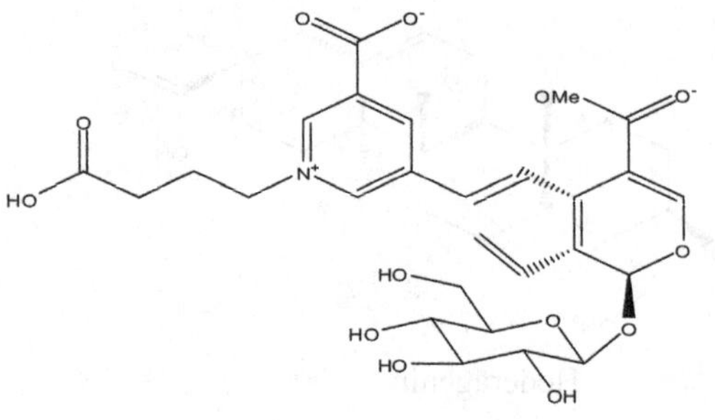

Lonijaposide A

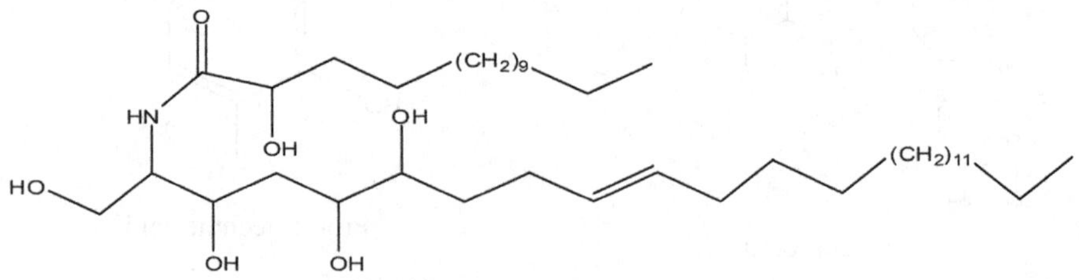

Lonijaposide A1

Fig. 2 (*Continued*)

1988, Kawai et al. (1988) firstly studied the saponins of the aerial parts of *Lonicera japonica*. 15 chemical compounds were found, and four compound were named, which were 3-O-α-l-arabinopyranosyl-28-O- [β-D-glucopyranosyl (1 → 6) -β-D-glucopyranosyl] oleanolic acid (101), 3-O- [α-l-rhamnopyranosyl (1 → 2) -α-L-arabinopyranosyl]-28-O-β-D-glucopyranosyl hederagenin (102), 3-O- [α-L-rahmnopyranosyl(1 → 2) -α-L-arabinopyranosyl] - 28-O- [β-D-glucopyranosyl(1 → 6) -β-D-glucopyranosyl] oleanolic acid (103), 3-O- [α-L-rhamnopyranosyl (1 → 2) -α-L-arabinopyranosyl] - 28-O- [6-acetyl-β-D-glucopyranosyl (1 → 6) -β-D-glucopyranosyl] hederagenin (104). At the same time, the pharmacology test proved that monodesmosides showed strong hemolytic activity, but bisdesmosides showed weak hemolytic activity. In 1996, Lou et al. (1996) has isolated three triterpenes compounds from the flower buds of *Lonicera japonica*. 3-O-α-L-rhamnopyranosyl- (1 → 2) -α-L-arabinopyranosy hederagenin 28-O-β-D-xylpyranosyl (1 → 6) -β-D-glucopyranosyl ester (105), 3-O-α-L-arabinopyranosy hederagenin 28-O-α-D-rhamnopyranosyl (1 → 2) [β-D-xylpyranosyl (1 → 6) -β-D-glucopyranosyl ester (106), 3-O-α-L-rhamnopyranosyl- (1 → 2) -α-L-arabinopyranosy hederagenin 28-O-α-D-rhamnopyranosyl (1 → 2) [β-D-xylpyranosyl (1 → 6) -β-D-glucopyranosyl ester (107). Then in 2000, 3-O-β-D-glucopyranosyl- (1 → 4) -β-L-glucopyranosyl (1 → 3) -α-L-rhamnopyranosyl (1 → 2) -α-L-arabinopyranosy hederagenin 28-O-β-D-glucopyranosyl (1 → 6) -β-D-glucopyranosyl ester (108), hederagenin-3-O-α-L-rhamnopyranosyl (1 → 2) -α-L-arabinopyranoside (109), 3-O-β-D-glucopyranosyl- (1→ 3) -α-L-rhamnopyranosyl (1 → 2) -α-L-arabinopyranosyl hederagenin 28-O-β-D-glucopyranosyl- (1 → 6) -β-D-glucopyranosyl ester (110), 3-O-α-L- rhamnopyranosyl - (1 → 2) - α - L - arabinopyranosy hederagenin 28 - O - β - D - glucopyranosyl (1 → 6) -β-D-glucopyranosyl ester (107) have been identified (Chen et al., 2000).

Then in 2003, a new triterpenoid saponin, loniceroside C (114) was isolated from the aerial parts, which showed *in vivo* antiinflammatory activity at the doses of 50, 100, 200 mg/kg (p.o.) against mouse ear edema provoked by croton oil with 15%, 31% and 28.7% inhibition rates, respectively. The positive drug aspirin (100 mg/kg, p.o.) was 15% inhibition (Kwak et al., 2003). And in 2008, two new triterpeniod saponins, loniceroside D (115) and loniceroside E (116), were isolated from the dry flowers and buds of *Lonicera japonica* (Lin et al., 2008).

3.6 Other compounds

From *Lonicera japonica*, 15 trace elements were found, such as Fe, Mn, Cu, Zn, Ti, Sr, Mo, Ba, Ni, Cr, Pb, V, Co, Li, Ca (Wu, 1988). In 2006, Kumar et al. identified six novel chemical compounds, lonijaposides A1-A4 (129-132), and lonijaposides B1-B2 (129-130) (Kumar et al., 2006). And the name of lonijaposide A1 was similar with lonijaposide A, but the chemical structures of them were different (Fig. 2). In 2008, Li isolated a new compound—shuangkangsu (141) with the marked anti-viral activity against influenza B virus ($P<0.5$), influenza A3 virus ($P<0.5$), and respiratory syncytial virus (RSV) ($P<0.005$), respectively (Li, 2008a). (+) -N- (3-methybutyryl-β-D-glucopyranoyl) -nicotinate (142), (+) -N- (3-methybut-2-enoyl-β-D-glucopyranosyl) -nicotinate (143) and three nucleosides also have been found from flower buds of *Lonicera japonica* (Song, 2008).

4 EFFECTS OF CRUDE EXTRACT

4.1 Anti-inflammatory activity

Recently, more and more experiments, *in vivo* and *in vitro*, showed different extracts of *Lonicera japonica* can inhibit various inflammatory reactions, and suppress various inflammatory factors.

In 1998, Lee et al. (1998) evaluated the anti-inflammatory activity of the n-butanol (BuOH, 4.2% based on the dry weight) fraction of *Lonicera japonica*. At oral doses of 400 mg/kg, it showed significant anti-inflammatory activities against AA ear edema, croton-oil ear edema, CGN-paw edema, rat cotton pellet granulomatic and AIA-inflammation models in mice and rats, the inhibitions were 27%, 23%, 26%, 18% and 42%, respectively. And the inhibition rate of the positive drug, aspirin (100 mg/kg) were 27%, 13%, 13%, 0% and 58%.

The anti-inflammatory properties of aqueous extracts from *Lonicera japonica* flower were evaluated in A549 cells. This extract directly inhibited both COX-1 and COX-2 activity, the expression of IL-1β-induced COX-2 protein and mRNA. But higher dose of the extract was required to suppress the expression of mRNA than protein. This result indicated that the extract acted translationally or post-translationally at lower doses and transcriptionally or post-transcriptionally at higher doses (Xu et al., 2007). Kang et al. examined the effect of water fraction of *Lonicera japonica* on trypsin-induced mast cell (HMC-1) activation. After stimulated with trypsin (100 μM), *Lonicera japonica* could inhibit TNF-α secretion, tryptase mRNA expression and trypsin-induced ERK phosphorylation at a dose-dependent manner. However, it did not affect the trypsin activity even with 1 000 μg/mL. These all indicated that *Lonicera japonica* might inhibit trypsin-induced mast cell activation through the inhibition of ERK phosphorylation than the inhibition of trypsin activity (Kang et al., 2004). On the other hand, the anti-inflammatory effects of water extract in proteinaseactivated receptor 2 (PAR2) -mediated mouse paw edema has been investigated. The results indicated that at doses of 50, 100 and 200 mg/kg o.p., it showed significant inhibition of both change in paw thickness and vascular permeability induced by PAR2, the inhibition rates were 41.8%, 69.1%, 70.9%, and 40.2%, 69.7%, 68.8%, respectively. The water extracts (100 mg/kg) also significantly inhibited PAR2 agonists-induced myeloperoxidase (MPO) activity and TNF-α expression in paw tissue. (Tae et al., 2003)

Su et al. (2006) used the supercritical CO_2 extraction process, the temperature 35 .C, the pressure 12 MPa, the CO_2 flowing 4.0 kg/h and time 1.5 h, to obtain 1.08% volatile oil from *Flos lonicerae*. The pharmacological studies suggested the potent anti-inflammatory effect of the volatile oil on the ear swelling model in mice.

All of these reports supported the traditional use of *Lonicera japonica*, and suggested it be a safe and mild anti-inflammatory agent for treating various inflammatory disorders.

4.2 Antiviral activity

Since 1980s, the antiviral activity of *Lonicera japonica* has been studied and proved, such as anti-RSV, anti-HIV, anti-HSV, anti-PRV and anti-NDV. At the same time, as important traditional medicines, *Lonicera japonica* had been used to treat some viral diseases in China.These pharmacology activities were supported to the traditional uses and drug's nature of *Lonicera japonica* in TCM.

Firstly, Ma et al. (2002) selected forty-four medicinal herbs, which are used for the treatment of respiratory tract infectious diseases in China, and tested the antiviral activities against respiratory syncytial virus (RSV) by means of the cytopathologic effect (CPE) assay.*Lonicera japonica* showed potent antiviral activities against RSV, the 50% inhibition concentration (IC_{50}) was 50. 0 g/mL, and selectivity index (SI) was more than 20. 0.Chang et al. (2003) applied a syncytia formation inhibition assay to study the anti-HIV agents of 80 MeOH extracts of Korean plants. This test based on the interaction between the HIV-1 envelope glycoprotein gp120/41 and the cellular membrane protein CD_4 of T lymphocytes.Flower of *Lonicera japonica* showed an inhibition of 5. 8± 1. 7% at a concentration 100 μg/mL.And in 2008, Xi (2008) suggested that *Lonicera japonica* extraction showed an obvious therapeutic action on the influenza A virus infected Pneumonia mouse. The lung indexes of *Lonicera japonica* group and the ribovirin group were lower than the model group with the significance difference ($P<0.01$), but no significance difference ($P>0.05$) between these two groups. *Lonicera japonica* could reduce the histopathological changes, the viral duplication, and the contents of influenza virus nucleic acid ($P<0.01$), compared to the model group).At the same time, the TNF-α, IL-1β expressions of *Lonicera japonica* group and the ribovirin group are lower than the model group with significant difference ($P<0.01$).

Then in 2009, Chen et al. (2009) proved that *Lonicera japonica* extracts and chlorogenic acid had the significant anticytomegalovirus activity, and the 0% toxic dose, minimum effective concentration (MEC) and therapeutic index (TI) of these two composites for human cytomegalovirus were 3000 μg/mL, 3 000 μg/mL, 1, and 100 μg/mL, 1 μg/mL, 100, respectively.*In vitro* tests, *Lonicera japonica* showed 104 and 72 times of TI for anti-HSV (Herpes simplex virus) -1F and anti-HSV-1HS-1 to acyclovir (ACV).But *in vivo* tests, the anti-viral activity of *Lonicera japonica* was closed to ACV (Wang, 1999).As to the caviid beta herpesvirus 1, *Lonicera japonica* showed significant inhibition of the duplication of guinea pig cytomegalovirus in cell level.TI and the inhibitory duplication index (100 and 2. 61) (Wang et al., 2005).The anti-virus (H9N2) activity and anti-AIV (LED = 3. 90 mg/mL, *in vitro*) of the flavones from flower buds of *Lonicera japonica* were also found (Li et al., 2001; Wang et al., 2006).At the same time, as the main composition, *Lonicera japonica* was used widely in TCM prescriptions, which were published by State Administration of TCM of China, to prevent and control SARS coronavirus in 2003 (http://www.satcm.gov.cn).And from the statistics of Beijing Youan hospital of capital medical university, the practical frequency of *Lonicera japonica* to treat influenza A virus subtype H1N1 was second in 113 species between May and November, 2009 (Zheng, 2010).

In Vero cells, three different extracts of *Lonicera japonica* (Hunan province), including volatile oil (P1), chlorogenic acids extracts (P2) and flavones extract (P3), were tested for the antiviral activity to the Pseudo rabies virus (PRV) and Newcastle disease virus (NDV).At the dose of 232. 7 μg/mL, 116. 35 μg/mL, 58. 18 μg/mL, 29. 09 μg/mL, the interdiction rates of P1 for PRV and NDV were 40. 13%, 17. 83%, 13. 16%, 2. 24%, and 75. 40%, 32. 01%, 12. 05%,

2.34%, in CPE respectively, and the LED of P1 (Least Effective Dose) were 232.7 g/mL and 232.7 g/mL. The interdiction rates of P2 (3.125 mg/mL, 1.563 mg/kg, 0.781 mg/mL and 0.391 mg/mL) for PRV and NDV were 63.74%, 46.27%, 13.10%, 3.51%, and 65.23%, 36.71%, 32.61%, 28.96% in CPE respectively. The interdiction rates of P3 (1.954 mg/mL, 0.977 mg/kg, 0.489 mg/mL and 0.244 mg/mL) for PRV and NDV were 94.00%, 78.42%, 42.30%, 3.36%, and 78.07%, 27.63%, 16.37%, 6.73%, respectively. LED against PRV and NVD of P2 and P3 were 0.997 mg/mL, 3.097 mg/mL ($P<0.05$), and 0.781 mg/mL, 1.563 mg/mL. These studies suggested that the extracts decrease CPE lesions and neutralize virus in dosages dependent, behave in inhibiting virus directly and promoting cell antivirus (Wang, 2008c).

As another kind of antiviral agents, several tannins of *Lonicera japonica* were investigated. 3,5-di-O-caffeoylquinic acid (6) and methyl 3,5-di-O-caffeoylquinic acid (30) had a strong inhibitory effect on HIV-1 RT and HDNAP-α. The ratio of IC_{50} of these two compounds for HIV-1 RT and HDNAP- was 2.0 and 2.2. While 3,4-di-O-caffeoylquinic acid (8) and methyl 3,4-di-O-caffeoylquinic acid (31) exhibited higher inhibitory effects on HDNAP- than HIV-1 RT (Chang et al., 1995). Meanwhile other 13 caffeoylquinic acids, caffeic acid (3) and caffeic acid methyl ester (32) isolated from *Lonicera japonica* also presented antiviral activities against respiratory viruses (Ma et al., 2005).

4.3 Antibacterial activity

Antibacterial activity, as another important property of *Lonicera japonica*, has been comprehensively studied. In 2009, Rahman et al. evaluated the antibacterial potential of essential oil from flowers and ethanol extracts from leaf. A remarkable antibacterial effect of the oil and extracts has been revealed against *Listederia monocytogenes ATCC 19116*, *Bacillus subtilis ATCC 6633*, *Bacillus cereus SCK 111*, *Staphylococcus aureus* (*ATCC 6538* and *KCTC 1916*), and *Salmonella enteritidis KCTC 12021*, *Salmanella typhimurium KCTC 2515*, *Enterobacter aerogenes KCTC 2190* and *Escherichia coli ATCC 8739*. The diameters of inhibitions zone were 20.3, 17.8, 15.2, 16.3, 14.1, 15.3, 14.0, 12.4, 12.1, and 16.2, 15.4, 14.0, 15.0, 14.1, 14.1, 14.2, 12.2, 10.3 mm, respectively. MIC values were 62.5, 62.5, 250, 125, 250, 125, 250, 500, 500, and 125, 125, 250, 125, 125, 250, 250, 500, 500 μg/mL. These findings suggested that the oil and extracts from *Lonicera japonica* may be a potential source of preservatives for the food or pharmaceutical industries (Rahman and Kang, 2009). And the antibacterial activities, against *Bacillus cereus* and *Staphylococcus aureus*, of the floral bud from *Lonicera japonica* were found by agar-well diffusion method *in vitro*. The diameters of inhibition zone were 6.3 and 7.2 mm, and this activity may be closely associated with the existence of phenolic constituents (Shan et al., 2007). Meanwhile *Lonicera japonica* also has showed the marked antibacterial activity against fourteen strains, including *Staphylococcus aureus*, *Streptococcus hemolyticus*, *Escherichia coli*, *Bacillus dysenteriae*, *Bacillus comma*, *Bacillus typhosus*, *Bacillus paratyphosus*, *Pseudomonas aeruginosa*, *Klebsiella pneumoniae*, *Bacillus tuberculosis*, *Streptococcus mutans*, *Bacillus adhaerens*, *Bacteroides melanogenicus and Haemophilus actinomycetemcomitans*, and so on. And the antibacterial study against different serotypes of Streptococcus mutans demonstrated that the extracts of *Lonicera japonica* could inhibit 87.5% strains with MIC 25 mg/mL (Sun, 2002; Song et al., 2003).

Then, the antimicrobial activity of water and alcohol extract from *Lonicera japonica* was investigated. MIC and MBC of water extract on *Staphyloccocu saureus* were 19.25% and 38.50%; MIC and MBC of alcohol extract on *Salmonella and Staphyloccocu aureus* were 9.80%, 19.60%, and 19.60%, 39.20%, respectively (Kang et al., 2010). Meanwhile the water decoctum from *Lonicera japonica* showed the reasonable eliminating effect on R plasmid from *Pseudomonas aeruginosa in vivo*. The elimination rate was 8% (Wang et al., 2000).

Finally, Tang et al. indicated that flavonoids from *Lonicera japonica* also have a strong antibacterial action, especially for Methicillin Resistant *Staphylococcus aureus* (MIC≤5 mg/mL) (Tang, 2008).

4.4 Antioxidant activity

Chio et al. has evaluated the antioxidant effects of *Lonicera japonica* flowers in 2007. The EtOAc fraction exhibited marked scavenging/inhibitory activities, as follows: IC_{50} values of 4.37, 27.58± 0.71, 0.47±0.05, and 12.13±0.79 μg/mL in the 1, 1-diphenyl-2-picrylhydrazyl (DPPH) radical, total reactive oxygen species (ROS), hydroxyl radical (^-OH), and peroxynitrite ($ONOO^-$) assays, respectively. And as the main compounds of the EtOAc fraction, luteolin, caffeic acid, protocatechuic acid, isorhamnetin-3-O-β-D-glucopyranoside, quercetin 3-O-β-D-glucopyranoside, and luteolin 7-O-β-D-glucopyranoside, also evidenced marked scavenging activities, with IC_{50} values of 2.08–11.76 μM for DPPH radicals, and 1.47–6.98 μM for ONOO. (Choi et al., 2007). And the results of other antioxidant tests indicated that the Trolox equivalent antioxidant capacity (TEAC) values and total phenolic content for methanolic extracts of the flora bud from *Lonicera japonica* were 589.1 mol Trolox equivalent/100 g dry weight (DW), and 3.63 gallic acid equivalent/100 g DW. These studies suggested that *Lonicera japonica* might be potential natural antioxidants and beneficial chemopreventive agent (Cai et al., 2004).

Then, antioxidant activity of polysaccharides with different molecule weights separated from *Flos lonicerae* by ultra-filtration was also studied. The reducing power of the polysaccharides has a direct correlation between antioxidant activity and concentration of certain plant extracts, and ultra-filtration fraction has a significant inhibit effect on superoxide radicals generated in a PMS/NADA/NBT system. Administered to rats, crude polysaccharides extracts (50–400 mg/kg) could reduce lipid peroxidate malony dialdehyde (MDA) content, improve glutathione peroxidase (GSH-Px) and catalase (CAT) activity, and enhance significantly superoxide dismutase (SOD) activity in serum and tissue (Li, 2008b).

Lan et al. (2007) used the biochemical method to investigate the activity and scavenging capability on free radicals of the membrane-protecting enzyme, which was extracted from flower bud, leaf and wattle of *Lonicera japonica* in different harvest time. The activity of the enzyme was increased since April and reached the highest level in June. The activity of the enzyme extracted from leaf was higher than those from flower bud and wattle, and the scavenging capacity of the extract from the flower bud on OH and H_2O_2 free radicals was stronger than those from leaf and wattle. Meanwhile the scavenging capacities on free radicals of five medicinal plants of *Lonicerae* from several provinces of China have been investigated. All the samples showed the scavenging capacities on three kinds of free radicals (O_2^-, ^-OH, H_2O_2) at some extent, and the samples from Henan and Shandong showed stronger scavenging capacities than those from Jiangsu province. The descending order of scavenging capacities from different species is *Lonicera macranthoides*, *Lonicera japonica*, *Lonicera similis*, *Lonicera fulvotomentosa* and *Lonicera hypoglauca* (Li et al., 2002b).

4.5 Hepatoprotective

In the dimethylnitrosamine (35 mg/kg o.p.7 days) induced liver fibrosis rats, 75% ethanol extract of Lonicera japonica showed significantly hepatoprotective effect by pathological analysis. 19 compounds, such as 8-phenyl-8-azbicyclo [4, 3, 0] non-3-ene-7, 9-dione, 3- [1,3] dioxolan-2-yl-4-methoxy-6-nitro-benzo [d] isoxazole, (E) -2-(5,5,5-Yip richloro-3-penten-1-yl) -1,3-dioxolane, 3,3-diphenylcyclopropene, 2- (methoxy-imino) -hexanedioic acid, gammalactone-2-methoximinegluconic acid, 2-(6-heptynyl) -1,3-dioxolane, and bis (o-methyloxime) -4-ketoglucose, have been isolated and identified by GC - MS method from 75% ethanol extract (Sun et al., 2010). But, Hu et al. suggested that the total flavones of *Lonicera japonica* have significantly protective effect on immunological liver injury in mice. Administered ig* 10d, the total flavones (100, 200, 400 mg/kg) could significantly decrease the raised liver and spleen indexes, and improve the aggravated liver histopathologic changes. In addition, this extract could reduce the high levels of NO and iNOS in liver homogenate and inhibit the expression of TNF-α in liver tissue. This mechanism possibly was due to diminishing the inflammation mediators (Hu et al., 2008).

4.6 Anti-tumor activity

The mechanisms of apoptosis induced by photodynamic therapy (PDT) in lung CH27 carcinoma cells, cultured with alcohol extract from *Lonicera japonica* as photosensitizer, have been explored. This extract exhibited significant photocytotoxicity in CH27 cells at a concentration range of 50-150 μg/mL, with 0.4-1.2 J/cm^2 light dose. The apoptosis induced by PDT combined with *Lonicera japonica* extract was accompanied by DNA condensation, externalization of phosphatidylserine and formation of apoptotic bodies, it showed to be caspase-3-independent apoptosis via activation of AIF. P38-associated pathway may be involved in apoptosis induced by PDT with *Lonicera japonica* in CH27 cells. These demonstrated that they induced CH27 cells apoptosis was probably related to its ability to change the protein expression and distribution of heat shock protein 27 (Leung et al., 2008). When the aqueous extract of *Lonicera japonica* (100 μg/mL) was applied to the HepG2 cells, the lysates of the treated cells were associated with increased stimulatory phosphorylation of JNK and p38 as compared to the basal value, similarly to the MAPK activation profile of PCA. Further experiments showed that this aqueous extract also decreased the viability of HepG2 cells to 50%. So Yip thought that the aqueous extract of *Lonicera japonica* could trigger HepG2 cell death in a JNK-dependent manner (Yip et al., 2006).

4.7 Insecticidal and acaricidal activities

95% methanol extracts from leaves and twigs of *Lonicera japonica* were tested at 10 000 ppm for evaluating their insecticidal and acaricidal activities against *Tetranychus urticae* Koch, *Aphis gossypii* Glover, *Myzus persicae* Sulzer, *Trialeurodes vaporariorum* (Westwood), and *Panonychus citri*

(McGregor). 24 and 48 h after application, the mortality (%) of adult *Tetranychus urticae* were 31.6 and 37.0, and the control group was 3.3 and 3.3. After 24, 48 and 72 h application, the number of laid eggs against the reproduction and repellent indexes were 2.2, 2.7, 4.0 and 91.1, 80.0, 55.6 gh, and the control group was 6.1, 6.4, and 11.7. At the same time, survival rates of *Panonychus citri* after 5 and 10 days application of the extracts were 41.7 and 31.3 (Kim et al., 2005).

Then, the repellent activities against *Aedes albopitus* of methanol extracts from different parts of *Lonicera japonica* have been studied on mouse skin. The results showed that the mortalities of n-hexane, ethyl acetate, n-butanol and water fractions of methanol extracts were 97%, 80%, 97%, 97% (stem), 97%, 80%, 0, 0 (leaf) and 77%, 90%, 50%, 80% (flower) (Yoon and Kyung, 2002). These results suggested that *Lonicera japonica* had the marked insecticidal and acaricidal activities.

4.8 Anti-pregnant activity

The good anti-pregnant effect of *Lonicera japonica* extract has been found in mice, dogs and monkeys. The extract has marked inhibitory effect on the deciduoma of pseudopregnant in mice, with a significant decrease on the level of plasma progesterone in pregnant rats after administration for 24 h. These results implied that the interruption effect of extract was likely concerned with the decrease on the level of plasma progesterone and/or prostaglandinlike action (Cao et al., 1986a, b).

4.9 Antihyperlipidemic and antithrombotic activities

In 1998, Pan et al. has found that *Lonicera japonica* could inhibit the increase of blood sugar and the content of high density lipoprotein cholesterin (HDL-C) in blood, reduce the level of serum cholesterol and accumulation index atherosclerosis in mice-induced alloxan (ALX) (Pan et al., 1998). Meanwhile Fan et al. observed the antithrombotic effects of the organic acid compounds in *Lonicera japonica* on the oxidative injury of HUVEC (Human Umbilical Vein Endothelial Cells). The IC_{50} of isomeric compounds of chlorogenic acids (2), caffeic acid, and isochlorogenic acid (3) were 0.0286 mg/mL, 1.707 mg/mL, 2.411 mg/mL, 0.026 mg/mL, 0.328 mg/mL, and 0.539 mg/mL respectively. The protect effect on the oxidative injury of HUVEC by H_2O_2 indicated a dosage depended manner. Caffeic acid and isochlorogenic acid have evident resistance to the oxidative injury, and chlorogenic acid has the preventive protect effect (Fan et al., 2007).

4.10 Anti-lipase activity

By screening for 75 medicinal plants on new pancreatic lipase (triacylglycerol lipase, EC 3.1.1.3), Sharma et al. suggested that 80% methanolic extract of whole plant from *Lonicera japonica* (0.2 mg/mL) present the marked anti-lipase activity, and the inhibition was 40.9% (Sharma et al., 2005).

5 APPLICATION ON VETERINARY AND AGRICULTURE

5.1 Application on veterinary

As a new medical feed additive, the effect of two Chinese medicinal herbs (*Astragalus membranaceus* and *Lonicera japonica*) and boron on non-specific immune response of Nile tilapia (*Ore-

ochromis niloticus) was investigated. After four weeks of feeding (contained 0.1% *Astragalus*, 0.1% *Lonicera* and a mixture of the two herbs with 0.05% boron), fish were infected with *Aeromonas hydrophila* and mortalities were recorded. Feeding tilapia with two herbs alone or in combination significantly enhanced phagocytic and respiratory burst activity of blood phagocytic cells, reduced the mortality following *Aeromonas hydrophila* infection. The lowest mortality was observed in the group fed with the combination of both herbs and boron. So both extracts and boron supplementation added to fish feed can act as immunostimulants and enhance the immune response and disease resistance of cultured fish (Ardóet al., 2008).

The fish (mean body weight of ca.110 g) were fed diets with a 0.1% supplement of *Astragalus radix*, *Lonicera japonica* or a mixture of these herbs for 8 weeks. Statistically significant intergroup differences were noted in the value of the hepatosomatic index, hepatocyte size, nucleus and nucleus/cytoplasm diameter ratio, and the appearance of the hepatic parenchyma and the protein content of the whole fish body. The analysis of the proximal composition of the fish viscera indicated significant differences in the fat content ($P<0.05$). Among the analyzed group of fatty acids (saturated - SFA, monoenoic - MUFA, polyenoic - PUFA) contained in the whole fish, the fillets and the viscera, significant intergroup differences were noted with regard to SFA (viscera) and MUFA (whole fish) ($P<0.05$). The total PUFA content was stable, although significant intergroup differences were noted with regard to a few of the acids that belong to this group ($P<0.05$) (Zdzislaw et al., 2008).

At the same time, the volatile oil extract and flavones extract promoted proliferation of chicken splenic lymphocytes significantly by MTT method ($P<0.05$), which were induced by ConA or LPS, and the effects were related to concentration of the extracts (Wang, 2008c).

5.2 Cadmium-hyperaccumulator

Phytoremediation using hyperaccumulators is a promising technique of removing soil pollutants. The growth responses, cadmium (Cd) accumulation capability and physiological mechanisms of *Lonicera japonica* under Cd stress were investigated. Exposed to 5 and 10 mg/L Cd, the plants did not show any visual symptoms. Furthermore, the height, dry biomass of leaves, roots and total and the chlorophyll content were obtained different grade increase. When the concentration of Cd was up to 50 mg/L, the height, dry biomass of leaves and roots had not significant differences compared with the control. The indexes of tolerance were all above 0.8. The maintenance of high superoxide dismutase and catalase activities was observed along with the increased Cd concentration, suggesting strong internal detoxification mechanisms inside plant cells. After 21 days exposure to 25 mg/L Cd, stem and shoot Cd concentrations reached (344.49±0.71) μg/g and (286.12±9.38) μg/g DW, respectively and the plant had higher bioaccumulation coefficient and translocation factor. According to these results, Liu et al. suggested that *Lonicera japonica* has strong tolerance and accumulation capability to Cd, and it is a potential Cd-hyperaccumulator (Liu et al., 2009).

6 PREPARATIONS, THE QUALITATIVE AND QUANTITATIVE ANALYSIS

In 2010, more than 12 preparations, in which *Lonicera japonica* was the main and active compositions, were listed in Chinese Pharmacopoeia and used to clear away the heat-evil and expel superficial evils, and cure fever, cough and pharyngalgia and the swell of throat, constipation, conjunctival congestion, etc., such as *Shuang Huang Lian Ke Li*, *Xiao Yin Pian*, *Ying Huang Kou Fu Ye* (Table 1).

In 2005, luteolin was added in Chinese Pharmacopoeia with chlorogenic acid to control the quality of medical materials by TLC and HPLC methods. The content of luteolin and chlorogenic acid in the *Flos Lonicera* should be more than 0.1% and 1.5%. An efficient microwave - assisted extraction (MAE) technique has been developed to extract chlorogenic acid from flower buds of *Lonicera japonica*. The yield of chlorogenic acid rapidly reached 6.14% within 5 min under the optimal MAE conditions, i.e. 50% ethanol as extraction solvent, 1 : 10 (w/v) of the solid/liquid ratio and 60℃ of extraction temperature (Zhang et al., 2008). Today, with the development of modern separation and identify techniques, it is widely accepted that the quality cannot be measured by mono-content. As the main and effective components of *Lonicera japonica*, essential oils and flavones were paid attention to qualitative and quantitative analysis *Lonicera japonica*. In 2008, Wang et al. (2008d) used normal HPLC.MS to separate and analyze the essential oils. 192 compounds were identified, and this technology provided the better method to analyze the chemical compounds of *Lonicera japonica* and should be beneficial for the study of the contents and active compounds.

Otherwise, a fast liquid chromatography method with diodearray detection (DAD) and time-of-flight mass spectrometry (TOF-MS) has been developed for analysis of constituents in *Flos Lonicera*. By accurate mass measurements within 4 ppm error for each molecular ion and subsequent fragment ions, as well as the 'full mass spectral' information of TOF-MS, a total of 41 compounds including 13 iridoid glycosides, 11 phenolic acids, 7 saponins, and 10 flavonoids were identified in a methanolic extract (Qi et al., 2009). In order to differentiate the sources and comprehensively control the quality of this medicinal plant, Chen et al. (2007) thought coupled with principal component analysis would be a well acceptable strategy. At the same time, capillary electrophoresis with electrochemical detection (CE-ED) also was employed to analyze active ingredients of *Lonicera japonica*. Operating in a walljet configuration, a 300 mm diameter carbon-disk electrode was used as the working electrode, which exhibits a good response at +0.90 V (vs. saturated calomel electrode) for four analytes. Under the optimum conditions, the analytes were baseline separated within 20 min in a 50 mmol/L borax buffer (pH 8.7). Notably, excellent linearity was obtained over two orders of magnitude with detection limits (S/N = 3) ranging from 0.1 to 0.5 mg/L for all the analytes (Peng et al., 2005).

Chemical fingerprint analysis has been introduced and accepted by WHO (1991), SFDAC (2000) and other authorities as a strategy for quality assessment of herbal medicines. It has been recognized as a rapid and reliable means for the identification and qualification of herbal medicines. Li et al. (2006) has employed the binary chromatographic profiling in fingerprint analysis of *Lonicera japonica*, Correlation analysis showed that six chromatographic peaks in ethanol extract were positively correlated with *in vitro* bacteriostasis activity. Two standard fingerprints were developed with 10 genuine samples of *Lonicera japonica*. Similarity analysis with a limited number of samples showed a fair consistence in the chromatographic profiling of Lonicera japonica from various sources and two harvests, and significant differences from other species.Combination use of the two fingerprints demonstrated confirmative identification and quality assessment of *Lonicera japonica*.

7 ACUTE AND SUBACUTE TOXICITY

Thanabhorn et al. (2006) has evaluated the acute and subacute toxicity of the ethanol extract from the leaves of *Lonicera japonica*.The single oral dose of the ethanol extract at 5000 mg/kg did not produce mortality or significant changes in the general behavior and gross appearance of the internal organs of rats.In subacute toxicity study, the ethanol extract was administered orally at a dose of 1000 mg/kg/day for a period of 14 days.The satellite group was treated with the ethanol extract at the same dose and the same period and kept for another 14 days after treatment.There were no significant differences in the body and organ weights between the control and the treated group of both sexes.Hematological analysis and clinical blood chemistry revealed no toxicity effects of the extract. Pathologically, neither gross abnormalities nor histopathological changes were observed. Then, Chang et al. (2003) assessed the toxicological assessment on food safety of the flower bud of *Lonicera japonica*.Acute toxicity test showed $LD_{50}>15$ g/kg Bw.on mice orally.And micronucleus test of bone marrow cell and *Salmanella typhimurium*/mammals microsomal enzyme test showed it was safe without mutagenesis.Meanwhile no toxicity test on mice and in antifertility test on SD female mouse. In summary, the extract was found to be fairly nontoxic when oral acute and subacute toxicities in rats were performed.Chronic toxicity study is needed for further support the safe use of this plant.

At the same time, Zhang (2007) used MTT method to study the toxicity of *Lonicera japonica* against THP-1, MT-2 cell *in vitro*.The result showed that compare to control group, 50 mg/mL could marked induce the cell death ($P<0.05$).

8 CONCLUSION

As the above said, *Lonicera japonica* was used and planted in China and East Asia.In 2003, *Lonicera japonica* was used as the most popular TCM to treat SARS coronavirus in China.But in Argentina, Australia, Brazil, Mexico, New Zealand and much of the United States, *Lonicera japonica* was thought as a major nuisance, and restricted in parts of North America and New Zealand (Starr et al., 2003).So, there are more potential utilization and development of *Lonicera japonica* out of Asia, especially out of China.

Obviously, the chemical components and pharmacology activities of *Lonicera japonica* have been studied, and many active compounds were isolated and identified.Of these, due to the good antibacterial, anti-inflammatory and anti-tumor activities, chlorogenic acid and luteolin were used as the indicator compound to characterize the quality of *Lonicera japonica* and the preparations in Chinese Pharmacopoeia.But recently, chlorogenic acid and luteolin were found in other medicinal plants, and even the contents were higher than *Lonicera japonica* (Committee for the Pharmacopoeia of PR China, 2010).So employing chlorogenic acid and luteolin to control the quality of *Lonicera japonica* lacked specificity.At the same time, Wu suggested that the antibacterial of the flavones of fresh flower buds were 4 times of chlorogenic acid, and iridoids were also presented the marked biology activities (Wu et al., 2001).The different habitat, harvest's time, medicinal parts, extraction methods, fresh flowers and dry flowers would result in the different chemical compositions and the quality of *Lonicera japonica*. So, how to exclusively and accurately control the quality of *Lonicera japonica* in TCM and preparations should be studied further.

In Chinese Pharmacopoeia (Committee for the Pharmacopoeia of PR China, 2010), only the flowers and flower buds have been officially used as the active parts to treat diseases. But the results of phytochemical researches indicated that some active compositions and compounds also existed in the leaves, vines and stems, such as essential oils, flavones and iridoids. So the intrinsically active compositions and the mechanisms of action of *Lonicera japonica* were also ambiguous.

In Chinese clinical practice, the traditional medicines with clearing away the heat-evil and expelling superficial evils effects are usually used to treat various infectious diseases. And the pharmacological studies proved that these medicines also have the anti-inflammatory, antiviral, antibacterial commonly. From the development of *Lonicera japonica*, the relationship between TCM and modern pharmacy also has been embodied.

In a word, phytochemical and pharmacological studies of *Lonicera japonica* have received much interest, more and more extracts and active compounds have been isolated and proved which has the anti-inflammatory, antiviral, antibacterial, antioxidant and enhance the immune response effects, etc. But until now, poor quality control, and fail development of *Lonicera japonica* always existed in. Further research especially on *in vitro/in vivo* antibacterial and antiviral effects should have priority.

ACKNOWLEDGEMENTS

This work was financed by The Special Fund of Chinese Central Government for Basic Scientific Research Operations in Commonweal Research Institutes (no.1610322011009). The authors would also like to express their gratitude to Lanzhou University PhD English writing foreign teacher Allan Grey who thoroughly corrected the English in the paper.

References

[1] Ardó, L., Yin, G.J., Xu, P., Váradi, L., Szigeti, G., Jeney, Z., Jeney, G., Chinese herbs (*Astragalus membranaceus* and *Lonicera japonica*) and boron enhance the nonspecific immune response of Nile tilapia (*Oreochromis niloticus*) and resistance against *Aeromonas hydrophila*. Aquaculture, 2008, 275: 26-33.

[2] Bassoli, B.K., Cassolla, P., Borba-Murad, G.R., Constantin, J., Salgueiro-Pagadigorria, C.L., Bazotte, R.B., da Silva, R.S.d.S.F., de Souza, H.M., Chlorogenic acid reduces the plasma glucose peak in the oral glucose tolerance test: effects on hepatic glucose release and glycaemia. Cell Biochemistry and Function, 2008, 26: 320-328.

[3] Cai, Y.Z., Luo, Q., Sun, M., Corke, H., Antioxidant activity and phenolic compounds of 112 traditional Chinese medicinal plants associated with anticancer. Life Sciences, 2004, 74: 2 157-2 184.

[4] Cao, C.P., Huang, Z.N., Qian, P.L., Pan, X.X., Studies on the anti-pregnant effect of the *Lonicera japonica* thunb extract. Pharmaceutical Industry, 1986a, 17, 115-117 (in Chinese).

[5] Cao, C.P., Huang, Z.N., Qian, P.L., Yan, D.P., Shen, Y.P., Pan, X.X., Pharmacological study of pregnancy termination effect of *Lonicera japonica* thunb. Pharmaceutical Industry, 1986b, 17: 319-321 (in Chinese).

[6] Chang, C.W., Lin, M.T., Lee, S.S., Karin, C.S., Liu, C., Hsu, F.L., Lin, J.Y., Differential inhibition of reverse transcriptase and cellular DNA polymerase-a activities by lignans isolated from Chinese herbs, *Phyllanthus myrtifolius* Moon, and tannins from *Lonicera japonica* Thunb and *Castanopsis hystrix*.Antiviral Research, 1995,27: 367-374.

[7] Chang, J., Sheng, P.P., Zhang, X.M., The toxicological assessment of *Lonicera japonica* on food safety.The Chinese Academic Medical Magazine of Organisms , 2003,2:63-64 (in Chinese).

[8] Choi, C.W., Jung, H.A., Kang, S.S., Choi, J.S., Antioxidant constitutes and a new triterpenoid glycoside from *Flos lonicerae*.Archives of Pharmacal Research, 2007,30:1-7.

[9] Chen, C.X., Wang, W.W., Ni, W., Chen, N.Y., Zhou, Jun, Triterpenoid glycosides from the *Lonicera japonica*.Acta Botanica Yunnanic, 2000,22,201-208 (in Chinese).

[10] Chen, C.Y., Qi, L.W., Li, H.J., Li, P., Yi, L., Ma, H.L., Tang, D., Simultaneous determination of iridoids, phenolic acids, flavonoids, and saponins in *Flos lonicerae* and *Flos lonicerae japonicae* by HPLC-DAD-ELSD coupled with principal component analysis.Journal of Separation Science, 2007,30:318-3 192.

[11] Chen, J.J., Fang, J.G., Wan, J., Feng, L., Zhang, Y.W., Zhao, J.H., Zhao, J.J., Wang, N., Chen, S.H., An in vitro study of the anti-cytomegalovirus effect of chlorogenic acid.Herald of Medicine, 2009,28: 1 138-1 140.

[12] Chen, X.C., Utilization of *Lonicera japonca*.Forest By-Product and Speciality in China, 2008,92: 37-39 (in Chinese).

[13] Committee for the pharmacopoeia of PR China, Pharmacopoeia of PR China, Part I.Chemical Industrial Press, Guangdong Science and Technology Publishing House (in Chinese), 1995.

[14] Committee for the Pharmacopoeia of PR China, Pharmacopoeia of PR China, Part I.China Medical Science and Technology Press, PR China, 2010 (in Chinese).

[15] Du, H.F., Zhang, Y., Wen, D.Q., Yang, D.J., Identify the contents of fresh flower of *L.japonica* using GC-MS with the different extraction methods.TCM Research of Chongqing, 2009,60:13-15 (in Chinese).

[16] Ehrman, T.M., Barlow, D.J., Hylands, P.J., In silico search for multi-target anti-inflammatories in Chinese herbs and formulas.Bioorganic & Medicinal Chemistry, 2010,18: 2 204-2 218.

[17] Fan, H.W., Li, Y.B., Sun, M., Zhu, Q., Antithrombus effect of honeysuckle and its organic acid compounds in vitro.Pharmacology and Clinics of Chinese Materia Medica, 2007,23:34-36 (in Chinese).

[18] Feng, Y., Study on the quality of *Lonicerae japonica* flos from different habitats.Chinese Medicine Journal Research Practice, 2010,24:34-35 (in Chinese).

[19] Gao, Y.M., Mu, H.J., Studies on the chemical constitutes of *Japanese Honeysuckle* (*Lonicera Japonica*).Chinese Traditional and Herbal Drugs, 1995. 26: 568-656 (in Chinese).

[20] Gu, G.G., et al., Sheng Nong Ben Cao Jing.Harrbing Publishing House, Harrbing, 2007 (in Chinese).

[21] He, S.Q., Hu, Q.F., Yang, G.Y., Research of honeysuckle.Yunnan Chemical Technology , 2010, 37:72-75 (in Chinese).

[22] http://www.satcm.gov.cn.State Administration of Traditional Chinese Medicine of the People's Republic of China(in Chinese).

[23] http://www.zysj.com.cn (in Chinese).

[24] Hu, C.M., Jiang, H., Liu, H.F., Li, R., Li, J., Effects of LJTF on immunological liver injury in mice.Anhui Medical and Pharmaceutical Journal, 2008,12: 295-296 (in Chinese).

[25] Hu, K.J., Sun, K.X., Wang, J.L., Inhibited effect of chlorogenic acid on virus in vitro.Journal of Harrbing Medical University, 2001,35: 430-432 (in Chinese).

[26] Huang, L.Y., Lv, Z.Z., Li, J.B., Studies on the chemical constitutes of *Japanese Honeysuckle*.Chinese Traditional and Herbal Drugs, 1996,27: 645-647 (in Chinese).

[27] Huo, X.F., Tang, Y.P., Zhang, Q.L., Luo, S.H., Effect of chlorogenic acid on the macrophages induced by lipoplysaccharide.Acta Academiae Medicine Zunyi, 2003,26: 507-508 (in Chinese).

[28] Ikeda, N., Ishihara, M., Tsuneya, T., Kawakita, M., Yoshihara, M., Volatile components of honeysuckle (*Lonicera japonica* Thunb.) flowers.Flavour and Fragrance Journal, 1994,9: 325-331.

[29] Iwanhashi, H., Negoroy, I., Inhibation by chlorogenic acid of haematincatalysed retinoic acid 5, 6-epoxidation.Journal of Biochemistry, 1986,239:641-646.

[30] Jeong, J.J., Ha, Y.M., Jin, Y.C., Lee, E.J., Kim, J.S., Kim, H.J., Seo, H.G., Lee, J.H., Kang, S.S., Kim, Y.S., Chang, K.C., Rutin from *Lonicera japonica* inhibits myocardial ischemia/reperfusion-induced apoptosis in vivo and protects H9c2 cells against hydrogen peroxide-mediated injury via ERK1/2 and PI3K/Akt signals in vitro.Food and Chemical Toxicology, 2009,47: 1 569-1 576.

[31] Ji, L., Pan, J.G., Xu, Z.L., The GC-MS analysis of volatile oil from *Lonicera japonica* thunb.Chinese Journal of Chinese Materia Medica, 1990. 15: 680 (in Chinese).

[32] Jiao, S.G., 2009.Research and comprehensive utilization of *honeysuckle*. Qilu Pharmaceutical Affairs 28: 487-489 (in Chinese).

[33] Kakuda, R., Imai, M., Yaoita, Y., Machida, K., Kikuchi, M., Secoiridoid glycosides from the flower buds of *Lonicera japonica*.Phytochemistry, 2000,55, 879-881.

[34] Kang, O.H., Choi, Y.A., Park, H.J., Lee, J.Y., Kim, D.K., Choi, S.C., Kim, T.H., Nah, Y.H., Yun, K.J., Choi, S.J., Kim, Y.H., Bae, K.H., Lee, Y.M., Inhibition of trypsin-induced mast cell activation by water fraction of *Lonicera japonica*. Archives of Pharmacal Research, 2004,27: 1 141-1 146.

[35] Kang, X., Li, D.S., Deng, C., Yuan, J.L., Study on antimicrobial activity of extracts from *Lonicera japonica* Thunb.Journal of Anhui Agricultural Science, 2010, 38: 14 935- 14 936 (in Chinese).

[36] Kawai, H., Kuroyanngi, M., Umehara, K., Ueno, A., Satake, M., Studies on the saponins of *Lonicera japonica* Thunb.Chemical & Pharmaceutical Bulletin, 1988,36:4 769-4 775.

[37] Kim, D.I., Park, J.D., Kim, S.G., Kuk, H., Jang, M.S., Kim, S.S., Screening of some crude plant extracts for their acaricidal and insecticidal efficacies.Journal of Asia-Pacific Entomology, 2005, 8:93-100.

[38] Kotanidou, A., Xagorari, A., Bagli, E., Kitsanta, P., Fotsis, T., Papapetropoulos, A., Roussos, C., Luteolin reduces lipopolysaccharide-induced lethal toxicity and expression of proinflammatory molecules in mice. American Journal of Respiratory and Critical Care Medicine, 2002, 165: 818-823.

[39] Kumar, N., Singh, B., Bhandari, P., Gupta, A.P., Uniyal, S.K., Kaul, V.K., Biflavonoids from *Lonicera japonica*.Phytochemistry, 2005,66: 2 740-2 744.

[40] Kumar, N., Singh, B., Gupta, A.P., Kaul, V.K., Lonijaposides, novel cerebrosides from *Lonicera japonica*.Tetrahedron, 2006,62: 4 317-4 322.

[41] Kwak, W.J., Han, C.K., Chang, H.W., Loniceroside C, an antiinflammatory saponin from *Lonicera japonica*.Chemical & Pharmaceutical Bulletin, 2003,51:333-433.

[42] Lan, H., Giao, D., Shen, H., Chen, Z.D., Studies on the activity of antioxidant enzyme and capability of scavenge free radicals extracted from *Lonicera Japonica* Thunb.Journal of Chinese Institute of

Food Science and Technology, 2007,7: 27-29 (in Chinese).

[43] Lee, E.J., Kim, J.S., Kim, H.P., Lee, J.H., Sam Sik Kang phenolic constituents from the flower buds of *Lonicera japonica* and their 5-lipoxygenase inhibitory activities. Food Chemistry, 2010, 120: 134-139.

[44] Lee, S.J., Son, K.H., Chang, H.W., Kang, S.S., Kim, H.P., Antiinflammatory activity of *Lonicera japonica*. Phytotherapy Research, 1998,12: 445-447.

[45] Lee, S.J., Shin, E.J., Son, K.H., Chang, H.W., Kang, S.S., Kim, H.P., Antiinflammatory activity of the major constituents of *Lonicera japonica*. Archives of Pharmacal Research, 1995a, 18: 133-135.

[46] Lee, S.J., Choi, J.H., Son, K.H., Chang, H.W., Kang, S.S., Kim, H.P., Suppression of mouse lymphocyte proliferation in vitro by naturally-occurring biflavovoids. Life Sciences, 1995b, 57:558.

[47] Leung, H.W.C., Hour, M.J., Chang, W.T., Wud, Y.C., Lai, M.Y., Wang, M.Y., Lee, H.Z., P38-associated pathway involvement in apoptosis induced by photodynamic therapy with *Lonicera japonica* in human lung squamous carcinoma CH27 cells. Food and Chemical Toxicology, 2008, 46: 3 389-3 400.

[48] Li, D.Z., Study on the synthesis and bioactivities of analogs of shuangkangsu-an anti-virus principle of Jinyinhua. Chinese Academy of Medical Science & Peking Union Medical College. Doctoral thesis, 2008a (in Chinese).

[49] Li, E.C., Studies on the isolation, purification and bioactivity of polysaccharides from *Flos lonicerae*. Shaanxi Normal University. Master thesis, 2008b (in Chinese).

[50] Li, F.M., Yuan, B., Xiong, Z.L., Lu, X.M., Qin, F., Chen, H.S., Liu, Z.G., Fingerprint analysis of Flos *Lonicerae japonicae* using binary HPLC profiling. Biomedical Chromatography, 2006,20: 634-641.

[51] Li, H.J., Zhang, C.Y., Li, P., Comparative study on volatile oils in flower and stem of *Lonicera japonica*. Journal of Chinese Medicinal Materials, 2002a,25: 476-477 (in Chinese).

[52] Li, H.J., Li, P., Zhang, C.Y., Zhao, M.Q., Scavenging capacities on radicals of *Flos Lonicerae* by chemiluminescence. Journal of China Pharmaceutical University, 2002b, 33: 496-549 (in Chinese).

[53] Li, H.J., Li, P., Wang, M.C., Ye, W.C., A new secoiridoid glucoside from *Lonicera japonica*. Chinese Journal of Natural Medicines 2003, 9:132-133 (in Chinese).

[54] Li, S.Z., Ben Cao Gang Mu. People's Medical Publishing House, Beijing, 1979 (in Chinese).

[55] Li, Y.M., Li, L., Bo, C., Li, D., Wang, T.Z., Study on the anti-adenovir effects of *L.japonica*. West China Journal of Pharmaceutical Sciences, 2001,16: 327-329.

[56] Lin, K., Analysis of essential oil from *lonicera japonica* in Fujian. Acta Agriculturae Jiangxi 21: 102-104, 2009 (in Chinese).

[57] Lin, L.M., Zhang, X.G., Zhu, J.J., Gao, H.M., Wang, Z.M., Wang, W.H., Two new triterpenoid saponins from the flowers and buds of *Lonicera japonica*. Journal of Asian Natural Products Research, 2008,10: 925-929.

[58] Liu, Z.L., He, X.Y., Chen, W., Yuan, F.H., Yan, K., Tao, D.L., Accumulation and tolerance characteristics of cadmium in a potential hyperaccumulator-*Lonicera japonica* Thunb. Journal of Hazardous Materials, 2009,169: 170-175.

[59] Lou, H.X., Lang, W.J., Lu, M.J., Water soluble constituents from *Lonicera japonica*. Chinese Traditional and Herbal Drugs, 1996,27: 195 (in Chinese).

[60] Ma, S.C., Du, J., But, P.P.-H., Deng, X.L., Zhang, Y.W., Ooi, V.E.-C., Xu, H.X., Lee, S.H-S., Lee, S.F., Antiviral Chinese medicinal herbs against respiratory syncytial virus.Journal of Ethnopharmacology, 2002,79: 205-211.

[61] Ma, S.C., Paul But, P.H., Vincent Ooi, E.C., Spencer Lee, H.S., Lee, S.F., Determination of the antiviral caffeoyl quinic acids isolated from *L.japonica* thunb.Chinese Journal Pharmaceutical Analysis, 2005,25: 751-753 (in Chinese).

[62] Machida, K., Asano, J., Kikuchi, M., Caeruleosides A and B, bis – iridoid glucosides from *Lonicera caerulea*.Phytochemistry, 1995,39: 111.

[63] Machida, K., Sasaki, H., Iijima, T., Kikuchi, M., Studies on the constituents of *Lonicera* species XVII.New iridoid glycosides of the stems and leaves of Lonicera japonica Thunb.Chemical & Pharmaceutical Bulletin, 2002,50: 1 041-1 044.

[64] Myung, W.B., Cheorun, J., Tae, W.J., Cheul, H.H., Effects of gamma irradiation on color characteristics and biological activities of extracts of *Lonicera japonica* (Japanese honeysuckle) with methanol andacetone.Lebensmittel-Wissenschaft & Technologie, 2004,37: 29-33.

[65] Pan, J.K., Liu, H.C., Liu, G.G., Chen, L.X., Qiu, Z.Q., *L.japonica* could reduce the level of blood sugar and blood fat of mice.Guangzhou Medicine, 1998,29: 59-61 (in Chinese).

[66] Park, E., Kum, S., Wang, C., Park, S.Y., Kim, B.S., Schuller-Levis, G., Antiinflammatory activity of herbal medicines: inhibition of nitric oxide production and tumor necrosis factor-alpha secretion in an activated macrophage-like cell line.The American Journal of Chinese Medicine, 2005,33: 415-424.

[67] Peng, L.Y., Mei, S.X., Jiang, B., Zhou, H., Sun, H.D., Constituents from *Lonicera japonica*. Fitoterapia, 2000,71: 713-715.

[68] Peng, Y.Y., Liu, F.H., Ye, J.N., Determination of phenolic acids and flavones in *Lonicera japonica* Thumb.By capillary electrophoresis with electrochemical detection. Electroanalysis, 2005, 17: 356-359.

[69] Qi, L.W., Chen, C.Y., Li, P., Structural characterization and identification of iridoid glycosides, saponins, phenolic acids and flavonoids in *Flos Lonicerae* Japonicae by a fast liquid chromatography method with diode-array detection and time-of-flight mass spectrometry.Rapid Communications in Mass Spectrometry , 2009,23: 3 227-3 242.

[70] Rahman, A., Kang, S.C., In vitro control of food-borne and food spoilage bacteria by essential oil and ethanol extracts of *Lonicera japonica* Thunb.Food Chemistry , 2009,116: 670-675.

[71] Ren, M.T., Song, J.C.Y., Sheng, L.S., Li, P., Qi, L.W., Identification and quantification of 32 bioactive compounds in *Lonicera* species by high performance liquid chromatography coupled with time-of-flight mass spectrometry.Journal of Pharmaceutical and Biomedical Analysis, 2008, 48: 1 351-1 360.

[72] Robinson, W.E., Cordeiro, J.M., Abdel-Malek, S., Jia, Q., Chow, S.A., Reinecke, M.G., Mitchell, W. M., Dicaffeoylquinic acid inhibitors of human immunodeficiency virus integrase: inhibition of the core catalytic domain of human immunodeficiency virus integrase.Molecular Pharmacology, 1996,50: 846-855.

[73] Shan, B., Cai, Y.Z., Brooks, J.D., Corke, H., The in vitro antibacterial activity of dietary spice and medicinal herb extracts.International Journal of Food Microbiology , 2007,117: 112-119.

[74] Shang, X.F., Antinociceptive and anti-inflammatory activities of iridoid glycosides extract of *Lamiophlomis rotata* (Benth.) Kudo.Lanzhou University.Master thesis (in Chinese), 2010.

[75] Shanghai Institute of Pharmaceutical Industry, Study on the active compositions of *Lonicera japonica*

Thunb.Chinese Journal of Pharmaceuticals, 1975,11: 24 (in Chinese).

[76] Sharma, N., Sharma, V.K., Seo, S.Y., Screening of some medicinal plants for anti-lipase activity. Journal of Ethnopharmacology, 2005,97: 453-456.

[77] Shu, Y.B., Zhang, J.R., Ma, X.W., Zhang, J.S., Supercritical extraction of volatile oil in *honeysuckle* and its application in cigarettes flavoring.Anhui Agricultural Science Bulletin, 2008,14: 76-78 (in Chinese).

[78] Son, K.H., Park, J.O., Chung, K.C., Chang, H.W., Kim, H.P., Kim, J.S., Kang, S.S., Flavonoids from the aerial parts of *Lonicera japonica*. Archives of Pharmacal Research, 1992, 15: 365-370.

[79] Son, K.H., Jung, K.Y., Chang, H.W., Kim, H.P., Kang, S.S., Triterpenoid saponinsfrom the aerial parts of *Lonicera japonica*, 1994,35: 1 005-1 008.

[80] Son, M.J., Moon, T.C., Lee, E.K., Son, K.H., Kim, H.P., Kang, S.S., Son, J.K., Leet, S.H., Chang, H.W., Naturally occurring biflavonoid, ochanflavone, inhibits cyclooxygenases-2 and 5-lipoxygenase in mouse bone marrowderived mast cells.Archives of Pharmacal Research, 2006,29: 282-286.

[81] Song, H.Y., Qiu, S.C., Wang, Z.Q., Zhang, Q., Yang, X., Research on the in vitro growth inhibition effect of *Lonicera japonica* Thunb. (LJT) on bacteri.Lishizhen Medicine and Materia Research, 2003,14: 269 (in Chinese).

[82] Song, W.X., Study on the Water-soluble Chemical Constituents of the Flower Buds of *Lonicera japonica*.Chinese Academy of Medical Science & Peking Union Medical College, 2008 (in Chinese).

[83] Song, W.X., Li, S., Wang, S.J.W.Y., Zi, J.C., Gan, M.L., Zhang, Y.L., Liu, M.T., Lin, S., Yang, Y.C., Shi, J.G., Pyridinium alkaloid-coupled secoiridoids from the flower buds of *Lonicera japonica*.Journal of Natural Product, 2008,71: 922-925.

[84] Song, Y., Li, S.L., Wu, M.H., Li, H.J., Li, P., Qualitative and quantitative analysis of iridoid glycosides in the flower buds of Lonicera species by capillary high performance liquid chromatography coupled with mass spectrometric detector.Analytica Chimica Acta, 2006,564: 211-218.

[85] State Food Drug Administration of China, Technical Requirements for the Development of Fingerprints of TCM Injections.SFDA, Beijing,2000 (in Chinese).

[86] Starr, F., Starr, K., Loope, L., *Lonicera japonica*.Caprifoliaceae.United States Geological Survey-Biological Resources Division,2003.

[87] Su, X.P., Song, B.W., Chen, Z.H., Niu, Y.P., Study on the volatile oil of *Flos lonicerae* by supercritical CO_2 extraction and its anti-inflammatory activity.Natural Product Research and Development, 2006,18: 663-666 (in Chinese).

[88] Sun, C.H., Teng, Y., Li, G.Z., Yoshioka, S., Yokota, J., Miyamura, M., Fang, H.Z., Zhang, Y., Metabonomics study of the protective effects of *Lonicera japonica* extract on acute liver injury in dimethylnitrosamine treated rats.Journal of Pharmaceutical and Biomedical Analysis, 2010,53: 98-102.

[89] Sun, D.M., Study on activity of antioxidation and effect of restraining bacterium of extract components in *Lonicera japonica* Thunb's leaf.Henan Science, 2002,20: 511-513 (in Chinese).

[90] Tae, J., Han, S.W., Yoo, J.Y., Kim, J.A., Kang, O.H., Baek, O.S., Lim, J.P., Kim, D.K., Kim, Y.H., Bae, K.H., Lee, Y.M., Anti-inflammatory effect of *Lonicera japonica* in proteinase-activated receptor 2-mediated paw edema.Clinica Chimica Acta, 2003,330: 165-171.

[91] Tang, M., Study of isolation and biological effects of active flavonoid components extracted from *Lonicera Japonica*.The Third Military Medical University (China).Doctoral thesis,2008 (in Chinese).

[92] Tao, H.J., Ming Yi Bie Lu.People's Medical Publishing House,Beijing (in Chinese).

[93] Thanabhorn, S., Jaijoy, K., Thamaree, S., Ingkaninan, K., Panthong, A., 2006.Acute and subacute toxicity study of the ethanol extract from *Lonicera japonica* Thunb.Journal of Ethnopharmacology, 1986,107: 370–373.

[94] Wagner, W.L., Herbst, D.R., Sohmer, S.H., Colocasia.In: Manual of the Flowering Plants of Hawaii.University of Hawaii Press, Honolulu, Hawaii, 1999:1 356–1 357.

[95] Wang, L.J., The study progress of *Lonicera japonica*.Medical Information, 2010,8: 2 293–2 296 (in Chinese).

[96] Wang, J.M., Ji, Z.Q., Kang, W.Y., The volatile oil from different parts of *Lonicera japonica* Thunb. Fine Chemicals, 2008a,25: 1 075–1 077 (in Chinese).

[97] Wang, L.M., Li, M.T., Yan, Y.Y., Ao, M.Z., Wu, G., Yu, L.J., Influence of flowering stage of *Lonicera japonica* Thunb.on variation in volatiles and chlorogenic acid.Journal of the Science of Food and Agriculture, 2009,89: 953–957.

[98] Wang, L.Q., Studies on antiviral effect and immunopotentiating activity of *Lonicera japonica* Thunb.and flos lonicerae in vitro.Agricultural University of Henan.Mater thesis,2008c (in Chinese).

[99] Wang, X.R., Chen, S.H., Qiao, F.Y., Experimental study of *honeysuckle* flower against guinea pig cytomegalovirus in vitro.Journal on Maternal and Children Health Care of China, 2005.17: 2 241–2 243 (in Chinese).

[100] Wang, X.Y., Jia, W., Zhao, A.H., Wang, X.R., Anti–influenza agents from plants and traditional Chinese medicine.Phytotherapy Research, 2006,20: 335–341.

[101] Wang, Y., Yu, J., Xiao, Y., Cheng, L., Guan, X.Z., Studies on the elimination of resistance plasmid from *P.aeruginosa* in vitro and vivo with *Lonicera japonica*.Journal of Norman Bethune University of Medical Science, 2000,26: 139–141 (in Chinese).

[102] Wang, Y.P., Xue, X.Y., Zhang, F.F., Xu, Q., Separation and analysis of volatile oil from *Lonicera japonica* thunb.Modernization of Traditional Chinese Medicine and Materia Medica, 2008b,10: 45–55 (in Chinese).

[103] Wang, Y.P., Xue, X.Y., Zhang, F.F., Liang, X.M., Separation and analysis of volatile oil from *Lonicera japonica* thunb using normal phase liquid chromatography–gas chromatography hyphenated with mass spectrometry.World Science and Technology/Modernization of Traditional Chinese Medicine and Materia Medica, 2008d.10: 45–48 (in Chinese).

[104] Wang, Z.J., Study on the anti-HSV-1F and anti-HSV-1HS-1 activities of *L.japonica*.Journal of Shandong University of TCM, 1999,23: 377–379 (in Chinese).

[105] World Health Organization, Guidelines for the Assessment of Herbal Medicines.WHO, Munich, Geneva,1991.

[106] Wu, E.X., Analysis on the trace element between *Lonicera japonica* Thunb and *Lonicera macranthoides* Hand.-Mazz.Chinese Traditional and Herbal Drugs, 1988,19:45–47 (in Chinese).

[107] Wu, C.X., Wang, J.M., Kang, W.Y., Component analysis on the volatile oil in different medicinal part of *Lonicera japonica* Thunb from Henan Province.China Pharmacy, 2009,20: 1 412–1 414 (in Chinese).

[108] Wu, L., Extraction and antioxidant ability of chlorogenic acid from *flos lonicera*.Tianjin University of Science and Technology.Master thesis,2005 (in Chinese).

[109] Wu, X.F., Jiao, X.Q., Li, G.R., Extracting of medicinal components *Lonicera japonica* Thunb's leaf and studying of their restraining bacterium test. Natural Product Research and Development, 2001,23:448 (in Chinese).

[110] Xagorari, A., Papapetropoulos, A., Mauromatis, A., Economou, M., Fotsis, T., Roussos, C., Luteolin inhibits an endotoxin-stimulated phosphorylation cascade and proinflammatory cytokine production in macrophages. The Journal of Pharmacology and Experimental Therapeutics, 2001, 296: 181-187.

[111] Xi, Q., Experimental study on the intervention of active component of *Lonicerae* Powder on immune cytokines of influenza viral pneumonia in mice. TCM University of Liaoning. Master thesis, 2008 (in Chinese).

[112] Xing, J.B., Li, H.J., Li, P., Studies on the quality control of *Flos Lonicerae*-Determination of total flavonoids. Chinese Journal of Modern Applied Pharmacy, 2002, 19: 169-170 (in Chinese).

[113] Xu, Y.B., Oliverson, B.G., Simmons, D.L., Trifunctional inhibition of COX-2 by extracts of *Lonicera japonica*: direct inhibition, transcriptional and post-transcriptional down regulation. Journal of Ethnopharmacology, 2007, 111: 667-670.

[114] Xu, Y.C., Study on extraction and isolation of the chlorogenic acid in *Honeysuckle* flowers and its bacteticidal activity. Chongqing University. Master thesis, 2008 (in Chinese).

[115] Yang, M.L., Zhao, Y.G., A comparative study of composition of volatile oil of *Flos Lonicerae* of Yuanzhou in Ningxia in different months. Journal of Ningxia University (Natural Science Edition), 2007, 28: 140-142 (in Chinese).

[116] Yip, E.C.H., Chan, A.S.L., Pang, H., Tam, Y.K., Wong, Y.H., Protocatechuic acid induces cell death in HepG2 hepatocellular carcinoma cells through a c-Jun N-terminal kinase-dependent mechanism. Cell Biology and Toxicology, 2006, 22: 293-302.

[117] Yoon, Y.H., Kyung, S.H., Repellent activity of *n*-hexane, ethyl acetate, *n*-butanol and water extracts of Native Plants against *Aedes albopitus*. Korean Journal of Entomology, 2002, 32: 61-64.

[118] Zdzisław, Z., Kowalska, A., Demska-Zakeś, K., Jeney, G., Jeney, Z., 2008. Effect of two medicinal herbs (*Astragalus radix* and *Lonicera japonica*) on the growth performance and body composition of juvenile pikeperch [*Sander lucioperca* (L.)]. Aquaculture Research 39: 1 149-1 160.

[119] Zhang, B., Yang, R.Y., Liu, C.Z., Microwave-assisted extraction of chlorogenic acid from flower buds of *Lonicera japonica* Thunb. Separation and Purification Technology, 2008, 62: 480-483.

[120] Zhang, H.F., Bai, Y.J., Lan, Y.S., Study on the anti-bacterium effect in vitro of *Lonicera japonica*. Journal of East China Normal University (Natural Science), 2000, 1: 107-110 (in Chinese).

[121] Zhang, N., Extraction of the antiseptic of cosmetic from honeysuckle's leaves. Jiangnan University. Master thesis, 2008 (in Chinese).

[122] Zhang, T.D., Pan, R.E., Bi, Z.M., Li, H.J., Li, P., Flavonoides from aerial parts of *Lonicera cheysantha*. Chinese Pharmaceutical Journal, 2006, 41: 741-743 (in Chinese).

[123] Zhang, X.C., Inhibition effects of allicin and flos lonicerae on the cells apoptosis induced by VSV and the possible molecular mechanism involved. Zhejiang University. Master thesis (in Chinese), 2007.

[124] Zheng, Y.H., Investigation of prescription to treat influenza A virus subtype H1N1 in Beijing YouAn Hospital. Shanxi Medical Journal, 2010, 39: 897-898 (in Chinese).

(发表于《Journal of Ethnopharmacology》, IF: 2.410)

Antinociceptive and Anti-inflammatory Activities of *Phlomis umbrosa* Turcz Extract

SHANG Xiao-fei[1,2], WANG Jin-hui[3], LI Mao-xing[2],
MIAO Xiao-lou[1], PAN Hu[1], YANG Yao-guang[1], WANG Yu[1]

(1. Engineering & Technology Center of Traditional Chinese Veterinary Medicine of Gansu.Research Center of Clinical Veterinary Medicine of CAAS.Lanzhou Institute of Animal and Veterinary Pharmaceutics Sciences, Chinese Academy of Agricultural Science, Lanzhou 730050, China; 2. Department of Pharmacy, Lanzhou General Hospital of PLA, Lanzhou 730050, China; 3. Department of Pharmacy, The Affiliated Hospital of Gansu College of T.C.M, Lanzhou 730000, China)

Abstract: *Phlomis umbrosa* Turcz has been used as the traditional medicine for thousands of years in China.In this paper, the acetic acid-induced writhing test, the hot plate test, the carrageenan-induced paw edema test, the xylene-induced ear swelling test, and the acetic acid-induced Evans blue leakage and leukocyte infiltration test were used to investigate the antinociceptive and anti-inflammatory activities of the aqueous extract of this plant (25, 50 and 100 mg/kg i.p.).Good dose-dependent effects were obtained in most of these tests, except in the hot plate test and the acetic acid-induced Evans blue leakage test.TLC and HPLC analyses showed iridoid glucosides were the main compositions of this extract.These findings suggested that the aqueous extract of *P.umbrosa* has significant antinociceptive and anti-inflammatory activities.

Key words: Phlomis umbrosa Turcz; Aqueous extract; Antinociceptive activity; Anti-inflammatory activity

1 INTRODUCTION

Phlomis umbrosa Turcz (Caosu) is a perennial herb (family Labiatae) from 40 cm to 100 cm tall which was native and distributed in several countries of the Southeast Asia.In China, it grows in wet meadows and woods at the range of 300–1500 m altitudes and distributes in Liaoning, Neimenggu, Hebei, Shan-dong, Shanxi, Shannxi, Gansu, Sichuan, Guizhou, Guangdong and other provinces[1].Root of *P.umbrosa* has been used as folk drugs with the effects of dispelling wind-evil and activating collaterals, reducing edema and curing pains, eliminating phlegm and detoxicating for thousands of years, etc[2].In some areas of China, *P.umbrosa* always was adopted as the substitutes of S.orientalis L.to treat cold, chronic bronchitis, arthralgia, and was recorded in 'Herbs of

Diannan'[3].

Iridoids and phenyethanoid glycosides were two main components of the aerial parts and roots of P.umbrosa. 8-O-acetyl shanzhiside methylester, phloyoside I, phloyoside II, shanzhiside methylester, 6″-syringyl-sesamoside, 8-O-acetyl-1-epishanzhigenin methyl ester, sesamoside and other compounds as the main iridoid glucosides have been isolated and identified. At the same time, decaffeoylverbascoside, calcelarioside B, 3-hydroxy-4-methoxy-β-phenylethoxy-O- [2, 3-diacetyl-α-L, rhamnopyranosyl- (1→3)]-4-O-feruloyl- [β-D-apio-furapiofuranosyl- (1→6)]-β-D-glucopyranoside (2‴, 3‴-diacetyl-O-betonyoside D), and other phenyethanoid glycosides have been isolated and reported in recent years[4-6].

Many plants allied to P.umbrosa have been reported with the antinociceptive and anti-inflammatory activities. Such as the iridoid glycosides extract of *Lamiophlomis rotata* (400 mg/kg, i.v.) and the ethanol extracts of *Phlomis younghusbandii* (200 mg/kg, i.p.) could decrease the number of writhings of mice induced by acetic acid and inhibit the inflammatory production induced by some agents ($p < 0.01$)[7]. Considering the traditional uses in alleviating pains of P. umbrosa, we evaluated systematically the antinociceptive and anti-inflammatory activities of aqueous extract of P. umbrosa (AEP) and elucidated its possible mechanism in this work.

2 MATERIALS AND METHODS

2.1 Plant material and extraction

The herbal samples of P.umbrosa were collected in Xiaolong mountain of Tianshui city in Gansu province, China in July of 2008. The raw material was identified by Prof. Zhigang Ma, Pharmacy College of Lanzhou University, China. A voucher specimen with accession number 20080728 was submitted to the Herbarium of the Laboratory of Pharmacognosy Department (Lanzhou University, China).

The aqueous extract of P.umbrosa (AEP) was prepared as previously described[8]. Briefly, 50 g powder of herbal P.umbrosa (aerial parts) was decocted in 100℃ with 500 mL water for 3 times, and every decoction time was 1 h. After they were filtered and vacuum dried at 60℃, the aqueous extract of P.umbrosa (AEP) (16.5 g) was obtained, and yield rate was 33.3%.

2.2 Chemical component analysis of AEP

2.2.1 Thin layer chromatography (TLC) analysis of AEP

The aqueous extracts of the aerial parts and the roots of P.umbrosa were prepared and spotted on the silica gel-CMC plate (Qindao Haiyang Chemical Reagent Factory, China) with 30μL, and $CHCl_3$-MeOH (4 : 1) was used as the developing solvent to develop it. 10% sulphuric acid with ethanol solvent was used as the color developing reagent and 8-O-acetyl shanzhiside methylester as standard preparation.

2.2.2 High performance liquid chromatography (HPLC) analysis of AEP

RP-HPLC methods was used to analyze the aqueous extracts of the aerial parts and the roots of P. umbrosa on a Waters apparatus (two solvent delivery systems, model 600, and a Photodiode Array detector, model 996), the gradient solvent system were 0-5 min, 6% CH_3CN; 5-10 min, 6%-

12% CH_3CN; 10-40 min, 12% CH_3CN.Data acquisition and quan-tification were performed using Millenium 2.10 version software (Waters).Symmetry reversed-phase column (C_{18}, 4.6 mm×150 mm; particle size 5 μm, Waters, made in Ireland) was maintained at ambient temperature (20.0℃).The mobile phase was filtered through a Millipore 0.45 mm filter and degassed prior to use. The peaks were detected at 235 nm and sesamoside, 8-O-acetylshanzhiside methylester, shanzhiside methylester, as the main iridoid glucosides were detected by comparing with the corresponding chemical marks, which were isolated from this plant and identified with MS, ^{1}H NMR, and ^{13}C NMR (Fig. 1).

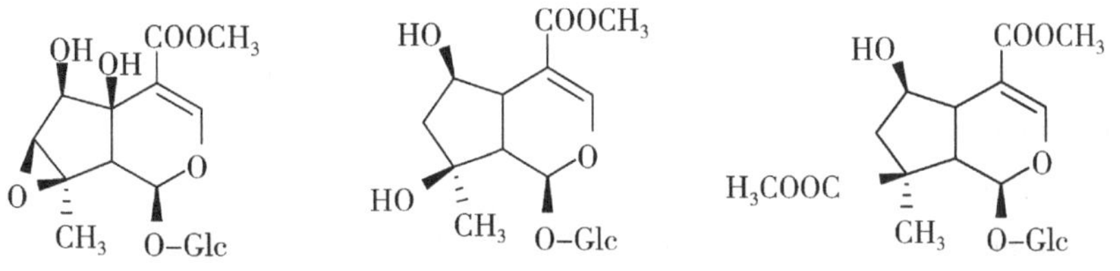

Fig. 1 The chemical structures of sesamoside (1), shanzhiside methylester (2), and 8-O-acetylshanzhiside methylester (3)

2.3 Experimental animals

Male or female Balb/C mice ((20±2) g, 2 months) were obtained from the Department of Animal Center, Lanzhou University (Lanzhou, China). They were kept in plastic cages at (22±2)℃ with free access to pellet food and water.Experimentscomplied with the rulings of Gansu Experimental Animal Center (Gansu, China) officially approved by the Ministry of Health, P.R. China in accordance with NIH guidelines[9].

2.4 Drugs and reagents

Acetic acid (Tianjing Chemical Reagent Company, China); xylene (Shanghai Chemical Reagent Company, China); aspirin (Hefei Jiulian Pharmaceutical Company, China); Morphine (The First Pharmaceutical Company Of Shenyang, China); crystal violet (Shanghai Xinzhong Chemical Reagent Company, China), carrageenan (Type I, Sigma Chemical Company, U.S.A.); Evans blue (E 2129, Sigma Chemical Company, U.S.A.).

2.5 Evaluation of antinociceptive and anti-inflammatory activities of AEP

2.5.1 Acetic acid-induced writhing response in mice

Experiments were carried out according to previously described method[10].AEP at doses of 25, 50 and 100 mg/kg was administrated by intraperitoneal injection to each mouse, the control group received normal saline (10 mL/kg, i.p.) and reference group received aspirin (100 mg/kg, i.p.).After 30 min treatment, 0.7% acetic acid (0.1 mL/10 g body weight) was injected intraperitoneally to mice.The mice were observed and counted for the number of abdominal constrictions and stretchings in a period of 0-30 min.

2.5.2 Hot plate test in mice

Experiments were carried out according to previously described method[11,12].AEP was administrated to the animals, the control group received normal saline (10 mL/kg, i.p.) and reference group received morphine sulphate (5 mg/kg, i.v.).30 min and 65 min after treatment, mice were individually placed on the heated plate which maintained at 55±1℃.The time for forepaw licking or jumping was taken as the latency time and the percentage of inhibition was determined.Before the experiment we screened and employed the mice which have the hot responses within 60 s.

2.5.3 Carrageenan-induced paw edema in mice

Experiments were carried out according to previously described method[13].Before injecting the test drugs, the initial paw thickness was determined with gauge calipers, after 30 min treatment, 25 μL 2% solution of carrageenan in normal saline was injected subcutaneously into the right hind paw.Paw thickness was measured at 0, 1, 3 and 5 h after injection, and the thickness of the edema was measured with gauge calipers and the percentage of inhibition was determined.

2.5.4 Xylene-induced ear swelling in mice

Experiments were carried out according to previously described method[13].After the test drugs were injected to animals 30 min, each animal received 30 μL of xylene on the anterior and posterior surfaces of the right ear lobe, the left ear was considered as control.One hour later, the animals were sacrificed by cervical dislocation and a diameter of 8 mm circular sections were taken from both ears with a cork borer, and weighed.The degree of ear swelling was calculated based on the weight of the left ear without applying xylene.

2.5.5 Leukocyte infiltration caused by intraperitoneal injection of 0.7% acetic acid

Experiments were carried out according to previously described method[14,15].AEP at doses of 25 mg/kg, 50 mg/kg and 100 mg/kg was administrated by intraperitoneal injection to each mouse, after 20 min 0.5% Evans blue solution (10 mL/kg) was injected by intravenous route; after 10 min mice were injected intraperitoneally with 0.7% acetic acid solution in normal saline and after another 20 min mice were sacrificed by cervical dislocation and the peritoneal cavity was washed by sterile saline and the volume collected with automatic pipettes.Finally the total volume of 5 mL and 100 μL of peritoneal fluid was mixed with 0.45 mL of Türk's solution (0.01% crystal violet in 3% acetic acid) and leukocyte cells were counted under microscope (Olympus Company, JAP).

2.5.6 Evans blue leakage caused by intraperitoneal injection of 0.7% acetic acid

Except 100μL of peritoneal fluid was employed to count the number of leukocyte cells.Other peritoneal fluids were centrifuged at 3000 rpm for 15 min and the absorbance of the supernatant was read at 610 nm with a HP-8453 UV spectrophotometer (HP Company, USA).The peritoneal capillary permeability induced by acetic acid is expressed in terms of dye (μg/mL), which leaked into the peritoneal cavity according to the standard curve of Evans blue dye[15].

2.6 Acute toxicity studies

The up-and-down or staircase method for acute toxicity testing was carried out as previously described[16,17].The dose was increased from 100 to 1000 mg/kg through the intraperitoneal route of administration.The animals were observed continuously for behavioral changes for the first 4 h and then observed for mortality if any 24 h after the drug administration.

2.7 Statistical analysis

The data obtained were analyzed using SPSS software program version 13.0 and expressed as a mean±S.E.M.Data were analyzed by a one-way ANOVA, followed by Student's two-tailed t-test for the comparison between test and control, and Dunnett's test when the data involved three or more groups.P-values less than 0.05 ($p<0.05$) were used as the significance level.

3 RESULTS AND DISCUSSION

Iridoid glycosides in some plants have been proved with the antinociceptive and anti-inflammatory activities[8]. So in this paper, firstly we have investigated the contents of iridoid glycosides which have been reported as the main components of *P.umbrosa*.TLC and HPLC analyses showed that the roots have more iridoid glycosides than the aerial parts. The contents of sesamoside, shanzhiside methylester and 8-O-acetylshanzhiside methyl ester in roots and the aerial parts were 13.67, 0.30, 8.47 mg/g and 3.29, 0.23, 0.37 mg/g, respectively. (Table 1) (Fig. 2 and Fig. 3).It is coincidental that the roots of this plant were more adopted in Chinese traditional uses.The antinociceptive and anti-inflammatory activities of these compounds were valuable to study further.

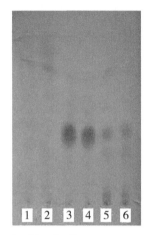

Fig. 2　The TLC chromatograms of 8-O-acetylshanzhiside methylester (3, 4), the aerial parts of *P. umbrosa* (1, 2) and the roots of *P.umbrosa* (5, 6)

Table 1　The contents of sesamoside, shanzhiside methylester, 8-O-acetylshanzhiside methylester, in the aerial parts and the roots of *P.umbrosa*

Sample	Sesamoside		Shanzhiside methyl ester		8-O-acetylshanzhiside methyl ester	
	Contents (mg/g)	RSD (%)	Contents (mg/g)	RSD (%)	Contents (mg/g)	RSD (%)
The aerial parts of *P.umbrosa*	3.29	2.35	0.23	5.73	0.37	5.01
The roots of *P.umbrosa*	13.67	4.51	0.30	5.53	8.47	3.85

Then, we have investigated the antinociceptive and anti-inflammatory activities of the aqueous extract of this plant with the two antinociceptive tests and four anti-inflammatory tests.Acetic acid causes an increase in peritoneal fluids of prosta-glandins such as prostaglandins, serotonin, and histamine involved in part, which was a model commonly used for screening peripheral analgesics[18].In our work, the results showed that AEP (25, 50 and 100 mg/kg) restrained the writhing response induced by intraperitoneal injection of 0.7% acetic acid in mice with inhibition rate of 27.81%, 73.13% and 81.56%, The decrease in the number of writhings was dosedependent and had the good effect of the peripheral analgesia.Aspirin (100 mg/kg, i.p.) produced a 79.66% reduction com-pared to the control (Table 2).But the writhing test may present many false positive tests, especially for neuromuscular block-ers. In the experiment, the abnormal locomotor activity was not found, but the motor performance and locomotor activity should be evaluated further.In order to evaluate the central antinociceptive activity of the extract, the hot-plate test was

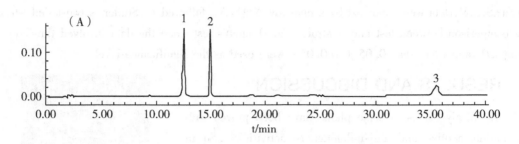
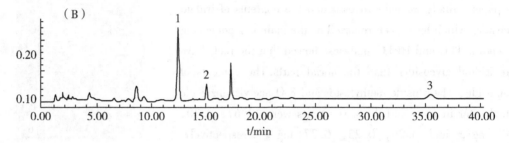
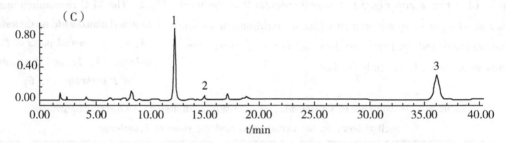

Fig. 3 The HPLC chromatograms at 235 nm of the reference substances (A), the aerial parts of *P.umbrosa* (B) and the roots of *P.umbrosa* (C). Sesamoside (1), shanzhiside methylester (2) and 8-O-acetylshanzhiside methylester (3)

carried out, because it had several advantages, particularly the sensitivity to strong analgesics and limited tissue damage[19]. Our results indicated that AEP could extend the animal's reaction time and show good analgesic activity against the hot plate, especially for the 100 mg/kg, the inhibition rate was 44.64%, 28.07% at 30 min and 65 min after treatment compared with 0 min (Table 3). From the results of two tests, we suggested that AEP has strong peripheral antinociceptive activity.

Table 2 Antinociceptive effect of *Phlomis umbrosa* extract on the writhing test

Group	Dose (mg/kg)	Number of writhings	Inhibition (%)
Control	—	32.0±13.08	—
AEP	25. i.p.	23.1±12.61	27.81
	50. i.p.	8.60±9.55	73.13
	100. i.p.	5.90±5.36	81.56
Aspirin	100. i.p.	6.51±7.34	79.66

Note: Each value represents the mean±S.E.M. of 10 mice.

* $P<0.05$ compared with control. ** $P<0.01$ compared with control.

Table 3 Antinociceptive effect of *Phlomis umbrosa* extract on the hot plate test

Group	Dose (mg/kg)	Reaction time (s)		
		0 min	30 min	65 min
Control	—	17.35±5.00	16.66±5.43	16.60±6.52
AEP	25 .i.p.	17.25±5.65	22.29±3.05#	19.29±5.15
	50 .i.p.	17.83±6.59	21.48±5.98	19.33±5.34
	100 .i.p.	17.55±5.00	31.79±7.22**##	24.40±6.31*#
Morphine	5. i.v.	17.88±6.06	51.59±2.18**##	49.10±3.46*##

Note: Each value represents the mean±S.E.M.of 10 mice.

** $P<0.05$, ** $P<0.01$ compared with Latency time in control group at each corresponding time.

$P<0.05$, ## $P<0.01$ compared with Latency time at 0 min in each corresponding group.

In anti-inflammatory tests, carrageenan-induced hind paw edema in mice is a biphasic phase of mediator release including histamine, serotonin and kinin in first phase.The inflammation reaches its maximum approximately 3 h post-treatment after which it begins to decline.The late phase, which is a complement-dependent reaction, has been shown to be due to overproduction of prostaglandin in tissues[20,21].The results showed that AEP at 1 h and 3 h significantly inhibited the release of histamine, serotonin and similar substances, but the effect of 5 h was weak (Table 4).Meanwhile xylene-induced neurogenous swelling, a common inflammatory model, was selected for evaluating vascular permeability which was partially associated with substance P[22]. The results showed that AEP at doses of 25, 50 and 100 mg/kg, i.p. dose-dependently suppressed xylene-induced ear swelling in mice, with 25.19%, 33.24% and 39.78% of the inhibition rate, respectively.Aspirin (100 mg/kg, i.p.) showed marked anti-inflammatory activity with a 54.81% reduction compared to the control ($p<0.01$), and indicted it might reduce the release of substance P or antagonize its action (Table 5).Evans blue leakage and leukocyte infiltration of mice could be induced by injected acetic acid, and the numbers of leukocyte and the contents of Evan blues would increase in the abdominal cavity.In our discussion in Sections 2.5.5 and 2.5.6, AEP (25, 50 and 100 mg/kg) can markedly reduce peritoneal leukocyte migration and the Evans blue leakage induced by acetic acid in mice with dose-dependent manner, and the inhibituon rates were 26.82%, 28.58%, 36.29%, and 30.48%, 30.00%, 33.33%, respectively (Tables 6 and 7).From the results of four tests, we suggested that AEP displays a significant anti-inflammatory effect.

Table 4 Anti-inflammatory effect of *Phlomis umbrosa* extract on the carrageenan-induced paw edema test

Group	Dose (mg/kg)	1 h		3 h		5 h	
		Swelling (%)	Inhibition (%)	Swelling (%)	Inhibition (%)	Swelling (%)	Inhibition (%)
Control	—	47.74±9.38	—	64.10±11.00	—	47.64±5.74	—
AEP	25	36.78±9.24*	22.06	52.22±13.11**	18.52	41.74±13.50	12.72
	50	35.33±10.12*	25.12	46.44±9.53**	27.32	41.42±11.34	13.44
	100	28.98±9.75**	39.06	43.21±9.40**	32.64	43.00±12.21	10.37
Aspirin	100	29.1±8.76**	38.3	42.50±4.75**	33.2	35.5±6.17**	29.7

Note: Each value represents the mean±S.E.M.of 10 mice.

* $P<0.05$ compared with control. ** $P<0.01$ compared with control.

Table 5 Anti-inflammatory effect of *Phlomis umbrosa* extract on the xylene-induced ear swelling test

Group	Dose (mg/kg)	Swelling (%)	Inhibition (%)
Control	—	45.25±3.86	—
AEP	25. i.p.	33.85±11.32	25.19
	50. i.p.	30.21±12.61*	33.24
	100. i.p.	27.25±9.62**	39.78
Aspirin	100. i.p.	20.45±7.12**	54.81

Note: Each value represents the mean±S.E.M. of 10 mice.
* $P<0.05$ compared with control. ** $P<0.01$ compared with control.

Table 6 Anti-inflammatory effect of *Phlomis umbrosa* extract on the acetic acid-induced leukocyte infiltration test

Group	Dose (mg/kg)	Total leukocytes ($\times 10^6$)	Inhibition (%)
Control	—	4.81±0.89	—
Model	—	25.02±5.19##	—
AEP	25 .i.p.	18.31±8.32	26.82
	50 .i.p.	17.87±6.57*	28.58
	100 .i.p.	15.94±7.02*	36.29

Note: Each value represents the mean±S.E.M. of 10 mice.
* $P<0.05$ compared with model.
** $P<0.01$ compared with control.

Table 7 Anti-inflammatory effect of *Phlomis umbrosa* extract on the acetic acid-induced Evans blue leakage test

Group	Dose (mg/kg)	OD_{610nm}	Evans blue dye (μg/mL)	Inhibition (%)
Control	—	0.23±0.48	2.76±6.56	—
Model	—	2.10±0.39	31.13±5.19##	—
AEP	25 i.p.	1.46±0.38	21.45±5.04*	30.48
	50 i.p.	1.47±0.69	21.60±9.75*	30.00
	100 i.p.	1.40±0.45	20.54±6.11*	33.33

Note: Each value represents the mean±S.E.M. of 10 mice.
* $P<0.05$ compared with model.
** $P<0.01$ compared with control.

Finally, we evaluated the acute toxicity of AEP (100–1000 mg/kg) which was administered by intraperitoneal injection. No mouse died or acute behavior changed during the observation periods. LD_{50} was estimated to more than 1000 mg/kg, i.p. AEP was safe at a given dose to mice. And further study on the chronic toxicity and physiological changes induced by AEP will be carried out.

4 CONCLUSION

Phlomis umbrosa Turcz has been used as the folk medicine to treat some diseases in China. In this paper, the present study demonstrated that the aqueous extract of P.umbrosa had the good antinociceptive and anti-inflammatory activities, and indirectly substantiated the traditional use of P. umbrosa in some inflammatory and pain disorders by the local folklore practitioners. At the same time, the results of chemical analysis showed that the main iridoid glucosides of P.umbrosa in the roots were more than in the aerial parts.So further studies should be carried to investigate whether the contents of iridoid glycosides of P.umbrosa contributed to its pharmacological activities.

ACKNOWLEDGEMENTS

This work wasfinanced by the National Scientific Research Institutes for basic R & D special fund (No.1610322011009).The authors would also like to express their gratitude to Lanzhou University PhD English writing foreign teacher Allan Grey who thoroughly corrected the English in the paper.

References

[1] Institute of botany, plants Chinese Academy of Sciences.The Picture Index of Senior China Plant (3). Beijing: Science and Technology Press, 1980: 650.

[2] Fu HZ, Liu SW, Lin WH.Acta Pharm Sin 1999, 34: 297-300.

[3] Jiangsu New Medical College. A Comprehensive Dictionary of Tradi-tional Chinese Materia Medica. Shanghai People's Press, 1986: 2 665.

[4] Li MX, Shang XF, Jia ZP, Zhang RX.Chem Biodivers, 2010, 7: 283-301.

[5] Amor IL, Boubaker J, Sgaier MB, Skandrani I, Bhouri W, Neffati A, et al.J Ethnopharmacol, 2009, 125: 183-202.

[6] Liu P, Takaishi Y, Duan HQ.Chin Chem Lett, 2007, 18: 155-157.

[7] Li MX, Shang XF, Zhang RX, Jia ZP, Fan PC, Ying Q, et al.Fitoterapia, 2010, 81: 167-172.

[8] Shang XF.Antinociceptive and anti-inflammatory activities of iridoid glycosides extract of *Lamiophlomis rotata* (Benth.) Kudo.Master thesis.Lanzhou University, 2010.

[9] NIH.Guidelines for Research Involving Recombinant DNA Molecules (NIH Guidelines).Bethesda, MD: Office of Biotechnology Activities, National Institute of Health, 2002.

[10] Koster R, Anderson M, Beer EJ.Fed Proc, 1959, 18: 418-420.

[11] Turner RA.In: Turner R, Hebborn P, editors.Analgesics: Screening Methods in Pharmacology.New York: Academic Press, 1965: 100.

[12] Shinde UA, Phadke AS, Nair AM, Mungantiwar AA, Dikshit VJ, Saraf MN.J Ethnopharmacol, 1999, 65: 21-27.

[13] Kou JP, Sun Y, Lin YY, Cheng ZH, Zheng W, Yu BY, et al.Biol Pharml Bull, 2005, 28: 1 234-1 238.

[14] Chen Q.Pharmacology Technology of Chinese Materia Medica, China.P.R.China: People's Medical Publishing House, 1994: 303.

[15] Lucena GMRS, Gadotti VM, Maffi LC, Silva GS, Azevedo MS, Santos ARS.J Ethnopharmacol, 2007, 112: 19-25.

[16] Bruce RD.Fund Appl Toxicol, 1985, 5: 151-157.

[17] Ghosh MN.Fundamentals of Experimental Pharmacology.Kolkata: Hilton and Company, 2005.

[18] Deraedt R, Jouguey S, Delevallee F, Flahaut M.Euro J Pharm, 1980, 5: 17-24.

[19] Ferreira MAD, Nunes ODRH, Fontenele JB, Pessoa ODL, Lemos TLG, Viana GSB.Phytomedicine, 2004, 11: 315-322.
[20] Di RM, Sorrentino L, Parente L.J Pharmac Pharmacol, 1972, 24: 575-577.
[21] Garćia MD, Fernαndez MA, Alvarez A, Saenz MT.J Ethnopharmacol, 2004, 91: 69-73.
[22] Luber-Narod J, Austin-Ritchie T, Hollins C, Menon M, Malhotra RK, Baker S, et al.Urol Res, 1997, 25: 395-399.

(发表于《Fitoterapia》, 院选 IF: 1.899)

A Novel Single Nucleotide Polymorphism in Exon 7 of LPL Gene and its Association with Carcass Traits and Visceral Fat Deposition in Yak (*Bos grunniens*) Steers

DING X.Z., LIANG C.N., GUO X., XING C.F.,
BAO P.J., CHU M., PEI J., ZHU X.S., YAN P.

Abstract: Lipoprotein lipase (LPL) is considered as a key enzyme in the lipid deposition and metabolism in tissues. It is assumed to be a major candidate gene for genetic markers in lipid deposition. Therefore, the polymorphisms of the LPL gene and associations with carcass traits and viscera fat content were examined in 398 individuals from five yak (*Bos grunniens*) breeds using PCR-SSCP analysis and DNA sequencing. A novel nucleotide polymorphism (SNP)-C→T (nt19913) was identified located in exon 7 in the coding region of the LPL gene, which replacement was responsible for a Phe-to-Ser substitution at amino acid. Two alleles (A and B) and three genotypes designed as AA, AB and BB were detected in the PCR products. The fre-quencies of allele A were 0.7928, 0.7421, 0.7357, 0.6900 and 0.7083 for Tianzhu white yak (WY), Gannan yak (GY), Qinghai-Plateau yak (PY), Xinjiang yak (XY) and Datong yak (DY), respectively. The SNP loci was in Hardy-Weinberg equilibrium in five yak populations ($P<0.05$). Polymorphism of LPL gene was shown to be associated with carcass traits and lipid deposition. Least squares analysis revealed that there was a significant effect on live-weight (LW) ($P<0.01$), average daily weight gain (ADG) and carcass weight ($P<0.05$). Individuals with genotype BB had lower mean values than those with genotype AA and AB for loin eye area and viscera fat

Key words: Yak LPL gene; Polymorphism; Carcass traits; Fat deposition

INTRODUCTION

Fat and fatty acids potentially make the significant con-tribution to the nutritional value of meat and to various aspects of meat quality[1]. Meat quality is largely influ-enced by i.m.fat (IMF) content, which affects sensory properties (i.e., tenderness and flavor) and nutritional val-ues of meats[2,3]. However, fat is an unpopular constit-uent of meat for consumers, as it is considered unhealthy[4]. Consequently, manipulation of the fat and fatty acid composition of muscle has been of great interest in recent years to produce meat with desirable nutritional and tech-nological qualities[3,5].

Lipoprotein lipase (LPL) is the rate-limiting enzyme for the import of triglyceride-derived fatty acids by muscle, for utilization, and adipose tissue, for storage in most species[6,7]. Regulation of LPL appears to be very complex and responds to dietary or hormonal changes and environmental conditions being clearly tissue-specific[7,8]. Furthermore, peroxisome proliferator-activated receptor γ (PPARγ), sterol regulatory element binding protein 1 (SREBP-1) and adipose fatty acids-binding protein (A-FABP), have recently been identified as genetic candidate genes being implicated in lipid metab-olism in adipose tissue by regulating gene expression of enzymes or proteins involved in lipid metabolism, or by transporting fatty acids[9-13].

As described in Genbank (Accession No. HM627756) the nucleotide sequences of the yak LPL cDNA in the coding regions are approximately 1.59 kb long and encode a protein of 479 amino acids (AA). Moreover, the inves-tigation of the levels of genetic diversity in the bovine LPL gene has led to the identification of Sau96I[14] and BsmAI[15] RFLP. Yak (*Bos grunniens*), is one of the world's most remarkable domestic animals and has developed special regulating mechanisms in adapting to the harsh environment of the Qinghai-Tibetan Plateau.[16,17]. Although yaks are multipurpose animals, they are raised mainly for meat production in the highland plains of China and yak meat and its products are widely consumed by local herders[18]. Thus, yak meat fills an important niche in Tibetan life and in developing the prosperity of the people in the vast mountainous regions. Modern knowl-edge indicates that yak meat has multifunctional effects on the health of Tibetans. However, information on LPL gene and its polymorphisms in domestic yak (*B.grunniens*) still remains very scarce. Hence, the objective of this study was to examine the relationship between polymorphisms of the LPL gene and it association with carcass traits, visceral fat deposition as well as meat quality in yak breeds.

MATERIALS AND METHODS

Animals and DNA samples

A total of 398 unrelated growing yak steers including 111 Tianzhu white yak (WY), 70 Qinghai-Plateau yak (PY), 50 Xinjiang yak (XY), 95 Gannan yak (GY) and 72 Datong yak (DY). 111 Tianzhu white yak steers at born-12 months old, 13-24 months old and 25-36 months old were selec-ted randomly and slaughtered to obtain the growth traits (live-weight and average daily gain) and carcass traits (carcass weight, viscera fat weight, loin-eye area) for sta-tistical analysis.

10 mL blood samples were collected from the jugular veins with 2 mL ACD for Genomic DNA extraction were quickly frozen in liquid nitrogen and stored at -80LC. Total DNA was extracted according the standard phenol-chloroform extraction protocol[19].

Primer design and PCR amplification

Primers used to amplify LPL gene exon 7 locus of yak were designed on the basis of available sequences of yak (GenBank accession numbers. NW_ 001495467). The primers for forward and reverse were 5'-AGTGCCTGC TTGTTTGTG-3' and 5'-TATGCCCTTTCTGTTCCT-3', respectively. The size of expected PCR products was 285 bp, including the whole exon 7 and partial sequence of 5' and 3' end. The PCR was carried out in 25μL of reaction volume containing 50 ng of genomic DNA template, 1.5 mM $MgCl_2$, 0.2 mM of each dNTP, 0.5 units of Taq DNA polymerase (TaKaRa, China) and 0.4 μM of each primer.10 cycles of amplification were carried out for each PCR.Each cycle consisted of an initial step at 94℃ for 5 min, denaturation step at 95℃ for 30 s, annealing tem-peratures for 40 s, and extension at 72℃ for 40 s.The final extension step was followed by a 10 min extension reac-tion at 72℃.

DNA sequencing and agarose electrophoresis analysis

2 μL PCR products were mixed with 6 μL denaturing solu-tion (95% formamide, 25 mM ED-TA, 0.025% xylenecya-nole and 0.025% bromophenol blue), heated for 10 min at 98℃ and chilled on ice.The samples were electrophoresed in 12% sodium dodecyl sulfate-polyacrylamide gel elec-trophoresis (SDS-PAGE).The electrophoresis was pro-grammed as 250 V, 40 mA (preelec-trophoresis) for 10 min for the first step, with the Silver Stain step of 150 V, 24 mA (Kucharczyk Techniki Elektroforetyczne) for 8 h after that.A refrigerated circulator was used to control the tem-perature (4℃) of gels.

Data analysis

Genotypes, allelic frequencies and population genetic indexes: He (gene heterozygosity), Ho (gene homozygos-ity), Ne (effective allele numbers) and PIC (Polymorphism Information Content) were calculated by using the POP-GENE software (ver.1.31).The Hardy-Weinberg equi-librium of the mutation was exam by χ^2-test.

The associations between SNP marker genotypes and growth traits in yak (WY, PY, GY, XY and DY) were analyzed by SPSS 13.0 software according to the following linear model:

$$Y_{ijk} = \mu + G_i + Age_j + marker_k + e_{ijk}$$

Where Y_{ijk} is observation for the carcass measurement traits, μ was the overall population mean, G_i is the fixed effect associated with ith genotype, Age_j is the effect of jth month of slaughte-ring, $marker_j$ is kth single SNP marker genotype, e_{ijk} is the random error.

RESULTS

Polymorphisms of LPL gene in five yak populations

The genetic polymorphism of the population was firstly detected by SSCP in the locus of exon 7 in yak steers (Fig. 1).The polymorphism of exon 7 was induced by C-T single nucleotide polymor-phism (SNP) in LPL gene (Fig. 2), in which C-T (nt19913) substitution led to an amino acid mutation (Phe>Ser).The genotypic and allelic frequencies of the locus were shown as in Table 1. Among the population, allele A was the predominant.The fre-quencies of allele A were 0.792 8, 0.742 1, 0.735 7, 0.690 0 and 0.708 3 for Tianzhu white yak (WY), Gannan yak (GY), Qinghai-Plateau yak (PY), Xinjiang yak (XY) and Datong yak (DY), respectively.Genotype frequencies, genetic indexes, and the equilibrium χ^2-test are shown in Tables 1 and 2. The SNPs loci were in Hardy-Weinberg equilibrium in the WY, GY, PY, XY and DY populations ($P > 0.05$).

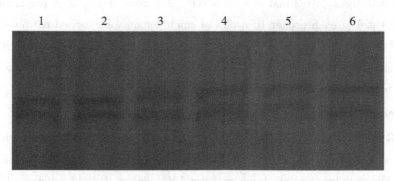

Fig. 1 The electrophoresis patterns of PCR-SSCP for LPL gene exon 7 in yak.
1, 2: AA genotype; 3, 4 and 5: AB genotype; 6: BB genotype

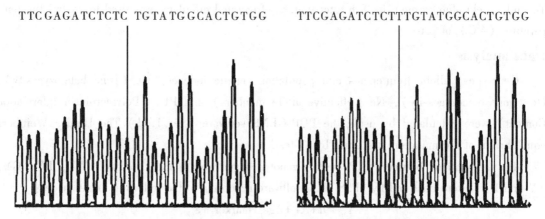

Fig. 2 Partial sequencing maps of exon 7 of LPL gene in the yak

Table 1 Genotypes distribution and allelic frequencies at the exon 7 of yak LPL gene in five populations

Breeds	Number of each genotype			Genotypic frequencies			Allelic frequencies		Equilibrium χ^2-test
	AA	AB	BB	AA	AB	BB	A	B	
DY	36	30	6	0.500 0	0.416 7	0.083 3	0.708 3	0.291 7	0.347 2
GY	55	31	9	0.578 9	0.326 3	0.094 7	0.742 1	0.257 9	0.254 6
WY	73	30	8	0.657 7	0.270 3	0.072 0	0.792 8	0.207 2	0.185 3
PY	38	27	5	0.542 9	0.385 7	0.071 4	0.735 7	0.264 3	0.295 0
XY	32	12	5	0.640 0	0.260 0	0.100 0	0.690 0	0.310 0	0.202 2

Table 2 Genetic diversity indexes at the exon 7 of yak LPL gene in five populations

Populations	He	Ho	Ne	PIC
DY	0.413 3	0.586 8	1.704 2	0.327 8
GY	0.382 8	0.617 2	1.620 2	0.309 5
WY	0.328 5	0.671 5	1.489 3	0.247 6
PY	0.388 9	0.611 1	1.636 4	0.312 7
XY	0.427 8	0.572 2	1.747 6	0.336 2

Note: He gene heterozygosity, Ho gene homozygosity, Ne effective allele numbers, PIC Polymorphism Information Content

Effect of the LPL gene on carcass and meat quality traits

Table 3 showed the comparison of the least square means and respective standard errors of carcass traits, viscera fat and meat quality, involving the genotypes of the LPL polymorphism.Correlation analysis indicated that in the born-12 months WY population, individuals with geno-type BB had more live-weight and ADG than those with genotype AA and AB in born-21 months and 13-24 months ($P<0.01$).However, the viscera fat weight and loin eye area did not find any associations among the populations.In contrast, Individuals with genotype BB had lower mean values than those with genotype AA and AB for loin eye area and viscera fat content in 25-36 months yak populations ($P<0.05$).

Table 3 Associations of SNP genotypes with carcass traits and fat content (LW%) in Tianzhu white yak breed

Ages (mon)	Genotypes	Traits				
		Live-weight (kg)	Average daily gain (g)	Carcass weight (kg)	Loin-eye area (cm²)	Viscera fat (LW%)
Born-12	AA	67.2±11.3[B]	157.8±13.31[b]	34.3±6.47[b]	29.2±2.34	0.9±0.04
	AB	69.5±13.6[B]	160.5±10.56[b]	35.4±5.40[b]	30.4±1.42	1.1±0.03
	BB	83.9±10.8[A]	192.2±11.98[a]	42.8±9.64[a]	28.7±1.86	1.1±0.01
13-24	AA	120.1±11.9[B]	152.3±10.64[b]	61.3±6.14[b]	31.2±2.28	1.3±0.15
	AB	126.0±7.1[B]	154.8±28.50[b]	64.3±5.48[b]	30.6±2.30	1.2±0.06
	BB	149.3±10.5[A]	167.3±25.80[a]	76.1±10.80[a]	29.8±0.69	1.4±0.11
25-36	AA	154.8±28.5	73.1±12.01[b]	78.9±6.91	35.4±1.67[b]	2.2±0.17[b]
	AB	152.1±5.4	71.5±12.18[b]	77.6±5.60	31.6±2.43[a]	1.6±0.31[a]
	BB	169.2±4.8	84.8±11.26[a]	86.2±8.06	30.3±2.72[a]	1.2±0.16[a]

Note: Values with different superscripts within the same line differ significantly at $P<0.01$ (A,B) and $P<0.05$ (a,b)

DISCUSSION

LPL is a key enzyme in lipoprotein metabolism by virtue of its capacity to hydrolyze triglycerides (TG) circulating in the form of lipoprotein particles, thereby delivering fatty acids to various tissues.LPL is synthesized by a variety of tissues, including adipose, skeletal and cardiac muscle, lactating mammary gland, and macrophages[20].There-fore, tissue LPL and its regulation became the subject of active investigation in rodents and humans[21] but also in ruminants[22]. The study of LPL is indeed of particular interest in tissues of meat-producing ruminants[23], since LPL controls TG partitioning between adipose tissues and muscles, thereby increasing fattening or providing energy in the form of fatty acids for muscle growth, which attracted much attention on the polymorphism and associ-ation[24].Research work from Chilliard's group[25-27] demonstrated that LPL also plays an important role in ruminant species.In adult sheep at maintenance levels, 55%-60% of the total amount of free fatty acids originates from hydrolysis of circulating triacylglycerols by LPL[28]. Hocquette et al.[25] demonstrated the levels of LPL transcripts are positively related to LPL activity in bovine tissues, including muscles and adipose tissues.

To our knowledge, there is currently no published lit-erature describing polymorphisms of LPL gene in yak breeds. In this study a novel single nucleotide polymor-phism (SNP) was detected among five yak populations exon 7 in the coding region of LPL gene. The mutations were firstly detected and classified by DNA sequencing and SSCP analysis. The nt19913 C>T mutation was found in all analyzed samples, and which lead to a p.Phe>Ser amino acid substitution. With almost 100 polymorphic sites[29], the levels of polymorphism observed in the human LPL gene are much higher than that reported for the bovine ortholog[24]. A clear trend that emerged from the locus showed intermediate polymorphism and the result of χ^2-test indicated that the locus in the population fitted with Hard-Weinberg equilibrium ($P<0.05$), which may be attributed to high genetic variation within the population.

Significant differences in LPL activity in ruminant species may suggest the existence of genetic factors that regulate the expression, localization, or activity of this enzyme[24]. With the objective of investigating the pos-sible functional role of the novel mutation that we char-acterized in the yak LPL gene, we performed an association analysis in different periods (born-12 months old, 13-24 months old and 25-36 months old) of WY population for which diverse carcass and fat deposition traits were registered (Table 3). Taking account into the slaughter cost and operation convenience[30], the 110 Tianzhu white yak (WY) were selected to analysis the carcass and fat deposition traits. Correlation analysis indicated that in the born-12 months WY population, individuals with genotype BB had more live-weight and ADG than those with genotype AA and AB ($P<0.05$). However, animals with genotype BB had lower mean values than those with genotype AA and AB for loin eye area and viscera fat content in 25-36 months yak popula-tions. As a consequence, comparisons of body weight and meat quality traits among the individuals with AA, AB and BB genotypes, individuals with genotypes BB had some superior economic traits. The results indicated that the mutations of CDS region in exon 7 of LPL gene in yak may simply be used as genetic markers linking to quantitative trait loci with effects on carcass traits, fat deposition as well as meat quality.

ACKNOWLEDGMENTS

We thank Y.J.Yue and J.B.Liu for instructive suggestions and stimulating discussions. This work is supported or partly supported by grants from National Beef Cattle Industry Technology & System (CARS-Beef Cattle System: CARS-38) and Special Fund for Agro-Scientific Research in the Public Interest (Contract No.201003061).

References

[1] Lee SH, Choi YM, Choe JH, Kim JM, Hong KC, Park HC, Kim BC (2010) Association between polymorphisms of the heart fatty acid binding protein gene and intramuscular fat content, fatty acid composition, and meat quality in Berkshire breed. Meat Sci 86: 794-800.

[2] Ruiz JA, Guerrero L, Arnau J, Guardia MD, Esteve GE (2001) Descriptive sensory analysis of meat

from broilers fed diets containing vitamin E or β-carotene as antioxidants and different supplemental fats. Poult Sci 80: 976-982.

[3] Wood JD, Richardson RI, Nute GR, Fisher AV, Campo MM, Kasapidou E, Sheard PR, Enser M (2003) Effects of fatty acids on meat quality: a review.Meat Sci 66: 21-32.

[4] Wood JD, Enser M, Fisher AV, Nute GR, Sheard PR, Richardson RI, Hughes SI, Whittngton FM (2008) Fat deposition, fatty acid composition and meat quality: a review.Meat Sci 78: 343-358.

[5] Forrest JC, Aberle ED, Hedrick HB, Judge MD, Merkel RA (1975) Principles of meat science.W. H.Freeman, San Francisco.

[6] Eckel RH (1987) In: Borensztajn J (ed) Lipoprotein lipase.Evener, Chicago, 79-132.

[7] Auwerx J, Leroy P, Schoonjans K (1992) Lipoprotein lipase: recent contributions from molecular biology.Crit Rev Clin Lab Sci 29: 243-268.

[8] Lu JZ, Li JX, Ji CN, Yu WY, Xu ZY, Huang SD (2008) Expression of lipoprotein lipase associated with lung adenocar-cinoma tissues.Mol Biol Rep 38: 131-137.

[9] Yin L, Zhang Y, Hillgartner FB (2002) Sterol regulatory element-binding protein-1 interacts with the nuclear thyroid hormone receptor to enhance acetyl-CoA carboxylase-α transcription in hepatocytes.J Biol Chem 277: 19 554-19 565.

[10] Stoeckman AK, Towle HC (2002) The role of SREBP-1c in nutritional regulation of lipogenic enzyme gene expression.J Biol Chem 277: 27 029-27 035.

[11] Brown MS, Goldstein JL (1999) A proteolytic pathway that controls the cholesterol content of membranes, cells, and blood.Proc Natl Acad Sci USA 96: 11 041-11 048.

[12] Grindflek E, Sundvold H, Lien S, Rothschild MF (2000) Rapid communication: physical and genetic mapping of the Peroxisome Proliferator Activated Receptor γ (PPAR γ) gene to porcine chromosome 13. J Anim Sci 78: 1 391-1 392.

[13] Elizabeth RS, Judith S (1999) The Adipocyte fatty acid-binding protein binds to membranes by electrostatic interactions.J Biol Chem 274: 35 325-35 330.

[14] Tank PA, Pomp D (1994) PCR-based Sau96I polymorphism in the bovine lipoprotein lipase gene.J Anim Sci 11: 3 032.

[15] Lien S, Gomez-Raya L, Vage DI (1995) A *Bsm*AI polymorphism in the bovine lipoprotein lipase gene.Anim Genet 26: 283-284.

[16] Wang DP, Li HG, Li YJ, Guo SC, Yang J, Qi DL, Jin C, Zhao XQ (2006) Hypoxia-inducible factor 1α cDNA cloning and its mRNA and protein tissue specific expression in domestic yak (*Bos grunniens*) from Qinghai-Tibetan plateau.Biochem Bio-phys Res Commun 348: 310-319.

[17] Wiener G, Han JL, Long RJ (2003) The yak, 2nd edn.Regional Office for Asia and the Pacific of the Food and Agriculture Organization of the United Nations, Bangkok, 154-156.

[18] Long RJ, Zhang DG, Wang X, Hu ZZ, Dong SK (1999) Effect of strategic feed supplementation on productive and reproductive performance in yak cows.Prev Vet Med 38: 195-206.

[19] Joseph S, David WR (2002) In: Huang PT (ed) Molecular cloning a laboratory manual, 3rd edn. Science Press, Beijing.

[20] Garfinkel AS, Schotz MC (1987) Lipoprotein lipase. In: Gotto AM Jr (ed) Plasma lipoproteins. Elsevier, Amsterdam, 335-357.

[21] Olivecrona T, Bengtsson-Olivecrona G, Chajek-Shaul T, Car-pentier Y, Deckelbaum R, Hultin M, Peterson J, Patsch J, Vilaro S (1991) Lipoprotein lipase.Sites of synthesis and sites of action.Atheroscler Rev 22: 21-25.

[22] Chilliard Y (1993) Dietary fat and adipose tissue metabolism in ruminants, pigs, and rodents: a re-

view.J Dairy Sci 76: 3 381-3 897.

[23] Ren MQ, Wegner J, Bellmann O, Brockmann GA, Schneider F, Teuscher F, Ender K (2002) Comparing mRNA levels of genes encoding leptin, leptin receptor, and lipoprotein lipase between dairy and beef cattle.Domest Anim Endocrinol 23: 371-381.

[24] Badaoui B, Serradilla JM, Tomas A, Urrutia J, Ares BL, Car-rizosa J, Sanche A, Jordana J, Amills M (2010) Identification of two polymorphisms in the goat lipoprotein lipase gene and their association with milk production traits.J Dairy Sci 90: 3 012-3 017.

[25] Hocquette JF, Graulet B, Olivecrona T (1998) Lipoprotein lipase activity and mRNA levels in bovine tissues.Comp Biochem Physiol B Biochem Mol Biol 121: 201-212.

[26] Bonnet M, Faulconnier Y, Flechet J, Hocquette JF, Leroux C, Langin D, Martin P, Chilliard Y (1998) Messenger RNAs encoding lipoprotein lipase, fatty acid synthase and hormone-sensitive lipase in the adipose tissue of underfed-refed ewes and cows.Reprod Nutr Dev 38: 297-307.

[27] Bonnet M, Leroux C, Faulconnier Y, Hocquette JF, Bocquier F, Martin P, Chilliard Y (2000) Lipoprotein lipase activity and mRNA are up-regulated by refeeding in adipose tissue and car-diac muscle of sheep.J Nutr 130: 749-756.

[28] Pethick DW, Dunshea FR (1993) Fat metabolism and turnover.In: Forbes JM, France J (eds) Quantitative aspects of ruminant digestion and metabolism. CAB International Press, Wallingford, 291-311.

[29] Merkel M, Eckel RH, Goldberg IJ (2002) Lipoprotein lipase: genetics, lipid uptake, and regulation.J Lipid Res 12: 1 997-2 006.

[30] Fan YY, Zan LS, Fu CZ, Tian WQ, Wang HB, Liu YY, Xin YP (2010) Three novel SNPs in the coding region of PPARc gene and their associations with meat quality traits in cattle.Mol Biol Rep 38: 131-137.

（发表于《Mol Biol Rep》，院选 IF：1.875）

Seasonal and Nutrients Intake Regulation of Lipoprotein Lipase (LPL) Activity in Grazing Yak (Bos grunniens) in the Alpine Regions around Qinghai Lake

DING X.Z., GUO X., YAN P., LIANG C.N., BAO P.J., CHU M.

(Lanzhou Institute of Husbandry and Pharmaceutical Sciences, Chinese Academy of Agricultural Sciences/Key Laboratory of Yak Breeding Engineering, Gansu Province, Lanzhou 730050, China)

Abstract: Lipoprotein lipase (LPL) is considered as a key enzyme in the lipid deposition and metabolism in tissues. To better understand fat cycling in grazing yak to adapt to the harsh environment on the Qinghai-Tibetan Plateau, and we have therefore explored seasonal and nutritional regulation of LPL in adipose tissue, liver and skeletal muscle. Sixty 3 year old growing yaks (BW.120.3±15.28 kg) were subdivided into six groups, each used to determine effects on chemical body composition and LPL activity in different tissues. The alpine pastures had the highest Crude protein (CP, 11.06%) contents and the gross energy (GE, 11.49 MJ/kg) in summer. Growth rates and body fat content were responsive to CP and GE intake regimen. Late spring up-regulation of LPL activity in the subcutaneous adipose tissue was consistent with a pronounced increase of body weight (BW) and whole body fat content. The highest LPL activity in the skeletal muscle was found in September (774±64.1 mU/g tissue), which may serve to cover the increased energy demands for compensatory growth and maintenance of adiposity in the coming cold season. Furthermore, the seasonal regulation of LPL involves some factors in addition to insulin and triglycerides (TG). These results suggest that yaks could rely, in part, on LPL activity to adapt to the harsh forage environment. During the growing season, an enhanced synthesis of LPL production in the adipose tissues along with mechanisms for the recycling of fat contributes toward the rapid recovery of body weight.

Key words: Grazing yak; Lipoprotein lipase; Adipose tissue; Nutritional regulation; Season

1 INTRODUCTION

Yak (*Bos grunniens*), the multipurpose herbivore, is one of the world's most remarkable domestic animals and exclusively living in alpine and subalpine regions at altitudes ranging from 3.000 to 5.000 m with a cold and harsh environment (Wang et al., 2006). About 15 million or more than

90% of the world's total yak population are currently herded in Chinese territories, which are the major source of livelihood for the nomadic Tibetans in the highland plains (Ding et al., 2010). Given the unique pattern of herbage mass distribution, the herds generally suffer from malnutrition for almost 8 months of the year. Performance of yaks is significantly affected by seasonal changes and they have adapted by displaying a circannual rhythm to better adapt to the harsh environment. A saying 'strong in summer, weighty in autumn, thin in winter and dead in spring' is and accurate description of the growing and survival pattern of yaks across the Qinghai-Tibetan plateau (Long et al., 1999). As a consequence of abundant grazing in summer and early autumn, protein mass and a layer of subcutaneous fat, as essential energy requirements are met through selective fat catabolism. These adaptive mechanisms make the yak an interesting model for the study of fat metabolism.

One enzyme key to adipocyte physiology is lipoprotein lipase (LPL), which plays a pivotal role in lipoprotein absorption and metabolism by catalyzing the hydrolysis of triglycerides (TG) transported in the bloodstream as very-low-density lipoproteins (VLDL) and intestinally derived chylomicrons (Mead et al., 2002). These triglycerides are hydrolysed into free fatty acids (FFA), which can then be re-esterified and as occurs in adipose tissue, or used as an energy source by peripheral tissues, mainly in muscle and heart of most species (Auwerx et al., 1992). Therefore, an important function of LPL activity in muscle tissues with regard to the total TG removal capacity of the body in both single-stomached and ruminant species (Hocquette et al., 1998). In addition, there is now evidence for a regulated LPL expression and activity under a number of physiological conditions that help to direct fatty acid utilization according to specific metabolic demands (Bonnet et al., 1998; Bonnet et al., 2000). Thus, LPL gene expression is regulated in a tissue-specific (adipose tissue, skeletal muscle and heart) in response to nutrient intake and body energy reserves (Bonnet et al., 2000).

The yak is a specialized species that has adapted to living in harsh environments. Thus, we tested the hypothesis that survival of the grazing yak on the Qinghai-Tibetan Plateau is highly dependent on the regulation of adipose accretion and metabolism. In addition, regulation of LPL in fat storage tissues liver and also in the skeletal muscle needs further functional and ecological adaptation studies.

2 MATERIALS AND METHODS

2.1 Experimental location

The experiment was carried out on natural rangeland typical alpine meadows dominated by herbage (sedge and grass) species of *Kobresia humilis*, *Kobresia pygmaea*, *Kobresia graminifolia*, *Elymus nutans*, *Polygonum viviparum*, and *Ana-phalis lacteal*, located in the area around Qinghai Lake (37°11' to 37°22'N, 100°32' to 101°26'E) of Qinghai Province, China. This area is 3200 to 4800 m above sea level, with a dry, cold winter climate. Alpine and sub-alpine herbage meadows in the region are an important pasture for yak production, with a seasonal migration between different grazing areas. Typically, transhumance farming defined by switching between differ-ent seasonal pasture sites is practiced, with spring and winter pastures belonging to one type, but being divided by fences as different seasonal pastures (Ding et al., 2008).

2.2 Experimental animals

A total of sixty 3 year old growing yak steers [initial mean BW. (120.3±15.28) kg] were randomly selected and appro-priately marked for identification, the animals remained with the rest of the herds throughout the year round.The tests were consecutive conducted bi-monthly starting from May and finishing on April of the next year, which was considered to correspond with the growing stages of the Plateau grasses, as well as the changes of the yak's nutrition status in a whole production year.The yaks were weighed during the exper-iment with an electronic balance.The slaughter group was subdivided into six groups of 5 yaks, each used to determine the seasonal and nutritional effects on chemical body composition and LPL activity in different tissues.

Blood samples were collected from the *Vena jugularis*, about 1 h prior to transport to the slaughterhouse.Plasma samples were stored at $-70°C$ after centrifugation at $3000 \times g$ for 20 min at $4°C$.After stunning with a shot device animals were killed by exsanguination.Samples of the liver, skeletal muscle were dissected from the sternomandibularis and subcutaneous, adipose tissue was sampled perianal and adjacent the caudal vertebrae.Then they were cut into pieces and were frozen less than 10 min post slaughter in liquid nitrogen and stored at $-80°C$ for subsequent analyses.

2.3 Herbage measurements and chemical composition

Herbage mass (HM) and daily intake at pastures on offer were determined in mid-month, on the forage samples cut within the frame of a 0.5 m×0.4 m quadrat.On each occasion, the quadrat was randomly positioned on 10-20 different places within each paddock, and the herbage were cut to ground level.Herbage samples were pooled, homogenized, weighed and two sub-samples were taken.Subsample 1 (500 g) was immediately oven-dried (60°C, 72 h) to determine dry matter (DM) contents and was subsequently used for HM estimations (kg DM ha^{-1}). Subsample 2 (300 g) was oven-dried (60°C for 72 h), ground to pass through a 1 mm screen, and later used for analyses of chemical compositions.The organic matter (OM) contents were calculated as the difference between DM and ash contents with ash being determined by combustion at 550°C overnight.N was determined by a Kjeldahl method using a Kjeltec 1030 autoanalyzer (Tecator Inc., Herndon., VI, USA).Neutral detergent fiber (aNDFom) and acid detergent fiber (ADFom) were determined according to the methods described by Van Soest et al. (1991) using α-amylase and sodium sulfite in the NDF procedure and correcting values for ash content.Gross energy was analyzed using an adiabatic bomb calorimeter (Gallenkamp Autobomb, Loughborough, Leics, UK). Intake on pasture was calculatedfor the experimental groupfromthe average difference in forage mass immediately pre- and post-grazing as measured on 3 to 5 consecutive days in each month (Ding et al., 2010).

2.4 Assay of LPL activity

Total tissues of LPL were released from frozen samples by homogenizing at 4°C in 9 mL per g of the following buffer: ammonia-HCl (25 mM), pH 8.2 containing EDTA (5 mM), Triton X-100 (0.8% w:v), SDS (0.01% w:v), heparin (5 000 IU·l^{-1}) and peptidase inhibitors [pepstatin A (1μg·mL^{-1}), leupeptin (10μg·mL^{-1}), and aprotinin (0.017 TIU·mL^{1})]. Insoluble mate-rial was removed by centrifugation (20 000×g for 20 min at 4°C).Activity of LPL was measured at 25°C with rat serum as activator and Intralipid ® (Pharmacia and Upjohn, Stockholm, Sweden) as substrate into which [^{3}H] triolein had been incorporated by sonification (Hocquette, et al., 1998). Activity of LPL was expressed in nmol fatty acid liberated per (min g tissue).

2.5 Plasma hormone assay

The determination of insulin concentration was performed by the porcine insulin RIA kit. Triglyceride was determined by Cleantech TG-S kit (Iatron Co. Ltd., Tokyo, Japan) following the standard protocol.

2.6 Statistics

Differences between groups were analyzed by one-way ANOVA analysis of variance, and the Student's t-test was used for comparisons between groups of animals, using the statistical program SPSS 12.0 (SPSS Inc, Chicago, IL, USA).

3 RESULTS

3.1 Herbage mass, chemical compositions and nutrients intake

The evolution of HM and daily intake at alpine meadows are shown in Fig. 1. The above ground HM of the native forages, in particular sedge, grass and forb, varied dramati-cally with the seasonal advances. Monthly biomass was highest in mid-summer and lowest in cold winter, ranged from 252 to 4578 kg DM/ha. The daily intakes of yak ranged from 1.80 to 8.50 kg DM throughout the grazing year. During the warm season (June to September), herbage provided by the native green forages was surplus to animal requirements, whereas, in the cold season (October to January) a relative surplus of more mature and dry forages is available to the yaks. With shortage of dry forages from February to May, yaks suffered malnutrition because intakes were far below those required for maintenance (NRC, 1989).

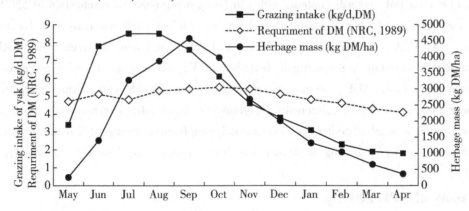

Fig. 1 Biomass production of above ground vegetation and grazing intake by yak during the year on the Qinghai-Tibetan Plateau

The CP and GE contents of plucked pasture samples were found highest in summer (GE, 21.31 MJ/kg; CP, 111.6 g/kg DM) and lowest (GE, 13.66 MJ/kg; CP, 48.2-54.1 g/kg DM) in spring and winter, whereas the aNDFom and the ADFom of the herbages on offer in summer (aNDFom, 497.4 g/kg DM; ADFom, 309.1 g/kg DM) were the lowest and highest in spring and winter (aNDFom, 580.3 – 598.8 g/kg DM; ADFom, 348.6 – 358.4 g/kg DM) (Table 1).

Table 1 Chemical compositions of herbages on offer and intake by yak in different grazing season (mean±S.E.)

	Spring	Summer	Autumn	Winter
Compositions (g/kg DM)				
Crude protein	48.2±6.93^c	113.6±8.72^a	90.1±3.99^b	54.1±9.81^c
Organic mater	943.9±1.37	938.2±4.02	927.5±2.81	827.8±4.50
ADFom	356.4±5.13^a	309.1±5.63^c	321.1±3.71^b	348.6±4.50^a
aNDFom	580.3±3.16^a	497.4±2.31^b	563.9±2.15^a	598.8±4.64^a
GE (MJ/kg)	14.63±0.68^c	21.31±1.07^a	17.86±1.03^b	13.66±0.88^c
Intake (kg/day)				
Dry matter	2.36±0.89^c	8.27±0.40^a	6.85±0.75^b	3.45±0.98^c
Crude protein	0.11±0.025^c	0.94±0.172^a	0.62±0.059^b	0.19±0.023^c
Organic matter	2.23±0.372^a	7.76±0.348^b	6.35±0.427^b	2.86±0.312^a
GE (MJ)	34.5±2.523^c	176.2±9.683^a	122.3±8.327^b	47.1±3.154^d

Note: Numbers within columns followed by different superscript letters are significantly different at the 0.05 level.

In the winter-spring and autumn, the CP intake of the grazing yak was 0.11–0.19 kg·d^{-1} and 0.62 kg·d^{-1}, respec-tively. The CP intake was found in summer, which is sufficient for the maintenance requirement of grazing yak. The GE intake of the grazing yak in summer, autumn, spring and winter was 176.2, 122.3, 34.5 and 47.1 KJ·d^{-1}, respectively.

3.2 Growth rate, body composition and plasma hormones

As would be expected, performance of the yak is highly influenced by the seasonal changes. In warm seasons, exper-imental animals showed ca.0.32 kg average daily gain. The weight loss over the winter was around 0.12–0.15 of the weight before the onset of winter. The maximum BW loss for all grazing animals appeared in March approximately to 0.25 of the weight (Fig.1). Typically, over the following summer and autumn yak regain their weight losses and may well double in weight before again losing, over the second winter of life, perhaps 0.34 of the maximum weight reached (Table 2). The cycle of weight gain and weight loss continues throughout life.

Table 2 Carcass measurements and plasma hormone levels in grazing yak in different season (mean±S.E.) ($n=5$)

	Warm sea			Cold sea		
	May	Jul.	Sept.	Nov.	Jan.	Mar.
Live weight	125.9±34.3^a	151.4±17.9^b	174.6±18.1^b	160.7±22.4^b	132.7±16.5^c	114.8±22.3^a
Body composition						
Moisture	65.2±8.7^a	76.3±10.3^c	72.8±6.5^c	73.7±7.4^c	69.8±3.9^b	68.8±6.2^b
Body fat (kg)	12.67±2.1^c	18.67±3.5^b	19.29±3.2^b	17.72±5.8^b	14.63±6.7^c	13.89±3.3^c
Plasma TG (mM)	12.67±2.3^c	14.63±6.6^c	13.89±5.9^c	18.67±8.2^a	19.29±7.4^a	17.72±5.6^a
Insulin level (μg/mL)	255.75±67.4^a	308.30±46.8^b	321.14±78.1^b	212.28±45.9^{ac}	250.55±48.2^a	197.33±34.3^c

Note: ^{a,b,c}Means with different superscripts within rows differ (P<0.05).

Body weight is the average value of triplicate in individual animal. The other values are the mean of 60 randomly selected grazing yaks.

Data on body composition of yak over different season of the year are presented inTable 2. The body fat content, in contrast, increased 0.16 during the warm season, and that reached the peak in September which was about 18.67 kg.The plasma insulin concentrations tended to be higher ($P<0.01$) than that in cold season.However, circulating levels of TG were higher in the cold season with the mean value of (18.56±0.79) nM ($P<0.05$) (Fig. 2).

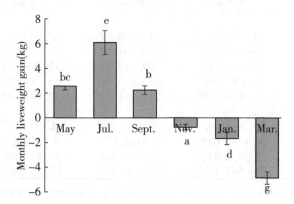

Fig. 2 Monthly gain in body weight gain of grazing yak cows (n=60) (mean4.E.) Different letters indicate significant differences (P<0.05)

3.3 Effect of nutrients intake on adipose tissue LPL activity

Average LPL activity over the entire grazing year revealed a graded in subcutaneous adipose tissue> skeletal muscle> liver with the average of 1609, 236 and 155 mU/g tissue, respectively.In adipose tissue, LPL activity was increased sharply ($P<0.05$) in tissue from yaks grazed on warm (3-14 fold) (Fig. 3A), and in March the lowest LPL activity (226 mU/g) was found. Conversely, in July, when ambient temperature was gradually rising, LPL activity increased and remained appreciable high (3130 mU/min g tissues) until the late October.In addition, LPL activity in the skeletal muscle was lowered in winter, increasing progressively and signif-icantly during spring and summer (Fig. 3B).The similar seasonal LPL pattern existed in liver, in which the lowest LPL activity was found in winter with a clear increase in earlier spring (Fig. 3C).

4 DISCUSSION

This paper describes for thefirst time, in domesticated yak species adipose lipoprotein lipase, a key enzyme in triglycer-ide storage, and the role of insulin in its seasonal activity.During the hydrolysis of circulating triglycerides by adipose LPL, liberated fatty acids are available as triglyceride for tissues as energy source storage for utilization (Hocquette et al., 1998).Generally, shift in both quantity and quality of Tibetan forages affects performance of yaks.In the present study, 0.21-0.29 live-weight of yak was reduced at early spring then recovered during following warm season when fed on pasture ad libitum, which coincided with the previously found in the alpine meadow (Long et al., 2005).Restricted food supply is severe across the seasons on the Qinghai-Tibetan Plateau regions.Refeeding after a restriction period results in a superior growth rate, and also in modification of fat storage and release by various tissues.

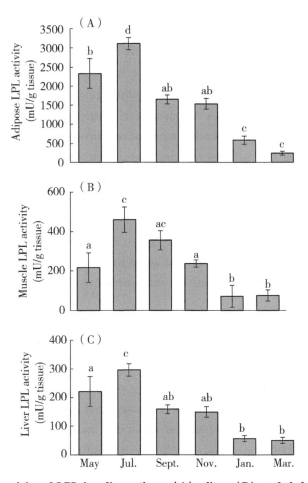

Fig. 3　Seasonal activity of LPL in adipose tissue (A), liver (B) and skeletal muscle (C)
Note: Data are represented as mean ± SE (n = 5). For each tissue, different letters indicate significant differences over sampling time (P<0.05).

In a similar manner, the previously reported decrease of feed intake and body fat content with the lower energy intake (Andersen et al., 1996) was related in this work with a lowered LPL activity in the adipose tissue during cold seasons when the forage was shortage. Until late spring, activity of LPL in the subcutaneous adipose tissue significantly increased (P<0.05), which may be part of the mechanisms involved in the replenishment of extra-hepatic lipid depots during compensatory growth (Hocquette, 2010; Hocquette et al., 1998), and changes that were likely associated with greater insulin secretion. Our results in yak species are in agreement with previous studies in sheep (Bonnet et al., 1998; Bonnet et al., 2000) and cows (Bonnet et al., 2004) that ruminant adipose tissue LPL activity is down-regulated by underfeeding and up-regulated by refeeding.

Greater and lower LPL activities occurred in adipose tissue and muscle tissue was seen in yak species. This may be attributed to the special physiological characteristics of yak, physiological characteristics, as the fat is located primarily on the outside of the carcass, so 95%–97% fat free in muscle. In ovine and other species the inverse relationship has been reported for different energy consumption (Bonnet et al., 2000; Doolittle et al., 1990). Additionally, expression and activity of

LPL are often inversely regulated in adipose tissue and muscle during fasting and feeding (Lithell et al., 1978; Mead et al., 2002; Ruge et al., 2005). Bonnet et al. (2000) demonstrated that LPL mRNA level is decreased by under-feeding and sharply increased by refeeding in the oxidative cardiac muscle and to a lesser extent in the oxidoglycolytic longissimus thoracis in sheep (Bonnet et al., 2000) and cows (Bonnet et al., 2004). This down-regulation may be specific to ruminants, thanks to their digestion particularities and low liver lipogenic capacity (Bonnet et al., 2000, 2004). Generally, to a great extent, underfeeding affects lipogenic activities in adipose tissue and refeeding restores those activities. How-ever, nutritional factors regulate sharply the expression and activities of LPL, moderately the activities of other lipogenic enzymes (glucose-6 phosphate dehydrogenase, malic en-zyme and glycerol-3-phosphate dehydrogenase) and to a lower extent the lipolytic activity of the hormone-sensitive lipase (Bonnet et al., 1998). However, reciprocal regulation of LPL between adipose tissue and skeletal muscle may be less distinct because starvation and realimentation did not cause truly opposite effects in these tissues in rats (Sugden et al., 1993). In the present study, it was shown that LPL activity was sensitive as a result of changes in season and nutritional conditions.

It is well known that insulin is a key hormone stimulating LPL activity, in tandem with a higher plasma insulin level in warm seasons is likely one of the reasons for higher body fat content in grazing yak (Table 2), which may trigger activation of the enzyme (Picard et al., 1999). As a consequence of the abundant grazing in summer and early autumn, yak are normally able to develop a layer of subcutaneous fat that also provide them with insulation from cold as well as an energy reserve during the period of nutritional deprivation over winter and spring.

Hocquette et al. (1998) demonstrated the levels of LPL transcripts are positively related to LPL activity in bovine tissues, including muscles and adipose tissues, which have been confirmed by the latter by the presence of substantial amounts of LPL transcripts in the fat depots of cattle. In adult sheep at maintenance levels, 55%–60% of the total amount of free fatty acids originate from hydrolysis of circulating triacylglycerols by LPL (Pethick and Dunshea, 1993), and the skeletal muscle mass and the heart together could utilize approximately 40% of the non-esterified fatty acid entry rate. It is also evident now for a tissue-specific regulation of LPL in muscle and adipose tissue in ruminants (Bonnet et al., 2000; Pethick and Dunshea, 1993). The fraction of VLDL is a minor component of plasma lipoproteins which reflects the lipid metabolism characteristics in the ruminant liver, and VLDL triglyceride secretion by the ruminant liver is considered to be very low (Bauchart et al., 1996; Emmison et al., 1991; Kleppe et al., 1988; Mamo et al., 1983; Mazur et al., 1992; Pullen et al., 1990). In addition, the LPL mRNA transcript in bovine liver was in amounts too low to detect. So, the LPL activity is that of the blood LPL transported from muscles and adipose tissue to the liver, not the liver LPL per se. In the current study, the liver also has the capacity to store appreciable amounts of TG in domestic yak, which in consistent with the previous observa-tion in ewes (Mazur et al., 1992). LPL activity in liver is up-regulated at late spring in the present study, which suggests that the LPL in the liver tissue is not causally related to the known autumn rise in liver fat content (Mingarro et al., 2002). This may be attributed to the fact that the hepatic TG entity is a dynamic fat depot derived from different sources (dietary chylomicrons, de novo fatty acid synthesis, and tissue-released NEFA) (Saera-Vila et al., 2005), which makes feasible the pivotal role of liver on the adjustment of lipid metabolic flow in response to changes in metabolic needs of the body (Albalat et al., 2006).

A clear seasonal pattern found in liver and adipose tissue has been reported infish by Saera-Vila et al. (2005), which is consistent with grazing yak in the present study. The highest LPL in muscle was observed in July and this would serve to increase energy availability over the course of summer growth spurt, which is also accompanied by the up-regulation of hepatic growth hormone receptors and circulating levels of insulin-like growth factor-I (IGF-I) (Calduch-Giner et al., 2003; Perez-Sanchez et al., 2002). In herbivores, the LPL transcript first in fat storage tissues and thereafter in skeletal muscle (Bonnet et al., 2000), this could be a major responsive factor for the observed decrease of circulating TG levels during the warm season.

5 CONCLUSIONS

The present study provides suitable evidence of LPL activity in grazing yak adipose tissue, muscle and liver, the tissue-regulation in response to seasonal shifts and nutrients intake of forages. Further, LPL appears to have evolved in a similar manner and the analysis of LPL activity is a useful tool to understand the regulation of lipid metabolism under different physiological conditions and to determine how dietary lipids are partitioned towards uptake, storage and mobilization. This could well point to an adaptive response of the yak to live at high altitude and to the nutritional deprivation that yak experience in winter and early spring. However, further quantification of LPL gen and mRNA expression is required, so as to assess the transcriptional regulation of LPL in different tissues.

ACKNOWLEDGMENTS

This work is partly supported by grants from National Natural Science Foundation of China (Grant No. 31101702), Central Public-interest Scientific Institution Basal Research Fund (Contract No.1610322011002), National Beef Cattle Industry Technology & System (MATS-Beef Cattle system: NYCYTX-38) and Special Fund for Agro-scientific Research in the Public Interest (Contract No.201003061). The authors thank the staff of San Jiao Cheng Sheep Breeding Farm for assistance with the fieldwork.

References

[1] Albalat, A., Sánchez-Gurmaches, J., Gutiérrez, J., Navarro, I., 2006. Regulation of lipoprotein lipase activity in rainbow trout (*Oncorhynchus mykiss*) tissues. Gen. Comp. Endocrinol. 146: 226–235.

[2] Andersen, M.K., Bailey, J.W., Wilken, C., Rule, D.C., 1996. Lipoprotein lipase and glycerophosphate acyltransferase in ovine tissues are influenced by growth and energy intake regimen. J. Nutr. Biochem. 7: 610–616.

[3] Auwerx, J., Leroy, P., Schoonjans, K., 1992. Lipoprotein lipase: recent contributions from molecular biology. Crit. Rev. Clin. Lab. Sci. 29: 243–268.

[4] Bauchart, D., Gruffat, D., Durand, D., 1996. Lipid absorption and hepatic metabolism in ruminants. Proc. Nutr. Soc. 55: 39–47.

[5] Bonnet, M., Faulconnier, Y., Fléchet, J., Hocquette, J.F., Leroux, C., Langin, D., Martin,

P., Chilliard, Y., 1998. Messenger RNAs encoding lipoprotein lipase, fatty acid synthase and hormone-sensitive lipase in the adipose tissue of underfed-refed ewes and cows. Reprod. Nutr. Dev. 38: 297-307.

[6] Bonnet, M., Leroux, C., Faulconnier, Y., Hocquette, J. F., Bocquier, F., Martin, P., Chilliard, Y., 2000. Lipoprotein lipase activity and mRNA are up regulated by refeeding in adipose tissue and cardiac muscle of sheep.J.Nutr.1 (30): 749-756.

[7] Bonnet, M., Faulconnier, Y., Hocquette, J. F., Bocquier, F., Leroux, C., Martin, P., Chilliard, Y., 2004. Nutritional status induces divergent variations of GLUT4 protein content, but not lipoprotein lipase activity, between adipose tissues and muscles in adult cattle.Br.J.Nutr.92: 617-625.

[8] Calduch-Giner, J. A., Mingarro, M., Vega-Rubin de Celis, S., Boujard, D., Perez-Sanchez, J., 2003. Molecular cloning and characterization of gilthead sea bream (*Sparus aurata*) growth hormone receptor (GHR).Assessment of alternative splicing.Comp.Biochem.Physiol.B 136: 1-13.

[9] Ding, L.M., Long, R.J., Shang, Z.H., Wang, C.T., Yang, Y.H., Xu, S.H., 2008. Feeding behaviour of yaks on spring, transitional, summer and winter pasture in the alpine region. Appl. Anim. Behav.Sci.111: 373-390.

[10] Ding, X.Z., Long, R.J., Kreuzer, M., Mi, J.D., Yang, B., 2010. Methane emissions from yak (*Bos grunniens*) steers grazing or kept indoors and fed diets with varying forage: concentrate ratio during the cold season on the Qinghai-Tibetan Plateau.Anim.Feed.Sci.Technol.162: 91-98.

[11] Doolittle, M.H., Benzeev, O., Elovson, J., Martin, D., Kirchgessner, T.G., 1990. The response of lipoprotein-lipase to feeding and fasting-evidence for posttranslational regulation. J. Biol. Chem. 265: 4 570-4 577.

[12] Emmison, N., Agius, L., Zammi, t V. A., 1991. Regulation of fatty acid metabolism and glucogenesis by growth hormone and insulin in sheep hepatocyte cultures.Effect of lactation and pregnancy.Biochem.J.274: 21-26.

[13] Hocquette, J.F., 2010. Endocrine and metabolic regulation of muscle growth and body composition in cattle.Animal.doi: 10. 1017/S1751731110001448.

[14] Hocquette, G.F., Graulet, B., Olivecrona, T., 1998. Lipoprotein lipase activity and mRNA levels in bovine tissues.Comp.Biochem.Physiol.B 121: 201-212.

[15] Kleppe, B.B., Aiello, R.J., Grummer, R.R., Armentano, L.E., 1988. Triglyceride accumulation and very low density lipoprotein secretion by rat and goat hepatocytes *in vitro*. J. Dairy Sci. 71: 1 813-1 822.

[16] Lithell, H., Boberg, J., Hellsing, K., Lundqvist, G., Vessby, B., 1978. Lipoprotein-lipase activity in human skeletal muscle and adipose tissue in the fasting and the fed states.Atherosclerosis 30: 89-94.

[17] Long, R.J., Zhang, D.G., Wang, X., Hu, Z.Z., Dong, S.K., 1999. Effect of strategic feed supplementation on productive and reproductive performance in yak cows.Prev.Vet.Med.38: 195-206.

[18] Long, R.J., Dong, S.K., Wei, X.H., Pu, X.P., 2005. The effect of supplementary feeds on the bodyweight of yaks in cold season.Livest.Prod.Sci.93: 197-204.

[19] Mamo, J.C., Snoswell, A.M., Topping, D.L., 1983. Plasma triacylglycerol secretion in sheep.Paradoxical effects of fasting and alloxan diabetes.Biochim.Biophys.Acta 753: 272-275.

[20] Mazur, A., Ayrault-Jarrie, r M., Chilliard, Y., Rayssiguier, Y., 1992. Lipoprotein metabolism in fatty liver dairy cows.Diabete Metab.18: 145-149.

[21] Mead, J.R., Irvine, S.A., Ramji, D.P., 2002. Lipoprotein lipase: structure, function, regulation, and role in disease.J.Mol.Med.80: 753-769.

[22] Mingarro, M., Vega-Rubin de Celis, S., Astola, A., Pendon, C., Valdivia, M. M., Perez-Sanchez, J., 2002. Endocrine mediators of seasonal growth in gilthead sea bream (*Sparus aurata*): the growth hormone and somato-lactin paradigm.Gen.Comp.Endocrinol.128: 102-111.

[23] NRC, 1989. Nutrient requirements of dairy cattle, Natl.Acad.Sci., Washing-ton, DC, 6th rev.ed..

[24] Perez-Sanchez, J., Calduch-Giner, J. A., Mingarro, M., Vega-Rubın de Celis, S., Gomez-Requeni, P., Saera-Vila, A., Astola, A., Valdivia, M.M., 2002. Overview of fish growth hormone family.New insights in genomic organization and heterogeneity of growth hormone receptors.Fish Physiol. Biochem.27: 243-258.

[25] Pethick, D.W., Dunshea, F.R., 1993. Fat metabolism and turnover.In: Forbes, J.M., France, J. (Eds.), Quantitative Aspects of Ruminant Digestion and Metabolism.CAB International Press, Wallingford: 291-311.

[26] Picard, F., Maimi, N., Richard, D., Deshaies, Y., 1999. Response of adipose tissue lipoprotein lipase to the cephalic phase of insulin secretion.Diabetes 48: 452-459.

[27] Pullen, D.L., Liesman, J.S., Emery, R.S., 1990. A species comparison of liver slice synthesis and secretion of triacylglycerol from nonesterified fatty acids in media.J.Anim.Sci.68: 1 395-1 399.

[28] Ruge, T., Svensson, M., Eriksson, J.W., Olivecrona, G., 2005. Tissue-specific regulation of lipoprotein lipase in humans: effects of fasting.Eur.J.Clin.Investig.35: 194-200.

[29] Saera-Vila, A., Calduch-Giner, J.A., Pérez-Sánchez, J., 2005. Duplication of growth hormone receptor (GHR) infish genome: gene organization and transcriptional regulation of GHR type I and II in gilthead sea bream (*Sparus aurata*).Gen.Comp.Endocrinol.142: 193-203.

[30] Sugden, M.C., Holness, M.J., Howard, R.M., 1993. Changes in lipoprotein lipase activities in adipose tissue, heart and skeletal muscle during continuous or interrupted feeding. Biochem. J. 292: 113-119.

[31] Van Soest, P.J., Robertson, J.B., Lewis, B.I., 1991. Methods for dietaryfiber, neutral detergent fiber and non-starch polysaccharide in relation to animal nutrition.J.Dairy Sci.74: 3 583-3 597.

[32] Wang, D.P., Li, H.G., Li, Y.J., Guo, S.C., Yang, J., Qi, D.L., Jin, C., Zhao, X.Q., 2006. Hypoxia-inducible factor 1α cDNA cloning and its mRNA and protein tissue specific expression in domestic yak (*Bos grunniens*) from Qinghai-Tibetan plateau. Biochem. Biophys. Res. Commun. 348: 310-319.

(发表于《Livestock Science》, 院选 IF: 1.294)

Effect of Several Supplemental Chinese Herbs Additives on Rumen Fermentation, Antioxidant Function and Nutrient Digestibility in Sheep[*]

QIAO G.H.[1], ZHOU X.H.[1], LI Y.[2], ZHANG H.S.[2], LI J.H.[2], WANG C.M.[2], LU Y.[2]

(1. Department of Prataculture and Fodder, Lanzhou Institute of Animal & Veterinarian Pharmaceutics Science, Chinese Academy of Agricultural Sci-ence, Lanzhou, China; 2. College of Food Science, Northeast Agricultural University, Harbin, China)

Abstract: Two experiments were carried out in this study. Experiment 1 was con-ducted to examine the effects of several supplemental Chinese herbs on antioxidant function and slaughtered body weight in sheep. Results indicated that Fructus Ligustri Lucidi supplementation improved the blood antioxidant function [higher concentration of glutathione reduc-tase (GR), superoxide dismutase and lower concentration of malondial-dehyde] and slaughtered body weight in sheep ($p < 0.05$). Experiment 2 was conducted to investigate the effect of Fructus Ligustri Lucidi extract (FLLE) on rumen fermentation and nutrient digestibility in sheep. Four levels of FLLE treatments, i.e. 0, 100, 300 and 500 mg/kg dry matter (DM), were used in this part. Addition of FLLE at 300 or 500 mg/kg DM increased total volatile fatty acid (VFA) concentration and propionate proportion, decreased ammonia-N concentration in the ruminal fluid, reduced blood urea nitrogen concentration at 2, 4, 6 and 8 h after morning feeding ($p < 0.05$). Addition of FLLE at all dosages had no effect on ruminal pH value and acetate concentration at all sampling time points in sheep ($p > 0.05$). Dynamic degradation coefficient c of maize DM was significantly increased by supplementing FLLE at 300 or 500 mg/kg DM ($p < 0.05$). Fructus Ligustri Lucidi extract addition had no effect on degradation coefficients a, b, c of DM and nitrogen of soy-bean meal; a, b of maize DM; a, b, c of maize nitrogen; and a, b, c of neutral detergent fibre (NDF) and acid detergent fibre (ADF) of Chinese wildrye ($p > 0.05$). Addition of FLLE at 300 or 500 mg/kg DM increased DM and organic matter digestibility of diet ($p < 0.05$). Fructus Ligustri Lucidi extract addition had no effect on digestibility of diet's NDF, ADF and crude protein ($p > 0.05$). From the aforementioned results, it is indicated that FLLE improved antioxidant status and slaughtered body weight. Fructus Ligustri Lucidi extract addition has capability to modu-late rumen fermentation, increase the maize

---

[*] Basal research fund of Chinese Academy of Agricul-tural Science (BRF100203).

degradation rate, total vola-tile fatty acid concentration and propionate proportion in sheep.

Key words: Chinese herbs; Rumen fermentation; Nutrient digestibility; Antioxidant function; Sheep

INTRODUCTION

The use of antibiotics in ruminant feeds is facing reduced social acceptance because of the appearance of residues and resistant strains of bacteria. The use of antibiotics has been banned in ruminant feeds in China. Now, animal nutritionists are studying the substitutes of antibiotics as green feed additives. There is a very long history of Chinese herbs and their extracts has been characterised as antiseptics and antimicrobials. Ma et al. (2004, 2005) reported that Fructus Ligustri Lucidi extract (FLLE) reduced lipid peroxides, superoxide dismutase (SOD) and glu-tathione peroxidase (GSH) content significantly in laying hens. Zhang et al. (2004, 2005) found that some Chinese herbs, Fructus Ligustri Lucidi, Radix Astragali, Radix Codonopsis, have capability to improve antioxidant function and milk yield in cows. In recent years, many sorts of plant extracts saponins, tannins and essential oils derived from yucca, chestnut, garlic, ginger, FLLE, etc. have been used in ruminant feeds to modulate rumen fermen-tation (Benchaar et al., 2006, 2008; Busquet et al., 2006; Cardozo et al., 2006; Lila et al., 2003; Avato et al., 2006). It was reported that addition of FLLE at 500 mg/kg dry matter (DM) improved nutrient deg-radation rate at 48 h in the rumen and increased total volatile fatty acid concentration in ruminal fluid in Northeast fine-wool sheep (Xu et al., 2008). The active components of FLLE are saponins and their derivatives (Che and Chen, 2009). But the action mechanism of FLLE in sheep production remains confusing. The aim of this study is to deter-mine the effects of Chinese herbs additives, Fructus Ligustri Lucidi, Radix Astragali, Radix Codonopsis, on antioxidant function and productive performance in sheep. Furthermore, herbs that improved produc-tive performance will be extracted. The extracts will be used as additives to investigate its action mecha-nism on rumen fermentation and nutrient digestibility in sheep.

MATERIALS AND METHODS

Animals and experimental procedure

In experiment 1, 16 sheep (Dorset sheep ♂ ×Small Tail Han ♀, initial body weight 22.30± 3.06 kg) at 4-month age were selected and fed 1 month in this part. Four sheep with close body weight were blocked into one treatment. At the end of the exper-iment, sheep was slaughtered and blood samples were removed. Basal diet was formulated as in Table 1. Bodies without head, hoof, leather and internal organs were weighted. Sheep was fed twice per day at 06:00 and 18:00 hours *ad libitum*. In experiment 2, four sheep (body weight 35±5 kg, 3.5 years old) with permanent ru-minal fistulas were used in the experiment. A 4×4 Latin square design (four animals; four periods) was carried out in this experiment. There were 28 days in each period (21 days for adaptation and 7 days for sample collection). Clean water was available at all time in the individual cage. Animals were fed twice per day at 06:00 and 18:00 hours *ad libitum*. The herb was mixed with a small part of concentrate in advance to ensure complete consumption of herb by sheep. The concentrate and forage of the diet were totally mixed by hand and offered to sheep. The diet formula is as in Table 1.

Table 1 Composition and nutrient level of the diets (DM basis, g/kg)

	Group 1	Group 2	Group 3	Group 4
Ingredients				
Chinese wildrye	400	400	400	400
Maize (cracked)	430	430	430	430
Soybean meal (sol)	140	140	140	140
Urea	10	10	10	10
Vitamin, mineral premix*	20	20	20	20
Control[†] (g/day)	0			
Fructus Ligustri Lucidi[†] (g/day)		6		
Radix Astragali[†] (g/day)			6	
Radix Codonopsis[†] (g/day)				6
Fructus Ligustri Lucidi extract[‡] (mg/kg)	0	100	300	500
Nutrient levels				
Gross energy (MJ/kg DM)	17.75	17.75	17.75	17.75
CP	133.6	133.6	133.6	133.6
NDF	313.7	313.7	313.7	313.7
ADF	184.6	184.6	184.6	184.6
Ca	6.4	6.4	6.4	6.4
P	4.2	4.2	4.2	4.2

Note: ADF, acid detergent fibre; DM, dry matter; NDF, neutral detergent fibre.

*Vitamin, mineral premix (g/kg): 250 bicalcic phosphate, 250 magnesium oxide, 25 zinc sulphate, 25 ferric sulphate, 10 manganese sulphate, 8.0 copper sulphate, 433 sodium bicarbonate, 0.5 of a mixture of vitamins A and D3, 1 vitamin E 50% and, (mg/kg): 0.5 sodium selenite, 2 potassium iodide, 2 cobalt sulphate.

[†] Represents treatments in experiment 1.

[‡] Represents treatments in experiment 2.

Chinese herbs additives

In experiment 1, Fructus Ligustri Lucidi (seed), Radix Astragali (root) and Radix Codonopsis (root) were collected in Anyue City, Sichuan Province in China. Fructus Ligustri Lucidi seed, Radix Asragali root and Radix Codonopsis root were washed by tap water until it is clean, then were put into 40℃ oven for 24 h. Dry seed and root of herbs were grounded to pass 1-mm screen for further usage. In experiment 2, Fructus Ligustri Lucidi seed was extracted in basic hydrolysis solution (2 N KOH in MeOH: H_2O) by rotary evaporator (RE5220, China; Tava et al., 1993).

The extraction procession of Fructus Ligustri Lucidi was performed in the Department of Traditional Chinese Veterinarian Pharmaceutics, Lanzhou Institute of Animal and Veterinary Pharmaceutics Science, Chinese Acad-emy of Agricultural Sciences.The solute containing extracts was freeze-dried and collected for further usage in the experiment.The main active compo-nents of FLLE extracted by the method of Tava et al. (1993) are oleanolic acid (pentacyclic triterp-enes) and saponins (ligustroside, 10-hydroxy ligu-stroside and nuezhenide) (Che and Chen, 2009).It was reported that the active components of extracts extracted by the method mentioned earlier are a group of steroidal glycosides.Herbs were added into diet at morning feed per day and ensured all herbs or FLLE supplemented were consumed by sheep.The dosages of three herbs in experiment 1 and FLLE in experiment 2 were according to Xu et al. (2008).

Ruminal and blood sampling

In experiment 2, approximately 100 mL rumen fluid samples were removed on day 26 of each period at 0, 2, 4, 6 and 8 h after morning feeding by vacuum pump.Each sample was strained immediately through four layers of cheesecloth.Ruminal pH was measured immediately after sampling using portable pH meter.The samples for VFA were acidified with 2 mL of 25% metaphosphoric acid and centrifuged at 4℃, 3000 g for 15 min.Supernatant was frozen (-20℃) for subsequent analysis. Rumen ammonia-N was determined by the hypochlorite-phenol proce-dure (Becher and Whitton, 1970).Volatile fatty acids were determined using a gas chromatograph (GC 2010, 30m× 0.32mm×0.25 mm capillary column, phase is diethylene glycol succinate and phosphoric acid, carrier gas is N_2), as described by Merchen et al. (1986).

Blood samples were removed from the jugular vein using 10 mL heparinised vacutainers at the same time when rumen fluid was sampled.The tubes were gently inverted a couple of times, then kept in an ice box and later centrifuged at 4℃, 2500 g for 15 min.The plasma was then transferred into storage tube and labelled with date and animal identification.The plasma samples were kept at) 20℃ until being analysed for blood urea nitrogen (BUN) using kit (Rongsheng Biochemical, Shanghai, China).All the blood antioxidant parameters of sheep in experiment 1 were determined using kits (Rongsheng Biochemical).

Feeds and faeces sampling

Feed intake was recorded at experimental period.Feeds were randomly collected and composited prior to analysis.Composited samples were ground to pass through a 1-mm screen and then analysed for DM, crude protein (CP), neutral detergent fibre (NDF), acid detergent fibre (ADF) and ash.Neutral deter-gent fibre contents of feed and faecal samples were determined using hot ethanol, 8M urea and heat-stable amylase (Sigma A 3306; Sigma Chemical, St.Louis, MO, USA; Van Soest et al., 1991).The repli-cas of each sample were eight.Faecal samples were totally collected from day 22 to 28 from sheep.Subs-amples (approximately 15% of total faeces) were taken and composited.The faeces were placed into an oven at 65℃ for 72 h, weighed and ground to pass through a 1-mm screen and then analysed for DM, ash, CP, NDF and ADF.

In situ degradability characteristics of feedstuffs

All feedstuff samples were ground to pass through a 1-mm screen for *in situ* degradability study.

Chemical components analysis of CP and DM in soybean meal and maize, and NDF and ADF in Chinese wildrye were analysed by the method of Van Soest et al. (1991). Ruminal degradation rate measurement was carried out using the nylon bag technique after 3 weeks adaptation period in each experimental per-iod. Five gram (fresh matter) of each test feed, maize, soybean meal and Chinese wildrye was accu-rately weighed into the nylon bag with a mean pore size of 53 μm (Shabi et al., 1998). Each feedstuff had three replicates. Bags plus samples were placed into the rumen at 30 min after the morning meal and retrieved after period of 2, 4, 6, 12, 24, 48 h for con-centrates and 2, 4, 6, 12, 24, 48, 72 h for Chinese wildrye respectively. Each feedstuff had three repli-cas. After removal from the rumen, bags were rinsed in pipe line fresh water and washed by hand under tap water until the bag became clear. The washed bags were placed into a hot air dry force oven at 65℃ for 48 h and weighed. To determine the con-tent of water-soluble material, bags representing 0 h degradation were also followed the same washing procedure as the incubated bags. Dried residues of each incubation time from each sheep were pooled. Dry matter and CP were ana-lysed. Dry matter and CP disappearance values were calculated as the difference between weight of nutrients before and after incubation of each sample. The degradability data obtained for DM and N for each feed were fit-ted to the equation $P = a + b(1 - e^{)ct})$ (Ørskov and McDonald, 1979), where P is the amount degraded at time t, a is the rapidly soluble fraction, b is the potentially de-gradable fraction and c is the degrada-tion rate of fraction b.

Statistical analysis

In experiment 1, randomized completely block design was used. Model sums of squares were sepa-rated into animal and treatment. The statistical model was as follows: $Y_{ijk} = \mu + a_i + \beta_j + \varepsilon_{ijk}$, where μ is the overall mean, a is the treatment effect, β is the animal effect and ε is the error term. The differ-ences between groups were established by means of Duncan's multiple range test at a signifi-cance level of $p<0.05$ or $p<0.01$. In experiment 2, a 4×4 Latin square design (four animals; four periods) was carried out in this experiment. All data obtained from the experiment were subjected to the general linear models (GLM) procedure for orthogonal poly-nomial analysis of SAS (SAS, 2003). Model sums of squares were separated into animal, period and treatment. The statistical model was as follows: $Y_{ijk} = \mu + a_i + \beta_j + \gamma_k + \varepsilon_{ijk}$, where μ is the overall mean, a is the treatment effect, β is the effect of the animal, γ is the effect of the period and e is the error term. The differ-ences between groups were estab-lished by means of Duncan's multiple range test at a significance level of $p<0.05$. Dry matter and CP degradation rate coefficients were calculated as described by Ørskov and McDonald (1979) using the NLIN procedure of SAS 8.2 (SAS Inst., Cary, NY, USA).

RESULTS

Effect of several Chinese herbs supplementation on blood antioxidant function and slaughtered body weight in sheep

The effect of several Chinese herbs supplementation on antioxidant function and slaughtered body weight gain in sheep is shown in Table 2. The initial body weight has no effect on all antioxidant parame-ters (p >0.05). Superoxide dismutase and GR in blood plasma were significantly increased in response to FLLE supplementation than the other Chinese herbs and control (p<0.05). The malondi-aldehyde (MDA) in blood plasma was significantly decreased in response to FLLE supplementation than the other Chinese herbs and control (p<0.05). Slaughtered body weight was significantly increased in response to FLLE supplementation than the other groups (p<0.05).

Table 2 Effect of Chinese herbs on SBW and blood antioxidant parameters in sheep

Items	Control	RC	RA	FLL	SEM	p-value Herbs	p-value IBW
SBW (kg)	13.92^a	13.39^a	14.14^a	16.89^b	0.36	<0.01	NS
GSH (mg/L)	173.33	173.64	172.70	174.48	0.44	NS	NS
GR (U/mL)	16.01^a	17.11^a	18.15^a	20.07^b	1.01	<0.05	NS
SOD (U/mL)	130.22^a	130.60^a	134.26^a	143.37^b	1.33	<0.05	NS
MDA (nmol/mL)	7.33^a	5.95^a	4.96^a	2.93^b	0.26	<0.05	NS

Note: GR, glutathione reductase; GSH, glutathione peroxidase; MDA, malondialdehyde; SOD, superoxide dismutase; RC, Radix Codonopsis; RA, Radix Astragali; FLL, Fructus Ligustri Lucidi; IBW, initial body weight; SBW, slaughtered body weight.

Values in the same row with different superscripts differ significantly (p<0.05 or p<0.01); NS represents not significant (p >0.05).

Effect of FLLE supplementation on rumen fermentation characteristics and BUN

Data of the rumen fermentation characteristics in sheep were shown in Fig. 1. Results indicate that ruminal fluid pH values were not changed at 0, 2, 4, 6 and 8 h after morning feeding by supplementing different dosages of FLLE (p >0.05, data were not shown). Ammonia nitrogen concentration of rumi-nal fluid peaked at 2 h after morning feeding, then went down at following sampling time points. Total volatile fatty acid concentration and propionate concentration were increased at 2, 4, 6 and 8 h in ruminal fluid of sheep in response to FLLE supple-mentation at dosages of 300 or 500 mg/kg DM after morning feeding (p<0.05). Acetate to propionate ratio was decreased at 2, 4, 6 and 8 h in ruminal fluid of sheep in response to FLLE supplementation at dosages of 300 or 500 mg/kg DM after morning feeding (p<0.05). Addition of FLLE at all dosages did not change acetate concentrations at each sam-pling time point (p >0.05). In response to addition of FLLE at 300 or 500 mg/kg DM, ruminal ammonia nitrogen concentration and blood urea nitrogen con-centration were decreased at 2, 4, 6 and 8 h after morning feeding (p<0.05; Fig. 2).

Effect of FLLE supplementation on nutrient degradation rate in sheep

Nutrient dynamic degradation rate of DM and nitro-gen in feedstuffs were shown in Table 3. Dry matter and nitrogen dynamic degradation coefficients (a, b) of maize, and a, b and c of soybean meal were not affected in response to addition of FLLE at all dos-ages. Only coefficient c value of maize DM was increased in response to addition of FLLE at 300 or 500 mg/kg DM (p<0.05).

Neutral detergent fibre and ADF dynamic degrada-tion coefficients a, b and c of Chinese wild-rye were not changed by supplementing FLLE at all levels (Table 4; p >0.05).

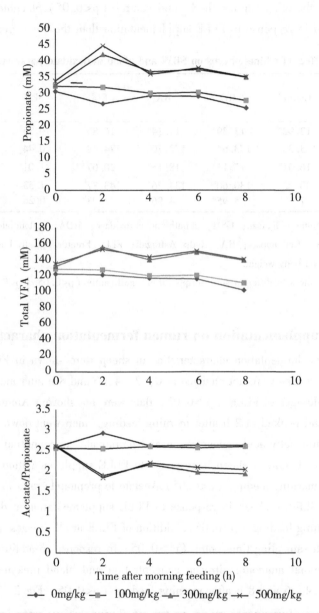

Fig. 1 Effect of supplemental Fructus Ligustri Lucidi extract on rumen fermentation parameters in sheep

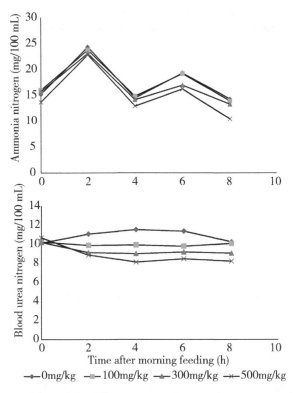

Fig. 2 Effect of Fructus Ligustri Lucidi extract supplementation on ammonia nitrogen concentration in ruminal fluid and blood urea nitrogen concentration in sheep

Table 3 Degradation coefficients of feedstuffs in sheep fed with different levels of FLLE

Feedstuffs	Coefficients	Treatments (mg/kg)				SEM	p-value
		0	100	300	500		
DM							
Maize	a	0.35	0.32	0.34	0.38	0.04	0.24
	b	0.63	0.66	0.63	0.60	0.07	0.27
	c per hour	0.023^a	0.025^a	0.029^b	0.036^c	0.06	0.01
Soybean meal	a	0.35	0.32	0.34	0.32	0.04	0.12
	b	0.63	0.66	0.61	0.64	0.04	0.37
	c per hour	0.045	0.057	0.042	0.046	0.07	0.19
N							
Maize	a	0.28	0.31	0.27	0.28	0.09	0.29
	b	0.44	0.46	0.44	0.42	0.03	0.82
	c per hour	0.052	0.058	0.046	0.055	0.06	0.12
Soybean meal	a	0.18	0.08	0.12	0.16	0.06	0.10
	b	0.74	0.88	0.83	0.85	0.07	0.65
	c per hour	0.047	0.046	0.048	0.044	0.11	0.11

Note: DM, dry matter; FLLE, Fructus Ligustri Lucidi extract; a, soluble fraction; b, slowly degradable fraction; c, fractional rate of disappearance of the b fraction; SEM, standard error of means.

Table 4 *In situ* degradation kinetics of Chinese wildrye incubated in the rumen of sheep fed diets supplemented with FLLE

Items	Treatments (mg/kg)				SEM	p-value
	0	100	300	500		
NDF						
a	0.07	0.07	0.07	0.06	0.12	0.14
b	0.77	0.72	0.73	0.74	0.14	0.10
c per hour	0.029	0.030	0.029	0.027	0.18	0.22
ADF						
a	0.02	0.03	0.02	0.02	0.14	0.17
b	0.83	0.76	0.75	0.79	0.11	0.28
c per hour	0.025	0.031	0.032	0.028	0.19	0.34

Note: ADF, acid detergent fibre; FLLE, Fructus Ligustri Lucidi extract; NDF, neutral detergent fibre; a, soluble fraction; b, slowly degradable fraction; c, fractional rate of disappearance of the b fraction.

Effect of supplemental FLLE on apparent nutrient digestibility in sheep

Effect of supplemental FLLE on nutrient digestibility in sheep was shown in Table 5. Results indicate that FLLE addition has no significant effect on diet's apparent digestibility of crude protein ($p>0.05$), NDF and ADF in sheep. Supplemental 500 mg/kg of DM in the diet increased apparent digestibility of DM and organic matter ($p<0.05$). Supplemental 300 mg/kg of DM to the diet increased apparent digestibility of organic matter ($p<0.05$). Dry matter intake (DMI) was not affected in response to addi-tion of FLLE at all dosages ($p>0.05$).

Table 5 Effect of supplemental FLLE on nutrient apparent total tract digestibility in sheep

Items	Treatments				SEM	p-value
	0 mg/kg	100 mg/kg	300 mg/kg	500 mg/kg		
DMI, g/day	109 0	110 0	108 7	109 5	4.56	0.25
Digestibility, %						
DM	63.8[a]	63.9[a]	65.4[a]	68.3[b]	0.58	0.04
OM	64.7[a]	64.0[a]	68.7[b]	69.5[b]	0.41	0.03
CP	62.3	61.1	59.8	60.4	1.45	0.15
NDF	55.7	57.1	54.6	57.3	1.52	0.19
ADF	50.5	49.2	47.7	50.7	1.23	0.22

Note: ADF, acid detergent fibre; DM, dry matter; DMI, dry matter intake; FLLE, Fructus Ligustri Lucidi extract; NDF, neutral detergent fibre.

Means within rows with different superscripts differ ($p<0.05$).

DISCUSSIONS

Effect of supplemental Chinese herbs on blood antioxidant function and slaughtered body weight in sheep

Under circumstances of poor antioxidant function status, unsaturated fatty acids in cell membrane will be oxidized, which leads to disruption of normal membrane structure and function, and production performance of animals will be decreased (Storey, 1996). Antioxidant enzymes such as SOD, GSH and GR present in blood are able to inhibit lipid oxida-tion of cell membrane and decrease end product (MDA) of cell membrane oxidation in blood. Anti-oxidant functional parameters and slaughtered body weight were significantly improved in response to FLLE supplementation, which indicates that FLLE has capability to improve antioxidant function of sheep. These results did support researches by some other scholars' who found that some Chinese herbs improved the antioxidant status of cows and laying hens. They found that milk yield was increased in lactating dairy cows and body weight were increased in laying hens by addition of FLLE (Ma et al., 2004; Zhang et al., 2004, 2005). Hu et al. (2006) reported that saponins from Medicago sp improved antioxidant status, nutrient digestibility and slaughtered body weight gain in sheep. Sub-stantial positive relationship between antioxidant status and production performance in sheep was declared in conclusion. In addition, nutrient digest-ibility of DM and organic matter (OM) was increased in response to FLLE supplementation in experiment 2. The improved blood antioxidant function and nutrient digestibility observed in these two experiments did support the idea of Hu et al.

Effect of supplemental FLLE on dry matter intake and nutrient digestibility

There are a few literatures reporting the effect of essential oils on DMI and nutrient digestibility in the *in vivo* experiments. It was reported that main essen-tial oil of Pseudostuga menziesii, limonene and pinene decreased the DMI in deer because of the organoleptic effects (Oh et al., 1968). Similarly, com-bination of cinnamaldehyde and eugenol reduced DMI in beef cattle (Cardozo et al., 2006). It was found that there was no effect of mixture of essen-tial oils (thymol, eugenol, vanillin, guaiacol and limonene) on DMI when lactating dairy cows were fed at a dose of 2 g/day (Benchaar et al., 2006). In this study, there is no effect of supplemental FLLE on DMI in sheep, which indicates that addition of FLLE has no capability to affect the palatability of the feed.

There is no effect of supplemental FLLE on crude-protein digestibility according to the result of this experiment. This result agrees with researches by some scholars' who reported that mixture of essential oils did not change the crude-protein digestibility (Benchaar et al., 2006; Ruppert et al., 2003). Degradation coefficients (a, b and c) of soy-bean meal was not changed by supplementing FLLE, which also approved constant crude-protein digestibility among all treatments in this study. The NDF and ADF digestibility was not affected by sup-plementing FLLE. This result disagrees with the result of Benchaar et al. (2006) who reported that NDF and ADF digestibility was increased by addition of mixture of essential oils. The result of Benchaar et al (2006) indicates that mixture of essential oils can stimulate the cellulose-degrading bacteria. In comparison, FLLE was not approved to have the capability to stimulate the activity of cel-lulose-degrading bacteria. According to the result of degradation coefficient of maize, dynamic DM deg-radation coefficient c was increased by supplement-ing FLLE, which indicates that FLLE can specifically stimulate the maize starch-degrading bacteria. A larger quantity of maize starch was fermented in the rumen of sheep in response to addition of FLLE. As a result, DM and organic matter digestibility were increased by supplementing FLLE.

Effect of supplemental FLLE on rumen fermentation characteristics

Saponins containing Yucca schidigera have been widely used at different dosages in ruminant diets. Protozoa in the rumen was inhibited, and ruminal ammonia nitrogen concentration was decreased (Busquet et al., 2006). The same effect of other essential oils (capsicum, carvacrol, carvone, cinnam-aldehyde, cinnamon oil and clove bud oil) on rumen nitrogen metabolism was also found (Busquet et al., 2006). In this study, addition of saponin containing FLLE at 300 or 500 mg/kg DM improved the pattern of rumen fermentation (i.e. less NH_3-N concentration and more total volatile fatty acid concentration). Diets supplemented with 300 or 500 mg/kg DM decreased ammonia nitrogen concentration of rumi-nal fluid at 4, 6 and 8 h after morning feeding (Fig. 2). The lower blood urea nitrogen concentration was also observed in sheep in response to addition of FLLE at dosages 300 or 500 mg/kg DM after morning feeding. It was reported that saponins reduced the growth of protein-fermenting ruminal bacteria *Streptococcus bovis*, *Prevotella bryanti* and *Ruminobacter amylophilus* (Wang et al., 2000). Therefore, reduced ammonia concentration by FLLE supplementation in this study may be due to the inhibition of these bac-teria. As a result, protein degradation or deamination of amino acids was inhibited. But, the lack of an effect of rumen *in situ* disappearance of N might sug-gest that less NH_3 is not due to this reason. The other explanation is that addition of FLLE improved incor-poration of ammonia-N in microbial protein. Further research needs to be performed to determine accu-rate pathway by which FLLE decreases ammonia-N concentration in ruminal fluid.

Effect of supplemental FLLE on ruminal degradation rate of feedstuffs

It was reported that there is no effect of feeding mix-ture of essential oils to lactating cows on ru-minal dynamic degradability of CP (i.e. soybean meal), fibre (grass silage) and starch (i.e. maize) (Benchaar et al., 2006; Molero et al., 2004). In this study, dynamic degradability coefficients of soybean meal (crude protein) and Chinese wildrye (NDF, ADF) did not change by supplementing FLLE, which agrees with the results of Benchaar and Molero (Table 4). But, dynamic degradability coefficient c of maize was significantly increased (Table 3). The results indicate that supplemental FLLE to diets may stimulate the amylolytic bacteria that play a positive rule to degrade maize starch in the rumen. The observed higher digestibility of diet's DM and organic matter in response to FLLE supplementation also approves that more maize starch was degraded in the rumen. Basically, maize starch was degraded into propio-nate. As a result, propionate concentration in rumi-nal fluid was increased, and total volatile fatty acid concentration was also increased in response to FLLE supplementation to the diet.

CONCLUSIONS

Addition of FLL improved blood antioxidant function and slaughtered body weight in sheep. Addition of FLLE has the capability to modulate rumen fermen-tation, increase the maize starch degradation rate, total volatile fatty acid concentration and propionate concentration in sheep. Addition of FLLE at dosages 300 or 500 mg/kg DM increased apparent digestibil-ity of organic matter and DM.

References

[1] Avato, P.; Bucci, R.; Tava, A.; Vitali, C.; Rosato, A.; Bialy, Z.; Jurzysta, M., 2006: Antimicrobial activity of sapo-nins from *Medicago* sp.: structure-activity relationship. *Phytother Research* 20:

454-457.

[2] Becher, G.R.; Whitton, B.K., 1970: Ammonia determination: reagents modification and interfering compounds.*Analytical Biochemistry*, 36: 243.

[3] Benchaar, C.; Petit, H.V.; Berthiaume, R.; Whyte, T.D.; Chouinard, P.Y., 2006: Effects of dietary addition of essential oils and monensin premix on digestion, ruminal fermentation characteristics milk production, and milk composition in dairy cows.*Journal of Dairy Science* 89: 4 352-4 364.

[4] Benchaar, C.; McAllister, T.A.; Chouinard, P.Y., 2008: Digestion, ruminal fermentation, ciliate protozoal populations, and milk production from dairy cows fed cinnamaldehyde, quebracho condensed tannin, or *yucca schidigera* saponin extracts.*Journal of Dairy Science* 91: 4 765-4 777.

[5] Busquet, M.; Calsamiglia, S.; Ferret, A.; Kamel, C., 2006: Plant extracts affect *in vitro* rumen microbial fermentation.*Journal of Dairy Science* 89: 761-771.

[6] Cardozo, P.W.; Calsamiglia, S.; Ferret, A.; Kamel, C., 2006: Effects of natural plant extracts on protein degradation and fermentation profiles in continuous culture. *Journal of Animal Science* 84: 2 801-2 808.

[7] Che, D.Y.; Chen, L., 2009: Chemical components and pathway advances of Fructus Ligustri Lucidi extract. *Journal of Modern Clinical Medicine* 5: 323-325.

[8] Hu, M.; Lu, D.X.; Niu, W.Y.; Ren, X.P.; Na, R.; Yi, J.; Bao, Q.J., 2006: Effect of saponins from medicage sp on nitrogen balance and production performance in sheep. *Chinese Animal Husbandry and Veterinary Medicine* In China 10: 21-24.

[9] Lila, Z.A.; Mohammed, N.; Kanda, S.; Kamada, T.; Itabashi, H., 2003: Effect of sarsaponin on ruminal fermentation with particular reference to methane production *in vitro*.*Journal of Dairy Science* 86: 3 330-3 336.

[10] Ma, D.Y.; Shan, A.S.; Li, Q.D., 2004: Effect of Chinese medical herbs on chickens growth and immunization.*Acta Zoonutrimenta Sinica* 2: 36-40.

[11] Ma, D.Y.; Shan, A.S.; Chen, Z.H., 2005: Influence of *Ligustrum lucidum*, *Schisandra Chinensis*, Sijunzitang and Daidzein on antioxidant status of laying hens under heat stress.*Acta Zoonutrimenta Sinica* 2: 23-27.

[12] Merchen, N.R.; Firkins, J.L.; Berger, L.L., 1986: Effect of intake and forage level on ruminal turnover rates, bacterial protein synthesis and duodenal amino acid flows in sheep.*Journal of Animal Science* 62: 216.

[13] Molero, R.; Ibara, M.; Calsamiglia, S.; Ferret, A.; Losa, R., 2004: Effects of a specific blend of essential oil compounds on dry matter and crude protein degradability in heifers fed diets with different forage to concentrate ratios.*Animal Feed Science and Technology* 114: 91-104.

[14] Oh, H.K.; Jones, M.B.; Longhurst, W.M., 1968: Comparison of rumen microbial inhibition resulting from various essential oils isolated from relatively unpalatable plant species.*Applied Microbiology* 16: 39-44.

[15] Ørskov, E.R.; McDonald, I., 1979: The estimation of protein degradability in the rumen from incubation measurements weighted according to rate of passage. *Journal of Agricultural Science* 92: 499-503.

[16] Ruppert, L.D.; Drackley, J.K.; Bremmer, D.R., 2003: Effects of tallow in diets based on maize silage or alfalfa silage on digestion and nutrient use by lactating dairy cows.*Journal of Dairy Science* 86: 593-609.

[17] SAS Institute, 2003: *SAS User's Guide.Statistics*, Version 6.12 ed.SAS Inst., Cary, NC.

[18] Shabi, Z.; Arieli, A.; Bruckental, L.; Aharoni, Y.; Zamwel, S.; Bor, A.; Tagari, H., 1998:

Effect of the synchroniza-tion of the degradation of dietary crude protein and organic matter and feeding frequency on ruminal fermentation and flow of digesta in the abomasum of dietary cows.*Journal of Dairy Science* 81: 1 991-2 000.

[19] Storey, K.B., 1996: Oxidative stress: animal adaptations in nature.*Brazilian Journal of Medical and Biological Research* 29: 1 715-1 733.

[20] Tava, A.; Oleszek, W.; Jurzysta, M.; Berardo, N.; Odoardi, M., 1993: Afalfa saponins and sapogenics: isolation and quantification in two different cultivars.*Phytochemistry Analysis* 4: 269-274.

[21] Van Soest, P.J.; Robertson, J.B.; Lewis, B.A., 1991: Methods for dietary fiber, neutral detergent fiber, and non-starch polysaccharides in relation to animal nutrition. *Journal of Dairy Science* 74: 3 583-3 597.

[22] Wang, Y.; McAllister, T.A.; Yanke, L.J.; Cheeke, P.R., 2000: Effect of steroidal saponin fromYucca schidigera extract on ruminal microbes.*Journal of Applied Micro-biology* 88: 887-896.

[23] Xu, Y.; Shan, A.S.; Qiao, G.H., 2008: Effect of *Ligustrum lucidum* extract on 48 h degradation rate of dietary nutrients in the rumen of northeast fine wool sheep.*The Journal of Northeast Agricultural University* 39: 25-28.

[24] Zhang, Q.R.; Li, J.G.; Li, X.M.; Ni, Y.D., 2004: Effect of Chinese herbs additives on immune function and milk yield in Chinese Holstein cows.*Journal of Heilongjiang Animal and Veterinary Medicine* In China 10: 22-24.

[25] Zhang, Q.R.; Li, J.G.; Li, X.M.; Ni, Y.D., 2005: Effect of Chinese herbs additives on endocrine function in Chinese Holstein cows.*Journal of Heilongjiang Animal and Veterinary Medicine* In China 4: 54-55.

(发表于《Journal of animal physiology and animal nutrition》，院选 SCI，IF：1.106)

Synthesis of Aspirin Eugenol Ester and its Biological Activity

LI Jian-yong, YU Yuan-guang, WANG Qi-wen, ZHANG Ji-yu,
YANG Ya-jun, LI Bing, ZHOU Xu-zheng, NIU Jian-rong,
WEI Xiao-juan, LIU Xi-wang, LIU Zhi-qi

(Key Lab of New Animal Drug Project of Gansu Province, Key Lab of Veterinary Pharmaceutics Discovery of Agricultural Ministry, Lanzhou Institute of Animal Science and Veterinary Pharmaceutics, Chinese Academy of Agricultural Sciences, Lanzhou 730050, China/The Academy of Life Science and Engineering, Lanzhou University of Technology, Lanzhou 730050, China)

Abstract: Aspirin, as starting precursor, was reacted with $SOCl_2$ to generate acyl chloride compound via esterifying eugenol to aspirin eugenol ester (AEE) with a yield of 65%. Tests of AEE for acute toxicity, anti-inflammatory, analgesic, and antipyretic effects were carried out in mice and rats. The results showed that toxicity of AEE was significantly reduced, about 50 and 3.7 times lower than aspirin and eugenol, respectively. Its strength of anti-inflammation, analgesia, and antipyresis was similar as aspirin and eugenol, but the effects lasted longer.

Key words: Aspirin eugenol ester (AEE); Synthesis; Toxicity; Anti-inflammation; Analgesic and antipyretic effects

Aspirin has been used as a treatment drug for inflammation and fever for more than a century. The biochemical mechanism of action of aspirin has been described previ-ously (Vane, 1971; Flower *et al.*, 1972; Vane and Botting, 1987). Aspirin produces its therapeutic (e.g., anti-inflammtory and analgesic) and side effect (e.g., gastroin-testinal ulcers) via inhibition of cyclooxygenase which is a key enzyme to catalyze prostaglandin formation (Vane, 1971). Recently, its use has been extended to prevention and treatment of cardiovascular disease based on its anti-thrombotic action in platelets since inhibition of cycloox-ygenase by aspirin blocks thromboxane A2 production which is crucial for blood clotting (Patrono, 1989). The anti-platelet effect of aspirin has been tested in various forms of coronary artery disease, pregnancy-induced hypertension, and preeclampsia in angiotensin-sensitive primigravida at low dosage and showed positive results in most of the reports (Schoemaker *et al.*, 1998; Wallenburg *et al.*, 1986).

Eugenol is the main component of volatile oil extracted from dry alabastrum of Eugenia Caryophyllata Thumb. Various therapeutic effects of eugenol have been demon-strated, including anti-virus, anti-bacteria, anti-pyresis, analgesia, anti-inflammation, anti-platelet aggregation, anticoagulation, anti-oxidation, anti-diarrhea, anti-hypoxia and, anti-ulcer, and inhibits intestinal movement, and ara-chidonic acid metabolism (Tragoolpua and Jatisatienr, 2007; Chami et al., 2005; Gill and Holley, 2004; Nagababu and Lakshmaiah, 1997; Hashimoto et al., 1988; Feng and Lipton, 1987; Raghavendra and Naidu, 2009). It is used to treat toothache, hepatopathy, and gastrointestinal diseases.

However, side effect of aspirin is very serious, such as gastrointestinal damage and eugenol is irritative and vul-nerable to oxidation. Chemically carboxyl group of aspirin and hydroxyl group of eugenol are responsible for these side effects and structural instability. Therefore, based on prodrug principal, aspirin and eugenol can be combined into aspirin eugenol ester (AEE) with reduced side effect and increased therapeutic effect and stabilization. AEE is supposed to be decomposed into aspirin and eugenol by the enzymes after absorption, which would show each original activity again.

We report here the synthesis of aspirin eugenol ester and some of its activities. Two synthetic methods of esterifi-cation for aspirin and eugenol were applied. When DCC (N, N'-dicyclohexylcarbodiimide) as catalysis reagent was used, yield of AEE was less than 10% with some by-products, such as (3', 4'-methylenedioxy) acetoxyphenol formed by the methyl group transfer during the esterifica-tion. In the second method, aspirin, as a starting precursor, was reacted with SOCl₂ to form acyl chloride compound, which esterified eugenol to AEE with a yield of over 60%. Therefore, the second method was selected (Scheme 1). The compound is confirmed with NMR, UV and EI-MS, etc. However, the melting point of title compounds was low, so it was kept at low temperature.

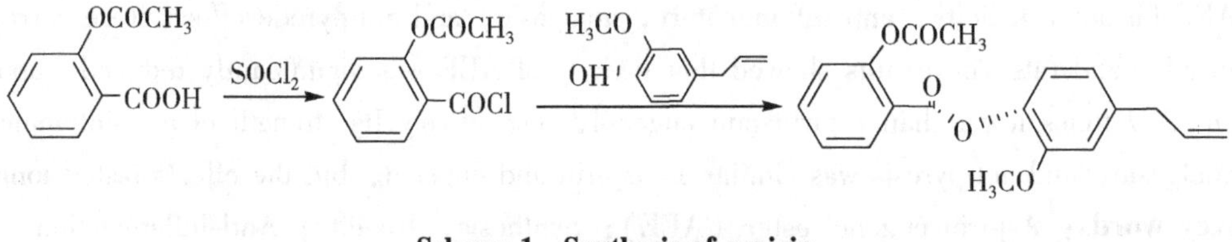

Scheme 1 Synthesis of aspirin

In order to test the toxicity and activities of AEE, the acute toxicity and anti-inflammatory, analgesic, antipyretic effect tests were carried out in mice and rats using aspirin and eugenol as controls. All animal experiments were carried out under the Guidelines for the care and use of laboratory animals as described by the U.S. National Institutes of Health. The results showed that toxicity of AEE was less than the controls, which was 0.02 times of aspirin and 0.27 times of eugenol. The anti-inflammatory, analgesic, and antipyretic effects of AEE were similar as aspirin and eugenol, but its effects lasted for a longer period.

EXPERIMENT

Melting points were determined using XT-4 binocular microscope melting point apparatus (Beijing Tech Instru-ment Co. Ltd.) without adjustment. Proton magnetic reso-nance (NMR) was determined at 400 MHz with a Varian INOVA 400 spectrometer. The chemical shift values were expressed in δ values related to the internal standard tet-ramethylsilane. A mass spectrum was recorded by electron impact (EIMS) method with a Finnigan Mat TSQ-700 spectrometer. Elementary analysis was carried out with an Elementary Vario EL III. The UV spectra were determined with an UV-240 Spectrophotometer (shimadzu).The IR spectra were determined with a Thermo Nicolet Nexus FTIR-470 Spectrometer for KBr pellets.

Preparation of 4'-allyl-2'-methoxyphenyl-2-acetoxybenzoate (aspirin eugenol ester, AEE)

Solution of aspirin (3.6 g, 0.02 mol) in DMF (dimethyl-formamide, 1 mL) was added with $SOCl_2$ (4.0 g, 0.03 mol) while stirring and allowed to reflux and stir for 3 h at 60-75℃. After the unreacted $SOCl_2$ was removed under pressure, the acyl chloride was dissolved into 8 mL benzene for next step.

Solution of eugenol (3.3 g, 0.02 mol) in 5 mL H_2O was added with 5% NaOH and 1.0 g PEG-1000 at 5-10℃ while mixing and allowed to be stirred for 0.5 h. After the above acyl chloride was added, the mixture continued to stir for 2 h and stood for substratification. After organic layer was separated and evaporated in vacuum, the solid was recrystallized in MeOH to give 4.24 g of AEE as a white solid (65%). Some parameters were as following. Mp: 71-72℃. ^1H NMR (DMSO-d_6) δ: 8.126, 7.750, 7.471, 7.306 (4H, m, C_6H_4), 7.071, 6.992, 6.813 (3H, m, C_6H_3), 5.979 (1H, m, -CH=), 5.057-5.147 (2H, m, =CH_2), 3.737 (3H, s, OCH_3), 3.387 (2H, d, -CH_2-), 2.215 (3H, s, OAc). ^{13}C-NMR (DMSO-d_6) δ: 169.116 (1C, C=O (OAc)), 162.142 (1C, C=O), 150.620 (1C, 2^0-C_6H_3), 150.505 (1C, 2-C_6H_4), 139.220 (1C, 4^0-C_6H_3), 137.435 (1C, -CH=), 137.198 (1C, 1^0-C_6H_3), 135.001 (1C, 4-C_6H_4), 131.727 (1C, 6-C_6H_4), 126.516 (1C, 5-C_6H_4), 124.341 (1C, 6^0-C_6H_3), 122.536 (1C, 3-C_6H_4), 122.159 (1C, 1-C_6H_4), 120.396 (1C, 5^0-C_6H_3); 116.070 (1C, =CH_2); 113.010 (1C, 3^0-C_6H_3); 55.714 (1C, OCH_3); 39.347 (1C, -CH_2-); 20.668 (1C, CH_3). UV (γ_{max}, nm) 210, 230, 279. IR (ν_{max} (KBr), cm^{-1}) 1762.9, 1729.3 (C=O), 1606.05 (C=C). MS (m/z) 326 (M^+), 164 ($C_{10}H_{12}O_2$), 163 ($CH_3COO-C_6H_4CO$), 121 (C_6H_5COO). Anal. Calc. for $C_{19}H_{18}O_5$: C 69.93, H 5.56, O 24.51. Found: C 69.93, H 5.53, O 24.54.

Acute toxicity test of AEE

Modified Kaber method (Li, 2003) was used to calculate LD_{50} of AEE in mice. There were seven groups with 10 mice in each group, including one control and 6 treatment groups as shown in Table 1. Ratio between male and female mice in each group was 1 : 1. Ratio of drug quantity between adjacent groups was 0.75. The maximum dosage is 20 g·kg^{-1} b.w. Before treatment, the animals were subject to absolute diet for 12 h with free access to water. The drug doses was divided into two and administrated intragastrically to mice for 6 h. The results were shown in Table 1.

Table 1 The results of oral acute toxic test of AEE in mice

Drug	Group	Dosage (g·kg^{-1})	Animal numbers	Death numbers	Death rate (p)	Survival rate (q)
AEE	1	20.00	10	10	1.0	0.0
	2	15.00	10	7	0.7	0.3
	3	11.25	10	4	0.4	0.6

(continued)

Drug	Group	Dosage (g · kg^{-1})	Animal numbers	Death numbers	Death rate (p)	Survival rate (q)
AEE	4	8.44	10	3	0.3	0.7
	5	6.33	10	2	0.2	0.8
	6	4.75	10	0	0	1.0
	7	—	10	0	0	1.0

$$LD_{50} = lg^{-1}[Xm - i(\Sigma p - 0.5)] \quad (I)$$

Based on Kaber formula (I), LD_{50} was 10.937 g · kg^{-1} (33.5 mmol · kg^{-1}) and 95% confidence interval was 9.309–12.850 g · kg^{-1} (28.6–39.4 mmol · kg^{-1}). It was well known that LD_{50} of aspirin and eugenol was 0.25 g · kg^{-1} (1.39 mmol · kg^{-1}) and 3.00 g · kg^{-1} (18.3 mmol · kg^{-1}), respectively (Cayman Chemical Company, 2005; WHO, 2006). Therefore, toxicity of AEE is less than that of aspirin and eugenol, only 0.02 times of aspirin and 0.27 times of eugenol in mass. According to the grade standard of acute toxicity for chemicals in European Union, AEE fall into non-toxic class. Reduced side effect for this aspirin derivative will be an advantage for clinical uses.

On the other hand, it was found that the mice appeared to die within 24 h after AEE was given and AEE toxicity action began approximately at 2 h following the drug administration. Dissection illustrated that hearts, livers, spleens, and kidneys of the dead mice were all normal, while their lungs bled slightly and a few of them were full of gases in their stomachs. In the survived mice over 7 days, all organs were normal. These results showed that AEE toxicity was predominantly to cause lung and stom-ach damages, which were similar to that of eugenol and aspirin, but with less severity.

Anti-inflammatory effects of AEE

Fifty mice were divided into such five groups as: control, aspirin 1.1 mmol · kg^{-1}, eugenol 1.1 mmol · kg^{-1}, AEE 1.1 and 0.56 mmol · kg^{-1}. Ratio between male and female mice in each group was 1:1. Control group was treated with 0.5% CMC-Na for 3 days while drug groups were administered intragastrically with 20 mL · kg^{-1} day^{-1} appropriate drug. All drugs were suspended in 0.5% CMC-Na solution. Xylene-induced ears swelling model was used as the acute inflammatory model to investigate the anti-inflammatory effects of AEE. The mice were anesthetized at 1 h by the end of administration and 50 ll xylene was smeared on both sides of its right earlap. After 1 h, the mice were put to death and the same parts of both ears were cut with a puncher for weighing. Then the degree and inhibition ratio of swelling were calculated as described previously (Li et al., 2009).

The results were shown in Table 2. Compared with the control group (7.05±1.19) mg, the xylene-induced ear swelling degree were significantly inhibited by AEE administration at 1.1 mmol · kg^{-1} (3.81±1.14) mg and 0.56 mmol · kg^{-1} (3.89±1.88) mg ($P<0.01$), and there were no significant differences of ear swelling degrees between AEE groups and aspirin and eugenol groups ($P<0.05$). This indicates that AEE and aspirin and eugenol possessed similar strength of anti-inflammation in this model. However, half dose of AEE produced similar strength of anti-inflammatory effect as aspirin and this suggests that lower dose of AEE can be selected to reduce the side effects for clinical uses.

Table 2 The effects of AEE on ear edema induced by xylene in mice ($\bar{x}\pm s$; $n=10$)

Group	Dosage (mmol·kg^{-1})	Animal numbers	Degree of ear swelling (mg)	Inhibition ratio (%)
Model group	—	10	7.05±1.19	—
Aspirin group	1.11	10	3.87±1.23**	45.19
Eugenol group	1.11	10	4.73±1.87*	32.93
High dosage of AEE	1.11	10	3.81±1.14**	45.94
Low dosage of AEE	0.56	10	3.89±1.88**	44.80

Note: Compared with model control group, * $P<0.05$, ** $P<0.01$

The analgesic effect of AEE

Fifty female mice were selected for pain threshold test. They were divided into five groups as control, aspirin 1.1 mmol·kg^{-1}, eugenol 1.1 mmol·kg^{-1}, AEE 1.1 and 0.56 mmol·kg^{-1}. Drugs formula and administration routine in each group were the same as described above. The analgesic effect of AEE was investigated using a Hot plate test. Hot plate pain threshold detector was set to (55±0.5)℃ for recording pain threshold, including time period of licking metapedes in mice. The pain thresholds for the mice were recorded at 30, 60, 90, 120, and 180 min following the administration according to the method reported previously (Nigade et al., 2010).

The results were given in Table 3. All drugs showed analgesic effect at different time points. The analgesic effect of aspirin reached peak at 1 h and it was very sig-nificant compared with control group. After 1 h, its effect was attenuated and not statistically significantly different compared with the control group although the threshold went up moderately ($P>0.05$). The analgesic effect of eugenol lasted longer than that of aspirin, in which at 2 and 3 h the thresholds in the mice receiving eugenol adminis-tration were higher than those of the control group. The analgesic effect of AEE lasted longer than both aspirin and eugenol, indicated by that the AEE groups had significantly higher thresholds than the control, aspirin and eugenol groups over a period of 3 h ($P<0.01$).

In summary, pain in mice caused by the thermal stim-ulation was significantly eased by the AEE treatment. Compared with the aspirin and eugenol, AEE treatment at same dose rate had similar analgesic effect for up to 2 h and stronger effect after 2 h, while AEE at half dose of aspirin and eugenol showed slower effect at 30 min after administration but stronger effect thereafter. This indicates AEE possessed longer effect than that of aspirin and eugenol. Therefore, AEE is a good candidate for devel-oping long-lasting analgesics.

Table 3 The analgesic effect of AEE in mice hot plate experiment ($\bar{x}\pm s$; $n=10$)

Group	Dosage (mmol·kg^{-1})	Pain threshold before administration (S)	Pain threshold after administration (S)			
			30 min	60 min	120 min	180 min
Model	—	20.35±6.19	24.88±5.28	21.28±3.00	19.06±4.55	17.67±5.78
Eugenol	1.1	19.85±5.00	30.76±6.18	28.01±8.16	24.86±7.56*	25.50±8.67*
High dosage of AEE	1.1	20.30±6.55	29.50±13.28	28.80±6.53*	28.74±9.90**	29.69±9.21**
Low dosage of AEE	0.56	22.72±5.40	24.45±7.27	28.81±7.56*	26.85±6.65**	29.13±7.66**

Note: compared with model control, * $P<0.05$ and ** $P<0.01$

The antipyretic effects of AEE

Seventy rats were divided into such seven groups that the mice received treatments of normal control, model control, aspirin 1.5 mmol·kg^{-1}, eugenol 1.5 mmol·kg^{-1}, or aspirin eugenol ester at 2.0, 1.5, or 1.0 mmol·kg^{-1} treatments. Ratio between male and female mice in each group was 1:1. Normal control group was given normal saline subcutane-ously at 10 mL·kg^{-1} and other groups were given 15% yeast fungus suspension subcutaneously at 10 mL·kg^{-1}. After 5 h, rat body temperatures were recorded and the rats with body temperature change less than 0.8℃ were replaced. Normal and model control groups were treated with 10 mL·kg^{-1} 0.5% CMC-Na solution and drug groups were administered intragastrically with 10 mL·kg^{-1} appropriate drug suspen-sion. All drugs were suspended in 0.5% CMC-Na solution. After intragastrical administration, body temperatures were recorded at 2, 4, and 6 h, respectively.

The results were shown in Table 4. The febrile pathol-ogy model was developed successfully in all testing mice and all drugs demonstrated antipyretic effect at different time points. The drug effects ranking from strong to weak were high dosage of AEE, aspirin, middle dosage of AEE, low dosage of AEE, and eugenol, while quantities of molecule of aspirin, middle dosage of AEE and eugenol group were equal. Three AEE treatments reduced response of rats to yeast-induced fever and the antipyretic effects were dose-dependent. The changes in body temperature after administration showed that the antipyretic effect of aspirin peaked at 2 h and attenuated thereafter and that of AEE was similar as aspirin at 2 h following administration and became stronger at 4 and 6 h.

The above results showed that AEE, aspirin, and euge-nol possessed similar acute toxicity and therapeutic effects, including anti-inflammation, analgesia, and antipyresis, but their durations and strength were different. In toxicity, AEE was less than aspirin and eugenol. In effects, AEE at same dose produced stronger and longer effects than aspirin and eugenol. This indicates that after absorption into the body, AEE should be effectively decomposed into aspirin and eugenol by the enzymes to produce their indi-vidual effects. The effects of these two compounds might act synergistically. This may partially explain why AEE had stronger and longer effect than aspirin. However, as AEE is a new compound, further studies on pharmacoki-netics and pharmacodynamics are required to fully under-stand its therapeutic and side effects and the mechanisms.

Table 4 The effect of AEE on yeast-induced fever in rats ($\bar{x}\pm s$; $n=10$)

Group	Dosage (mmol·kg^{-1})	Basal body temperature (℃)	5 h after fever (℃)	Body temperature change after administration (℃)		
				2 h	4 h	6 h
Blank	—	37.51±0.37	37.51±0.43	+0.17±0.22	+0.28±0.13	+0.40±0.22
Model	—	37.30±0.29	39.45±0.40	-0.35±0.30	-0.40±0.38	-0.75±0.48
Aspirin	1.5	37.04±0.23	39.21±0.38	-2.19±0.32**	-1.80±0.55**	-1.78±0.51**
Eugenol	1.5	37.50±0.20	39.26±0.29	-0.94±0.42*	-1.11±0.19*	-1.21±0.29*
High dosage of AEE	2	37.36±0.24	39.41±0.22	-2.09±0.45**	-2.46±0.47**	-2.49±0.49**
Middle dosage of AEE	1.5	37.37±0.36	38.94±0.53	-1.14±0.35*	-1.40±0.35*	-1.45±0.41*
Low dosage of AEE	1	37.26±0.18	39.34±0.35	-1.2±0.28*	-1.29±0.32*	-1.32±0.34*

Note: compared with model control, * $P<0.05$, ** $P<0.01$

ACKNOWLEDGMENTS

NMR, EIMS, and element analysis were carried out in the center of analysis and identification, department of chem-istry, Lanzhou University. The project was supported by Specific Program Funded by Basic Scientific Research Operating Expenses of Central Public Scientific Research Institutes of China (BRF060403), Key Project of Scientific and Technical Supporting Programs of Gansu Province of China (0804NKCA074).

References

[1] Cayman Chemical Company (2005) Aspirin. MATERIAL SAFETY DATA SHEET: 3.

[2] Chami N, Bennis S, Chami F, Aboussekhra A, Remmal A (2005) Study of anticandidal activity of carvacrol and eugenol in vitro and in vivo. Oral Microbiol Immunol 20: 106-111.

[3] Feng J, Lipton JM (1987) Eugenol: antipyretic activity in rabbits. Neuropharmacology 26: 1 775-1 778.

[4] Flower R, Gryglewshi R, Herbaczynska-cedro K, Vane JR (1972) Effects of anti-inflammatory drugs on prostaglandin biosynthe-sis. Nat New Biol 238: 104-106.

[5] Gill AO, Holley RA (2004) Mechanisms of bactericidal action of cinnamaldehyde against *Listeria monocytogenes* and of eugenol against *L. monocytogenes and Lactobacillus sakei*. Appl Environ Microbiol 70: 5 750-5 755.

[6] Hashimoto S, Uchiyama K, Maeda M, Ishitsuka K, Furumoto K, Nakamura Y (1988) In vivo and in vitro effects of zinc oxide-eugenol (ZOE) on biosynthesis of cyclo-oxygenase products in rat dental pulp. J Dent Res 67: 1 092-1 096.

[7] Li SQ (2003) Principal and method for animal toxicology, 2nd edn. Sichuan university publishing house, Chengdu, 56.

[8] Li J, Wang L, Bai H, Yang B, Yang H (2009) Synthesis, characterization, and anti-inflammatory activities of rare earth metal complexes of luteolin. Med Chem Res 18: 1-5. doi: 10.1007/s00044-009-9289-2.

[9] Nagababu E, Lakshmaiah N (1997) Inhibition of xanthine oxidase-xanthine-iron mediated lipid peroxidation by eugenol in lipo-somes. Mol Cell Biochem 166: 65-71.

[10] Nigade G, Chavan P, Deodhar M (2010) Synthesis and analgesic activity of new pyridine-based het-

erocyclic derivatives.Med Chem Res 19: 1-11. doi: 10. 1007/s00044-010-9489-9.

[11] Patrono C (1989) Aspirin and human platelets from clinical trials to acetylation of cyclooxygenase and back.TiPS 10: 453-458.

[12] Raghavendra RH, Naidu KA (2009) Spice active principles as the inhibitors of human platelet aggregation and thromboxane biosynthesis.Prostaglandins Leukot Essent Fatty Acids 81: 73-78.

[13] Schoemaker RG, Saxena PR, Kalkman EA (1998) Low-dose aspirin improves in vivo hemodynamics in conscious, chronically infarcted rats.Cardiovasc Res 37: 108-114.

[14] Tragoolpua Y, Jatisatienr A (2007) Anti-herpes simplex virus activities of *Eugenia caryophyllus* (Spreng.) Bullock & S.G.Harrison and essential oil, eugenol.Phytother Res 21: 1 153-1 158.

[15] Vane JR (1971) Inhibition of prostaglandin synthesis as a mechanism of action for aspirin-like drugs. Nat New Biol 231: 232-235.

[16] Vane JR, Botting R (1987) Inflammation and mechanism of anti-inflammatory drugs. FASEB J 1: 89-96.

[17] Wallenburg HC, Dekker GA, Makovitz JW, Rotmams P (1986) Low-dose aspirin prevents pregnancy-induced hypertension and pre-eclampsia in angiotensin-sensitive primigravide.Lancet 1: 1-3.

[18] WHO (2006) Summary of evaluations performed by the joint FAO/ WHO expert committee on food additives, Eugenol No.1529, WHO Food Additives Series 17, http: //www.inchem.org/ documents/jecfa/jecmono/v17je10. htm.

（发表于《Med Chem Res》，院选 SCI，IF：1.058）

Development of Gastric and Pancreatic Enzyme Activities and Their Relationship with Some Gut Regulatory Peptides in Grazing Sheep

LANG Xia[1,2]*, WANG Cai-lian[1]

(1. Faculty of Animal Science and Technology, Gansu Agricultural University, Lanzhou, 730070, China; 2. Lanzhou Institute of Animal Science and Veterinary Pharmaceutics Sciences, Chinese Academy of Agricultural Science (CAAS), Lanzhou, 730050, China)

Abstract: Forty-four Gansu Alpine Fine-wool lambs were used to study changes in the activities of three gastric and five pancreatic enzymes under grazing conditions between 0 and 56 days of age. The lambs were slaughtered on days 0, 3, 7, 14, 21, 28, 42 and 56, the abomasal contents, mucosa and pancreas were immediately removed and placed into liquid nitrogen and enzyme activities were determined. Gastric enzyme (chymosin, pepsin and pregastrc esterase) activities were relatively high at birth, especially chymosin, but decreased quickly between day 0 and 21. The activity of pepsin changed insignificantly with increasing age. There was no significant change in the pancreatic enzyme activities (trypsin, chymotrypsin, α-amylase, lipase and lactase). The activity of trypsin was relatively higher than that of the other pancreatic enzymes, and lactase activity was low. These ontogenic patterns might be under the control of many gut regulatory peptides, the plasma concentrations of which changed simultaneously. Some gastric and pancreatic enzymes were correlated with plasma concentrations of these gut regulatory peptides.

Key words: Digestive Enzymes; Plasma Regulatory Peptides; Grazing Lambs

INTRODUCTION

During the first post-natal months, the young ruminant is faced with three types of situations requiring physiological and digestive adaptation: adaptation to the extra uterine environment, maintenance in an extended preruminant stage and weaning (Thivend et al., 1980). The effects of age on pancreatic secretions have been more thoroughly investigated in monogastric animals (Corring et al., 1978; Kretchmer, 1985) than in ruminant species (Huber et al., 1961; Guilloteau et al., 1983, 1984, 1985). However, few studies report effects of age on gastric enzyme secretions in

---

* Corresponding Author: Lang Xia. E-mail: langxiax@163.com
Received August 26, 2010; Accepted November 12, 2010

young mammals (Walker, 1959; Hartman et al., 1961; Henschel, 1973; Garnot et al., 1977; Foltmann et al., 1981). There are some reports on digestive physiology for non-grazing lambs; for example, it is generally acknowledged that the pre-ruminant abomasum characteristically secretes large amounts of chymosin, which, with pepsin and hydrochloric acid, coagulates milk casein. At 2 days of age there is a threshold of development of enzyme secretion potentiality in lambs. Quantities of gastric enzymes in relation to empty live weight increase between birth and 2 days, but that of chymosin then decreases, whereas pepsin does not change significantly. The evolution of pancreatic enzyme activity was usually the reverse of that of chymosin; however, trypsin activity was low at birth (Guilloteau et al., 1983). But very few studies report on digestive physiology for grazing lambs, and especially in grazing lambs distributed over a plateau at 2 600 m above sea level.

In China, lambs usually stay with their mothers until weaning of non-grazing lambs at the age of two months and of grazing lambs usually at the age of four months. In our experiment, the animals were Gansu Alpine Fine-wool sheep, which are distributed over the northern slope of the east Qilian mountains at the border of Gansu and Qinghai (101°45′E, 37°53′N). The area is a cold alpine pastoral range, and belongs to sub-alpine meadow and mountain rangeland. The altitude ranges from 2 600 m to 3 500 m, even to 4 000 m, and the cold season is longer and there is shortage of pasture forage. Gansu Alpine Fine-wool sheep are a special breed grazing on the plateau at 2 600 m above sea level. The study indicated that grazing lambs had a lot of grass in the rumen at 7 days postpartum, which means that at 7 days of age the lambs can eat solid food but are still maintained in the pre-ruminant state until day 28. Solid food enters the rumen, and an increasing amount of microbial digestion is observed together with increased development of the forestomachs. At the end of weaning (4 months old), lambs are no longer pre-ruminant but ruminant. Depending on the feeding conditions, i.e., milk or solid food, substantial differences in the enzyme activities along the digestive tract occur between these two quite different physiological situations. The lamb, therefore, is an interesting model for investigating the effects of development on digestive functions, and studies can result in useful parameters on digestive physiology for grazing lambs on the plateau, regardless of the nutritional substrates involved.

The present work aimed to study the patterns of abomasal and pancreatic enzyme secretions in grazing lambs. The development of gastric and pancreatic digestive functions was tentatively correlated with changes in the plasma concentrations of five gut regulatory peptides.

MATERIALS AND METHODS

Study site description

The animals were from Huangcheng sheep breeding testing farm of Gansu, China. The farm was the first breeding farm for Gansu Alpine Fine-wool sheep which is located at Huangcheng District in the Sunan Yugu autonomous county of Gansu Province of China, which is on the northern slope of the east Qilian mountains at the border of Gansu and Qinghai (101°45′ E, 37°53′ N). The area is cold alpine pastoral range and belongs to sub-alpine meadow and mountain rangeland. The altitude is from 2 600 m to 3 500 m, even up to 4 000 m. The climate is variable over the year and 4 seasons are not clear. Plateau mainland climate, winds and dryness. The climatic characteristics during the year are drought and cold in winter, warm and humid in summer. Mean annual temperature is 0.6 to 3.8℃, the highest is 31℃ in July, and lowest is −29℃ in January. Annual sunshine hours are 2 272 h. The absolute frost-free period is 45 to 60 days. Mean annual rainfall is 361.6 mm, annual evaporation is 1 111.9 mm, mean annual relative humidity is 38% to 58%. The natural grass sprouts in April and dies in September, the wilt period of pasture is above 7 months. The soil type is kastanozem, mountain phaeozem and cinnamon soil. The soil is fertile, and water is abundant. The types of natural grassland are grass family, sedge family, weeds of grass family and sedge family and scrub weeds. The vegetation mainly consists of grass family and sedge family as well as a few pea family.

Animals and feeding

There were about 10 000 Gansu Alpine Fine-wool lambs in a breeding flock single born between April 22 and May 22 in 2007. Lambs suckled their dams and grazed on native grassland pasture, with no additional feed. Animals were penned at pasture at night without access to feed and water. There were no feed supplements, but a salt mineral mix was provided to the lambs throughout the year; animals had access to water only once a day. Lambs were not weaned until 4 months old, and therefore they remained with their dams throughout the lactation period. Forty-four single-born male lambs obtained at birth from the herd wererandomly distributed into eight groups (Table 1). The experimental period was from May to June, and grass samples were collected according to pasture type, one pasture type from five sampling spots and each sampling area was 1 m^2. Grass kinds included bush cinquefoil (*Potentilla fruticosa Linn.*), linearleaf kobresia (*Kobresia capillifolia* (*Decne.*) *Clarke*), larch needlegrass (*Stipagrandis P. Smirn*), bellard kobresia (*Kobresia bellardii* (*All.*) *Degl.*), and mountain willow (*Salix oritrepha Schneid*). The original samples were clipped. Then, the samples were divided into four equal parts and a randomly chosen part was used to analyze the composition. Nutrient composition of grazed pasture is presented in Table 2.

Table 1 Base status of the test lambs

Age (d)	Slaughter number (n)	Birth weight (kg)	Slaughter weight (kg)	Daily gain (g)	Withers height (cm)	Body length (cm)	Heart girth (cm)
0	6	4.02±0.175	4.02±0.175	—	36.00±2.828	29.83±1.472	34.50±2.739
3	8	3.97±0.283	4.57±0.488	229.58±118.127	36.38±2.973	32.88±1.553	36.50±1.690
7	5	3.86±0.498	5.11±0.244	178.86±49.54	38.20±1.095	35.20±1.643	38.60±2.302
14	5	3.59±0.397	6.84±0.996	232.14±66.047	39.80±1.924	36.60±1.673	43.40±2.191
21	5	4.19±0.394	7.01±0.415	134.19±27.376	41.80±1.924	38.80±0.837	42.40±1.517
28	5	4.14±0.397	8.86±1.841	168.57±72.464	39.00±3.674	40.20±1.789	47.40±3.362
42	5	3.83±0.303	10.20±0.837	151.67±21.716	47.20±1.924	44.20±2.588	49.40±2.074
56	5	4.11±0.292	16.44±2.674	220.18±46.73	52.60±2.793	51.40±2.302	61.60±3.715

Note: * Values are means±SEM. ** Statistics is Tukey' test under One-Way ANOVA in SPSS12.0.

Table 2 Nutrient component of pasture*

Month	DM	CP	EE	NDF	ADF	Ash	Ca
5	62.22	10.33	2.19	71.67	37.98	11.31	0.61
6	37.36	10.04	2.14	71.57	39.35	11.16	0.59

Note: * They were expressed on dry matter bases but dry matter.

** DM=Dry matter; CP=Crude protein; EE=Ether extract; NDF=Neutral detergent fiber; ADF=Acid detergent fiber.

Tissue sampling

On the day of slaughter at 3, 7, 14, 21, 28, 42 and 56 days of age the lambs were fed as usual; the lambs slaughtered at 0 d of age were fed nothing. Lambs were removed from their dams at approximately 09:00 h, transported to the research laboratory and final weights were obtained and recorded. Lambs were slaughtered by severing the jugular vein. The abdominal cavity was opened and the entire gastrointestinal tract was removed. The pancreas was carefully dissected free, cleaned of extraneous tissue, weighed, wrapped in pledget and immediately placed into liquid nitrogen. The abomasum was cut open, 100 to 200 g abomasal contents were collected and put into liquid nitrogen, then emptied of its contents, rinsed with ice-cold isotonic saline, gently blotted with filter paper, defatted, weighed and spread out onto a glass plate lying on ice. Finally, the abomasal mucosa of the cardiac, fundic and pyloric gland regions were scraped off with a glass slide, and put into liquid nitrogen. All the frozen samples were subsequently stored at −80℃ until use.

Sample preparation

The abomasal mucosa and pancreas were all minced with scissors, and homogenized in 5 volumes of ice-cold 0.4 M KCl (W/V=1 : 5) for 45 s in an ice-cold vessel (4℃ overnight). The homogenate was centrifuged at 15,000 g (4℃ for 20 min), and aliquots of the supernatant were stored at −80℃.

Enzyme analyses

The homogenate of abomasal mucosa was analyzed for chymosin activity using the method described by Arima (1967). Pepsin activity and pregastric esterase activity were determined by spectrophotometry using bovine haemoglobin and olive oil, respectively, as substrates. The activities of pancreatic trypsin and chymotrypsin were determined by spectrophotometry using benzoyl-L-arginine-p-nitroanilide and N-glutaryl-L-phenylalanine-p-nitroanilide, respectively, as substrates. Lipase activity was determined by spectrophotometry using olive oil as a substrate. The α-amylase and lactase activities were assayed using starch and ONPG (o-nitrophenyl-β-D-galactoside), respectively, as substrates.

The results are expressed as enzyme quantities per milligram of tissue protein content for all the enzymes.

Plasma gut regulatory peptide analysis

Blood was collected on EDTA (4%) from an external jugular vein at 9:00 h on the morning before slaughter. Plasma samples were stored at −20℃ and concentrations of five gut regulatory peptides were subsequently determined by ELISA methods.

Statistical analysis

Variance analysis was used to assess the effects of age, and Tukey's test was used to classify the means. Correlation coefficients were used to study the relationships between enzyme activities and plasma concentrations of the gut regulatory peptides. Values were considered to be significant at $p<0.05$.

RESULTS

Development of abomasum pH

pH either in abomasal contents or in gland region mucosa was highest at birth and then decreased with increasing age (except for a high value at 28 d in contents), but remained within the optimal range for enzyme activity. pH was generally similar in each gland region mucosa (Table 3).

Table 3 Changes of pH of abomasal contents and mucosa

Age (d)	Slaughter number (n)	Contents	Cardiac gland region mucosa	Fundic gland region mucosa	Pyloric gland region mucosa
0	6	4.67±0.681^a	4.74±0.896^a	4.43±0.794^a	4.36±0.745^a
3	8	3.39±0.461^{ab}	3.53±0.349^a	3.54±0.362^{ab}	3.49±0.326^a
7	5	3.36±0.474^{ab}	3.75±0.621^a	3.39±0.602^{ab}	3.27±0.815^a
14	5	3.37±0.538^{ab}	3.48±0.674^a	3.62±0.536^{ab}	3.52±0.488^a
21	5	3.29±0.896^{ab}	3.75±1.041^a	3.30±1.070^{ab}	3.35±0.905^a
28	5	4.09±1.331^{ab}	3.26±0.644^a	3.16±0.717^{ab}	3.32±0.750^a
42	5	3.59±0.736^{ab}	3.62±0.390^a	2.86±0.789^b	3.55±0.579^a
56	5	2.96±0.184^b	3.41±0.175^a	2.94±0.485^b	3.41±0.357^a

Note: * Values are means±SEM.

** Statistics is Tukey' test under One-Way ANOVA in SPSS12.0.

Abomasal enzymes

In lambs, the specific activity of chymosin in abomasal contents was highest at birth, but tended to decrease between 0 and 56 d, possibly due to dilution by the intake of milk and grass. There were more differences in enzyme activity of abomasal contents among individuals, possibly affected by enzyme secretion, milk and grass intake. At d28, d35 and d42 the enzyme activity was high for unknown reasons. Between days 0 and 14 the specific activity of chymosin was high in both contents and mucosa, and thereafter it decreased. Between days 0 and 14 the specific activity was lower in the mucosa of the fundic gland region than in the cardiac and pyloric gland regions, and was similar in all regions thereafter except for d28 in the cardiac gland region (Table 4).

Table 4 Developmental changes of abomasal chymosin (U/mg protein)

Age (d)	Slaughter number (n)	Contents	Cardiac gland region mucosa	Fundic gland region mucosa	Pyloric gland region mucosa
0	6	43.00±34.430^a	21.13±16.498^{ab}	18.32±10.377^{ac}	37.94±15.660^{ab}
3	8	6.92±2.867^a	52.41±16.556^a	39.02±16.270^a	58.60±28.920^a

(continued)

Age (d)	Slaughter number (n)	Contents	Cardiac gland region mucosa	Fundic gland region mucosa	Pyloric gland region mucosa
7	5	4.21±1.985^a	49.27±22.490^{ab}	55.60±27.100^{abc}	14.55±1.328^{ab}
14	5	6.64±2.991^a	29.80±11.575^{ab}	21.80±1.323^{ab}	29.74±18.168^{ab}
21	5	16.24±9.409^a	12.61±7.702^b	9.15±3.867^c	6.92±5.440^b
28	5	14.60±15.237^a	23.28±11.671^{ab}	8.57±3.117^c	11.89±8.037^{ab}
42	5	12.57±14.391^a	7.04±3.041^b	10.79±5.108^{bc}	10.23±10.745^b
56	5	5.35±3.601^a	5.22±3.229^b	3.43±2.327^c	4.07±4.910^b

Note: * Statistics is Tukey' test under One-Way ANOVA in SPSS12.0.

Compared to chymosin, the specific activity of abomasal pepsin was lower. Pepsin activity was high at birth and increased little with increasing age. There was a little difference in different gland region, in fundic gland region mucosa which was highest, and in pyloric gland region was lowest; and in contents which was lowest because of milk and grass diluted (0.003 to 0.009 U/mg protein) (Table 5).

Table 5 Developmental changes of abomasal pepsin (U/mg protein)

Age (d)	Slaughter number (n)	Contents	Cardiac gland region mucosa	Fundic gland region mucosa	Pyloric gland region mucosa
0	6	0.008±0.005^a	0.025±0.005^a	0.027±0.004^a	0.016±0.002^{ab}
3	8	0.003±0.001^a	0.025±0.002^a	0.031±0.003^a	0.017±0.003^a
7	5	0.003±0.001^a	0.027±0.003^a	0.033±0.004^a	0.013±0.002^{ab}
14	5	0.004±0.002^a	0.028±0.003^a	0.034±0.001^a	0.013±0.004^{ab}
21	5	0.005±0.003^a	0.032±0.006^a	0.032±0.006^a	0.011±0.003^{ab}
28	5	0.005±0.001^a	0.030±0.004^a	0.030±0.013^a	0.009±0.001^b
42	5	0.009±0.010^a	0.027±0.002^a	0.029±0.004^a	0.014±0.008^{ab}
56	5	0.006±0.002^a	0.026±0.004^a	0.031±0.003^a	0.013±0.004^{ab}

Note: * Statistics is Tukey' test under One-Way ANOVA in SPSS12.0.

Pregastric esterase activity in abomasal contents was high at birth, and decreased thereafter by milk and grass diluted. In d28 it tended to increase. The enzyme is secreted by the tongue root. It seemed that the secretion capacity had little change from day 0 to 56 (Table 6).

Pancreatic enzymes

The specific activity of trypsin was higher than that of α-amylase, chymotrypsin and lipase. Lactase activity was low. The specific activity of all the tested pancreatic enzymes fluctuated with increasing age except that of chymotrypsin which tended to increase after d14.

Table 6 Developmental changes of abomasal pregastric esterase (U/mg protein)

Age (d)	Slaughter number (n)	Pregastric esterase activity
0	6	0.083±0.041^a
3	8	0.034±0.008^a
7	5	0.037±0.004^a
14	5	0.040±0.010^a

(continued)

Age (d)	Slaughter number (n)	Pregastric esterase activity
21	5	0.036 ± 0.009^a
28	5	0.052 ± 0.013^a
42	5	0.057 ± 0.012^a
56	5	0.077 ± 0.019^a

Note: * Statistics is Tukey' test under One-Way ANOVA in SPSS12.0.

Plasma concentrations of gut regulatory peptides

In lambs, considerable changes in the plasma concentrations of gastrin, GIP, CCK, and sectin (Table 8) occurred with increasing age, especially between days 0 and 42, which tended to decrease. CCK concentration was sharply decreased by 1/15 between 3 and 42 d, but increased by 3-fold between d 42 and 56. The secretin and gastrin concentrations reached a peak at d 3, and thereafter which decreased and on d 42 were as 45% and 36% as of d 3. GIP concentration was high between 0 and 7 d and decreased at d 14 and d 28. At d 42 it decreased 9.2% of d 7. Increased plasma PP concentration was recorded between 0 and 42 d.

Table 7 Developmental changes of pancreatic enzymes (U/mg protein)

Age (d)	Slaughter number(n)	α-amylase	Lactase	Lipase	Trypsinase	Chymotrypsin
0	6	0.267 ± 0.016^a	0.006 ± 0.002^b	0.041 ± 0.007^a	1.607 ± 0.317^a	0.105 ± 0.032^b
3	8	0.244 ± 0.018^{ab}	0.010 ± 0.002^a	0.030 ± 0.004^a	1.475 ± 0.407^a	0.124 ± 0.031^{ab}
7	5	0.249 ± 0.017^{ab}	0.007 ± 0.001^{ab}	0.038 ± 0.010^a	1.325 ± 0.322^a	0.105 ± 0.012^b
14	5	0.259 ± 0.021^a	0.010 ± 0.002^{ab}	0.039 ± 0.004^a	1.644 ± 0.080^a	0.158 ± 0.032^{ab}
21	5	0.209 ± 0.037^b	0.008 ± 0.002^{ab}	0.037 ± 0.010^a	1.617 ± 0.129^a	0.157 ± 0.031^{ab}
28	5	0.261 ± 0.031^a	0.010 ± 0.003^{ab}	0.038 ± 0.006^a	1.534 ± 0.201^a	0.166 ± 0.028^a
42	5	0.259 ± 0.015^a	0.009 ± 0.002^{ab}	0.035 ± 0.009^a	1.556 ± 0.195^a	0.156 ± 0.024^{ab}
56	5	0.248 ± 0.013^{ab}	0.010 ± 0.006^{ab}	0.037 ± 0.004^a	1.579 ± 0.121^a	0.156 ± 0.040^{ab}

Note: * Statistics is Tukey' test under One-Way ANOVA in SPSS12.0.

Table 8 Development changes of plasma regulatory peptide concentrations

Age (d)	Slaughter number(n)	CCK (pmol/L)	Secretin (ng/mL)	PP (ng/mL)	Gastrin (ng/mL)	GIP (ng/mL)
0	6	28.128 ± 10.164^{ab}	36.440 ± 14.221^{abc}	0.223 ± 0.057^b	9.108 ± 1.921^{ab}	0.441 ± 0.125^a
3	8	53.801 ± 17.396^a	54.496 ± 7.731^a	0.308 ± 0.047^{ab}	12.494 ± 1.873^a	0.425 ± 0.192^a
7	5	51.624 ± 26.714^{abc}	52.522 ± 9.701^{ab}	0.342 ± 0.024^{ab}	11.872 ± 2.852^{ab}	0.467 ± 0.315^{ab}
14	5	27.985 ± 25.671^{abc}	36.895 ± 14.235^{abc}	0.427 ± 0.068^a	7.051 ± 3.513^{ab}	0.167 ± 0.084^{ab}
21	5	43.454 ± 32.693^{abc}	43.970 ± 17.876^{abc}	0.403 ± 0.127^{ab}	9.724 ± 4.906^{ab}	0.161 ± 0.116^{ab}
28	5	20.582 ± 20.512^{abc}	31.282 ± 12.29^{bc}	0.440 ± 0.087^{ab}	6.300 ± 3.388^{ab}	0.068 ± 0.051^b
42	5	3.543 ± 2.799^c	24.820 ± 8.593^c	0.585 ± 0.347^{ab}	4.476 ± 2.093^b	0.043 ± 0.051^b
56	5	10.032 ± 14.654^b	29.251 ± 8.349^{bc}	0.437 ± 0.098^{ab}	5.090 ± 2.320^b	0.062 ± 0.049^b

Note: * Statistics is Tukey' test under One-Way ANOVA in SPSS12.0.

** CCK=Cholecystokinin; PP=Pancreatic polypeptide; GIP=Gastric inhibitory peptide; same as following.

Correlation analyses

No significant correlation was found between plasma concentrations of CCK, secretin, PP, gastrin, GIP and chymosin activity in the contents (p>0.05). By contrast, significant positive correlations were observed between plasma concentrations of CCK, secretin, gastrin, GIP and the chymosin activity in the mucosa of the cardiac, fundic and pyloric gland regions ($p<0.01$ or $p<0.05$), while significant negative correlation was recorded between plasma concentration of PP and the chymosin activity in the mucosa of the fundic and pyloric gland regions (p<0.01 or p<0.05). Significant negative correlations were found between plasma concentrations of CCK, secretin, gastrin, GIP and pepsin activity in the contents (p<0.05). Only plasma concentration of GIP was positively correlated with the specific activity of pepsin in the mucosa of the pyloric gland region (p<0.05). The plasma concentration of PP was positively correlated with the specific activity of pepsin in the mucosa of the cardiac gland region while negatively correlated with the pyloric gland region (p<0.05). Significant negative correlations were recorded between the plasma concentrations of CCK, secretin, gastrin and pregastric esterase activity of the contents (p<0.05), and no significant correlation was observed between the plasma concentrations of PP, GIP and pregastric esterase activity (p>0.05).

Table 9 Correlation between plasma regulatory peptide concentrations and abomasal enzyme activities

		CCK	Secretin	PP	Gastrin	GIP
Chymosin	Cardiac gland region	0.633**	0.530**	-0.230	0.595**	0.607**
	Fundic gland region	0.616**	0.570**	-0.343*	0.598**	0.691**
	Pyloric gland region	0.495**	0.377*	-0.438**	0.500**	0.676**
Pepsin	Cardiac gland region	-0.061	-0.106	0.309*	-0.153	-0.222
	Fundic gland region	0.064	0.201	0.234	0.089	-0.050
	Pyloric gland region	0.217	0.199	-0.326*	0.271	0.471*
Pregastric esterase	Contents	-0.346*	-0.350*	0.018	-0.345*	-0.200

Note: * Statistics is Pearson in SPSS12.0.

The plasma concentrations of CCK, secretin, PP, gastrin and GIP were all positively or negatively correlated with the specific activity of chymotrypsin (p<0.01 or p<0.05). The plasma concentrations of CCK and GIP were negatively correlated with the activity of lactase (p<0.05). Moreover, negative correlation was observed between α-amylase activity and the plasma concentration of secretin (p<0.05). No significant correlation was found between the plasma concentrations of CCK, secretin, PP, gastrin, GIP and the lipase, trypsin activities (p>0.05).

DISCUSSION

Grazing condition and grazing pasture

Gansu Alpine Fine-wool sheep grazes on alpine cold pastoral range which belongs to sub-alpine meadow and mountain rangeland. The altitude is from 2 600 m to 3 500 m, even to 4 000 m. The climate is plateau mainland climate, winds and dryness. The climatic characteristics in a year are drought and cold in winter, warm and humidity in summer. Mean annual temperature is 0.6 to

3.8℃, the highest is 31℃ in July, and lowest is −29℃ in January. Annual sunshine hours are 2 272 h. The absolute frost-free period is 45 to 60 days. Mean annual rainfall is 361.6 mm, annual evaporation is 1 111.9 mm, mean annual relative humidity is 38% to 58%. The natural grass sprouts in April and dies in September, the wilt period of pasture is above 7 months. The soil type is kastanozem, mountain phaeozem and cinnamon soil. The types of natural grassland are grass family, sedge family, weeds of grass family and sedge family, scrub weeds. The vegetation mainly consists of grass family and sedge family as well as a few pea family.

Lambs suckle their mothers and graze on native grassland pasture, with no additional feed. Animals are penned at pasture at night without access to feed and water. There are no feed supplements, but a salt mineral mix is provided to lambs throughout the year, animals have access to water only once a day. Lambs are not weaned until 4 months old.

The abomasal pH

The pH changed from 2.96 to 4.7 in the abomasal contents and from 2.8 to 4.8 in the mucosa. The pH was relatively low in the mucosa of the fundic gland region. The pH of pancreas didn't change with increasing age. The results agree with previous findings in young ruminants (Gorrill et al., 1967). But these are the first for lambs on pasture in China.

Change of abomasal enzyme activities

All the enzyme activities recorded in the abomasum were high at birth, agree with the results reported by previous reports (Huber, 1969; Era Berinkovd, 1988). Pepsin activity was lower than that of chymosin. This finding confirmed the importance of chymosin in milk clotting, because its activity level was 100%−200% higher than that reported for pepsin (Raymond et al., 1973). Pepsin activity tended to increase with increasing age, but chymosin activity reversed. After d14 chymosin activity in the mucosa decreased sharply. The results agree with the most reports about chymosin of calves abomasal mucosa (Andren, 1980; Valles, 1980), contents (Garnot et al., 1977), gastric juice (Alais, 1963; Hill et al., 1970; Kirton et al., 1971; Hagyard and Davey, 1972) and lambs abomasal mucosa (Guilloteau, 1983), but disagree with Garnot (Garnot et al., 1977). The decreasing chymosin activity may be the increasing grass intake and decreasing milk intake after d14, which weaken the stimulus of abomasum resulting in the decrease of chymosin secretion. According to the previous reporting, milk (most probably its casein fraction) is responsible for the activation of chymosin secretion (Garnot et al., 1977).

Pepsin activity in abomasal contents was relatively high and decreased to the lowest at d3, possibly because of dilution by milk intake, and thereafter it began to increase. The development of secretion potentiality of abomasum appeared to depend on age and intake. In the testing, pepsin activity in the mucosa of the cardiac and fundic gland regions was higher than that in the pyloric gland region. As a whole, the changes of pepsin activity agree with previous findings in young ruminants (Guilloteau et al., 1984; Huber et al., 1961; Isabelle et al., 1992). Pregastric esterase activity was high at birth, at d 3 which decreased by dilution of milk intake and increased a little after d 28. It appeared that the secretion potentialities of root of tongue changed little between d 0 and 56.

Table 10 Correlation between plasma regulatory peptide concentration and pancreatic enzyme activities

	CCK	Secretin	PP	Gastrin
α-Amylase	−0.264	−0.298*	0.048	−0.219
Lactase	−0.305*	−0.196	0.228	−0.294
Lipase	−0.190	−0.238	0.221	−0.183
Trypsinase	0.008	0.016	0.147	−0.008
Chymotrypsin	−0.303*	−0.307*	0.514**	−0.363*

Note: * Statistics is Pearson in SPSS12.0.

Change of pancreatic enzyme activities

Trypsin, chymotrypsin are all secreted by pancreas. It also secretes carboxypeptidases, nucleic acid enzyme and so on. All those enzymes flow into small intestine through bile duct along with pancreatic liquid. It reported that pancreatic proteolytic enzyme contributes approximately 72% of the protein in pancreas. Trypsin and chymotrypsin activities were 6-fold of the summation of α-amylase, lactase as well as lipase, and erepsin secreted by small intestine epithelia, so the digestibility of protein was very strong. The pancreas was relatively mature at birth, but the relative weight (% body weight) of it was still growing, and the function was increasing after d 21. Trypsin activity changed little with increasing age, but chymotrypsin activity tended to increase after d 14. Lipase activity existed at birth but change little with age. Under grazing conditions, the dietary contain more fat only in the period of suckling. The fat in the grass is about 1% to 4%, which is hydrolyzed by rumen microorganism into digestible fatty acid (salt) and flows into small intestine. Lipase activity in pancreas and small intestine mucosa increased little with increasing age, but still digested the fat in the diet. Pancreatic lactase activity was high at birth, but was lower than that in small intestine mucosa. Disaccharidases (lactase and maltase) are mainly secreted by small intestine mucosa and epithelium.

α-Amylase is secreted by pancreas and small intestine. According to the reports, in young pigs and calves, α-amylase was low at birth and increased with age. The dietary of weaning animals doesn't add starch generally. In the present study, the animals were grazing lambs, starch in the dietary was very little, but α-amylase activity was observed in pancreas at birth. Pancreatic activity was greater than that in small intestine mucosa. Pancreatic α-amylase activity had insignificant change with increasing age.

Relationships with gut regulatory peptides

Significant change was observed in plasma concentrations of gut regulatory peptides. High concentrations of gastrin, CCK and secretin at d3, and GIP at 7 d in plasma were observed. Thereafter they decreased and at d 42 reached the lowest point. By contrast, the concentration of plasma PP was relatively low at birth and increased with age and by d42 it reached the highest. The changes of all plasma gut regulatory peptides tested were consistent with the change of the rumen function. Adaptation of chymosin activity to dietary changes under the grazing conditions may be influenced by the decrease of CCK, secretin, gastrin, GIP, or increase of PP concentrations. In the present study, all the tested plasma concentrations of gut regulatory peptides seemed to be involved in the regulation of abomasal mucosa chymosin activity, because it was highly correlated with abomasal mucosa chymosin activity. And the change of CCK, secretin, gastrin and GIP were consistent with the change of chymosin activity. All of them decreased with increasing age, while PP reversed. On the contrary, all the tested plasma concentrations of gut regulatory peptides did not seem to be involved in the regulation of pepsin and pregastric esterase activities, because they had little correlation with the specific activity of them. This disagreed with previous reporting (Daviccon et al., 1980; Pierzynowski et al., 1991), maybe influenced by the management system (grazing) or the time of blood collection. The plasma concentrations of all the tested plasma gut regulatory peptides may have decisive, but opposite, influences on pancreatic chymotrypsin is supported by the correlations observed with the pancreatic chymotrypsin. But they did not seem to be involved in the regulation of pancreatic trypsin and lipase, because they were little correlated with trypsin and lipase activities. Only secretin seemed to be involved in the regulation of pancreatic α-amylase because they negatively correlated with each other. CCK, GIP seemed to be involved in the regulation of pancreatic lactase because they were negatively correlated with each other (but we didn't find reports at present), because lactase always be deemed to be secreted by small intestine mucosa. But in the present study considerable lactase activity was observed.

CONCLUSION

pH was acidic in abomasum (relatively lower in fundic gland region) and alkaline and stable in pancreas. α-Amylase, lactase, lipase, trypsinase and chymotrypsin activities were observed in pancreas at birth, and changed little with increasing age. The concentrations of gastrin, GIP, CCK, and secretin in plasma tended to decreasing with increasing age, and PP tended to increase. The correlation between the concentration of pancreatic polypeptide and other regulatory peptide concentrations seemed to be negative. Gastrin, GIP, CCK, and secretin may play a positive regulatory function to chymotrypsin activity in abomasum, while pancreatic polypeptide had a negative regulatory function. The tested plasma regulatory peptides were insignificantly correlated with pepsin and pregastric esterase activities in abomasum. The plasma regulatory peptides correlated with chymotrypsin activity in pancreas positively or negatively; PP positively regulates α-amylase activity in pancreas; GIP and CCK negatively regulate the lactase activity in pancreas.

References

[1] Alais, C. 1963. Etude de la secretion d'enzymes coagulant dans la caillette de l'agneau. Ann. Biol. Anita. Biochim. Biophys. 3: 65.

[2] Andren, A., L. Bjorck and O. Claesson. 1980. Quantification of chymosin (rennin) and pepsin in bovine abomasum by rocket immuno-electrophoresis. Swed. J. Agric. Res. 10: 123.

[3] Arima, K. 1967. Milk-clotting enzyme from microorganism, part I, screening test and identification of the potent fungus. Agric. Biol. Chem. 31: 540-545.

[4] Corring, T., A. Aumaltre and G. Durand. 1978. Development of digestive enzymes in the piglet from birth to 8 weeks. I. Pancreas and pancreatic enzymes. Nutr. Metab. 22: 231-243.

[5] Davicco, M.J., J. Lefaivre, and J.P. Barlet. 1980. The endocrine regulation of exocrine pancreas in preruminant milk-fed calves. Ann. Rech. Vet. 11 (2): 123-132.

[6] Era Berinkovd, Jiri Sajdok, Pavel Rauch and Jan Kag.1988. Determination of chymosin and bovine pepsin content of bovine rennets by high performance liquid chromatography.Neth.Milk Dairy J.42: 337.

[7] Foltmann, B., A.L.Jensen, P.Lonblad, E.Smidt and N.H.Axelsen.1981. A developmental analysis of the production of chymosin and pepsin in pigs.Comp.Biochem.Physiol.68: 9.

[8] Garnot, P., R.Toullec, J.L.Thapon, T.P.Martin and M.T.Hoang. 1977. Influence of age, dietary protein and weaning on calf abomasal enzymic secretion.J.Dairy Res.Feb; 44 (1): 9-23.

[9] Garnot, P., R.Toullec, J.L.Thapon, P.Martin, M.T.Hoang, C.M.Mathieu and B.Ribadeau-Dumas. 1977. Influence of age, dietary protein and weaning on calf abomasal enzymatic secretion.J.Dairy Res. 44: 9.

[10] Guilloteau, P., T.Corring, R.Toullec and J.1984. Enzyme potentialities of the abomasum and pancreas of the calf.I.Effect of age in the preruminant.Reprod.Nutr.Dev.24 (3): 315-325.

[11] Guilloteau, P., T.Corring, Pascaline Garont, P.Martin, R.Toullec and G.Durand.1983. Effects of age and weaning on enzyme activities of abomasum and pancreas of the lamb.J.Dairy Sci.66: 2 373-2 385.

[12] Guilloteau, P.1983. Effects of age and weaning on enzyme activities of abomasum and pancreas of the lamb. J.Dairy Sci.66: 2 373-2 385.

[13] Guilloteau, P., T.Corring, P.Garnot, P.Martin, R.Toullec and G.Durand. 1983. Effects of age and weaning on enzyme activities of abomasum and pancreas of the lamb.J.Dairy Sci.66: 2 373-2 385.

[14] Guilloteau, P., T.Corring, R.Toullec and R.Guilhermet.1985. Enzyme potentialities of the abomasum and pancreas of the calf.II.Effects of weaning and feeding a liquid supplement to ruminant animals.Reprod.Nutr.Dev.25: 481-493.

[15] Guilloteau, P., T.Corring, R.Toullec and J.Robelin.1984. Enzyme potentialities of the abomasum and pancreas of the calf.I.Effect of age in the preruminant.Reprod.Nutr.Dev.24: 315-325.

[16] Hagyard, C.J.and Davey CL.Rennin yield.1972. The effects of bobby calf age and degree of starvation. New Zealand J.Dairy Sci.Technol.7: 140.

[17] Hartman, P.A., V.W.Hays, R.C.Baker, L.H.Neagle and D.V.Catron.1961. Digestive enzyme development in the young pig.J.Anim.Sci.20: 114.

[18] Henschel, M.J.1973. Comparison of the development of proteolytic activity in the abomasums of the preruminant calf with that in the stomach of the young rabbit and guinea-pig.Br.J.Nutr.30: 285.

[19] Hill, K.J., D.E.Noakes and R.A.Lowe.1970. Gastric digestive physiology of the calf and piglet.pp.166-179 in Physiology of digestion and metabolism in the ruminant (Ed.A.T.Phillipson).IIIth Int.Symp.Ruminant Physiol., Oriel Press Ltd., Newcastle upon Tyne.

[20] Huber, J.T., N.L.Jacobson, R.S.Allen and P.A.Hartman.1961. Digestive enzyme activities in the young calf.J.Dairy Sci.44: 1 494.

[21] Huber, J.T.1969. Calf Nutrition and rearing.Development of the digestive and metabolic apparatus of the calf.J.Dairy Sci.52: 1 303.

[22] Huber, J.T., N.L.Jacobson, R.S.Allen and P.A.Hartman.1961. Digestive enzyme activities in the young calf.J.Dairy Sci.44: 1 494-1 501.

[23] Isabelle Le Huerou-luron, Paul Guilloteau, Catherine Wicker-Planquart, Jean-Alain Chayvialle, John Burton, Aziz Mouats, Rene Toullec and Antoine Puigserver.1992. Gastric and pancreatic enzyme activities and their relationship with some gut regulatory peptides during postnatal development and weaning in calves.J.Nutr.122: 1 434-1 445.

[24] Kirton, A.H., D.J.Paterson and N.H.Clarke.1971. Slaughter information and rennin production from bobby calves.New Zealand J.Agric.Res.14: 397.

[25] Kretchmer, N.1985. Weaning: enzymatic adaptation.Am.J.Clin.Nutr.41: 391-398.

[26] Pierzynowski, S.G., R.Zabielsli, B.R.Weström, M.Mikolajczyk and W.Barej.1991. Development of the exocrine pancreatic function in chronically cannulated calves from the preweaning period up to early rumination.J.Anim.Physiol.Anim.Nutr.65: 165-172.

[27] Raymond, M. N., E. Ericas, R. Salesse, F. Gamier, P. Garnot and B. Ribadeau-Dumas. 1973. Proteolytic unit for chymosin (rennin) activity based on a reference synthetic peptide.J.Dairy Sci. 56: 419-425.

[28] Thivend, P., R.Toullec and P.Guilloteau.1980. Digestive adaptation in the preruminant.561-586 in Digestive physiology and metabolism in ruminants (Ed.Y.Ruckebusch and P.Thivend).Vth Int.Symp.Rumin.Physiol.M.T.P.Press Ltd., Lancaster.

[29] Valles E.1980. Les pepsins gastriques bovines utilisees en fromagerie.These Doct.Ing, Paris.

[30] Walker, D.M.1959. The development of the digestive system of the young animal.IV.Proteolytic enzyme development in the young lamb.J.Agric.Sci.53: 381.

(发表于《Asian-Australian Journal of Animal Science》，院选 IF: 0.4889)

4-Allyl-2-methoxyphenyl 2-acetoxy-benzoate

LIU Xi-Wang, LI Jian-Yong*, YANG Ya-Jun, ZHANG Ji-Yu

(Key Laboratory of New Animal Drug Project/Gansu Province Key Laboratory of Veterinary Drug Discovery, Ministry of Agriculture/Lanzhou Institute of Animal Science and Veterinary Pharmaceutics, Chinese Academy of Agricultural Sciences, Lanzhou 730050, China)

Key indicators: single-crystal X-ray study; T = 296 K; mean (C-C) = 0.005 Å; R factor = 0.051; wR factor = 0.142; data-to-parameter ratio = 14.0.

In the title compound, $C_{19}H_{18}O_5$, the ester group is twisted with respect to the acetylsalicylic acid and eugenol rings at dihedral angles of 22.48 (2) and 81.07 (1)°, respectively. The dihedral angle between the two benzene rings is 60.72 (1)°. The crystal packing exhibits no significantly short inter-molecular contacts.

RELATED LITERATURE

For background regarding the medicinal properties of eugenol, see: Feng & Lipton (1987); Dohi et al. (1989). For the synthesis of the aspirin eugenol ester and its biological activity, see: Li et al. (2011).

EXPERIMENTAL

Crystal data

$C_{19}H_{18}O_5$

$M_r = 326.33$

Monoclinic, $P2_1 = n$

$a = 10.60$ (2) Å

$b = 12.58$ (2) Å

$c = 13.23$ (2) Å

$\beta = 109.020$ (17)°

$V = 1667$ (5) Å^3

$Z = 4$

Mo Kα radiation

$\mu = 0.09$ mm^{-1}

$T = 296$ K

0.26 mm × 0.24 mm × 0.22 mm

Data collection

Bruker APEXII CCD diffractometer

Absorption correction: multi-scan (*SADABS*; Sheldrick, 1996)

$T_{min} = 0.976$, $T_{max} = 0.980$

8695 measured reflections

3089 independent reflections

1786 reflections with $I > 2\sigma$ (I)

$R_{int} = 0.054$

Refinement

$R[F^2 > 2\sigma F^2)] = 0.051$

wR $(F^2) = 0.142$

$S = 1.01$

3089 reflections

220 parameters

H-atom parameters constrained

$\Delta\rho_{max} = 0.49$ eÅ^{-3}

$\Delta\rho_{min} = -0.25$ eÅ^{-3}

Data collection: *APEX2* (Bruker, 2007); cell refinement: *SAINT* (Bruker, 2007); data reduction: *SAINT*; program (s) used to solve structure: *SHELXS*97 (Sheldrick, 2008); program (s) used to refine structure: *SHELXL*97 (Sheldrick, 2008); molecular graphics: *SHELXTL* (Sheldrick, 2008); software used to prepare material for publication: *SHELXTL*.

This work was supported by the earmarked fund for China Agriculture Research System (cars-38).

Supplementary data and figures for this paper are available from the IUCr electronic archives (Reference: EZ2244).

References

[1] Bruker (2007). APEX2 and SAINT.Bruker AXS Inc., Madison, Wisconsin, USA.

[2] Dohi, T., Terada, H., Anamura, S., Okamoto, H.& Tsujimoto, A. (1989).*Jpn. J. Pharmacol.*49: 535-539.

[3] Feng, J.D.& Lipton, J.M. (1987).*Neuropharmacology*, 26: 1 775-1 778.

[4] Li, J.-Y., Yu, Y.-G., Wang, Q.W., Zhang, J.-Y., Yang, Y.-J., Li, B., Zhou, X.-Z., Niu, J.-R., Wei, X.-J., Liu, X.-W.& Liu, Z.-Q. (2011).*Med. Chem. Res.* doi: 10.1007/s00044-011-9609-1.

[5] Sheldrick, G.M. (1996).*SADABS*.University of Göttingen, Germany. Sheldrick, G.M. (2008).*Acta Cryst.*A64: 112-122.

Supplementary Materials:
4-Allyl-2-methoxyphenyl 2-acetoxybenzoate

LIU X.W., LI J.Y., YANG Y.J., ZHANG J.Y.

COMMENT

Aspirin has been widely used as an analgesic and anti-inflammatory drug. As the major constituent of clove oil, eugenol also shows antipyretic activity (Feng & Lipton, 1987) and anti-inflammatory activity (Dohi et al., 1989). In this paper, we report the structure of the title compound, which was synthesized from the reaction of aspirin and eugenol in sodium hydroxide solution.

EXPERIMENTAL

The title compound was obtained according to the literature method (Li et al., 2011). Acetylsalicylic acid (0.025 mol) and thionyl chloride (2.5 mL) were mixed in 10 mL tetrahydrofuran (THF), and refluxed at 343 K for 2 h. The surplus thionyl chloride and THF were removed under reduced pressure. The target O-Acetylsalicylyl chloride was dissolved in 5 mL THF, added dropwise to an iced solution of eugenol (0.025 mol) and sodium hydroxide (0.04 mol) in 40 mL water. After stirring at room temprature for 3 h, the crude product was obtained by filtration. The crystals were obtained by recrystalization from methanol. Elemental analysis: calculated for $C_{19}H_{18}O_5$: C 69.93%, H 5.56%, O 24.51%; found: C 69.93%, H 5.53%, O 24.54%.

REFINEMENT

The positions of all H atoms were determined geometrically and refined using a riding model with C—H = 0.93–0.97 Å and U_{iso}(methyl H) = 1.5 U_{eq}(C) and 1.2 U_{eq} for other H atoms.

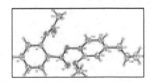

Fig. 1 The molecular structure of (I), with atom labels and displacement ellipsoids drawn at the 30% probability level.

4-ALLYL-2-METHOXYPHENYL 2-ACETOXYBENZOATE

Crystal data

$C_{19}H_{18}O_5$ $F(000) = 688$

M_r = 326.33

Monoclinic, $P2_1/n$

Hall symbol: P 21/n

a = 10.60 (2) Å

b = 12.58 (2) Å

c = 13.23 (2) Å

β = 109.020 (17)°

V = 1667 (5) Å^3

Z = 4

D_x = 1.300 Mg·m^{-3}

Melting point = 344–345 K

Mo $K\alpha$ radiation, λ = 0.71073 Å

Cell parameters from 2334 reflections

θ = 2.3–23.9°

μ = 0.09 mm^{-1}

T = 296 K

Block, colorless

0.26 mm×0.24 mm×0.22 mm

Data collection

Bruker APEXII CCD diffractometer

Radiation source: fine-focus sealed tube graphite

φ and ω scans

Absorption correction: multi-scan (*SADABS*; Sheldrick, 1996)

T_{min} = 0.976, T_{max} = 0.980

8695 measured reflections

3089 independent reflections

1786 reflections with $I > 2\sigma(I)$

R_{int} = 0.054

θ_{max} = 25.5°, θ_{min} = 2.3°

h = −11→12

k = −15→15

l = −16→15

Refinement

Refinement on F^2

Least-squares matrix: full

$R[F^2>2\sigma(F^2)]$ = 0.051

$wR(F^2)$ = 0.142

S = 1.01

3089 reflections

220 parameters

0 restraints

Primary atom site location: structure-invariant direct methods

Secondary atom site location: difference Fourier map

Hydrogen site location: inferred from neighbouring sites

H-atom parameters constrained

$w = 1/[\sigma^2(F_o^2)+(0.0532P)^2+0.6544P]$

where $P = (F_o^2 + 2F_c^2)/3$

$(\Delta/\sigma)_{max}$ < 0.001 $\Delta\rho_{max}$ = 0.49 e Å^{-3}

$\Delta\rho_{min}$ = −0.25 e Å^{-3}

Extinction correction: *SHELXL97* (Sheldrick, 2008),

Fc* = kFc[1+0.001xFc$^2\lambda^3$/sin(2θ)]$^{-1/4}$

Extinction coefficient: 0.044 (3)

Special details

Geometry.All e.s.d.'s (except the e.s.d.in the dihedral angle between two l.s.planes) are estimated using the full covariance mat-rix.The cell e.s.d.'s are taken into account individually in the estimation of e.s.d.'s in distances, angles and torsion angles; correlations between e.s.d.'s in cell parameters are only used when they are defined by crystal symmetry. An approximate (isotropic) treatment of cell e.s.d.'s is used for estimating e.s.d.'s involving l.s.planes.

Refinement. Refinement of F^2 against ALL reflections. The weighted R-factor wR and goodness of fit S are based on F^2, convention-al R-factors R are based on F, with F set to zero for negative F^2. The threshold expression of $F^2 > \sigma(F^2)$ is used only for calculating R-factors (gt) etc. and is not relevant to the choice of reflections for refinement. R-factors based on F^2 are statistically about twice as large as those based on F, and R-factors based on ALL data will be even larger.

Fractional atomic coordinates and isotropic or equivalent isotropic displacement parameters ($Å^2$)

	x	y	z	$U_{iso}*/U_{eq}$
C1	0.5429 (3)	0.8815 (2)	1.1109 (2)	0.0552 (7)
H1	0.5154	0.9490	1.0842	0.066*
C2	0.6740 (3)	0.8662 (3)	1.1758 (3)	0.0651 (9)
H2	0.7329	0.9233	1.1933	0.078*
C3	0.7161 (3)	0.7671 (3)	1.2139 (2)	0.0667 (9)
H3	0.8038	0.7566	1.2575	0.080*
C4	0.6290 (3)	0.6826 (3)	1.1880 (2)	0.0589 (8)
H4	0.6582	0.6150	1.2135	0.071*
C5	0.4980 (2)	0.6979 (2)	1.1242 (2)	0.0454 (6)
C6	0.4519 (2)	0.7981 (2)	1.0850 (2)	0.0445 (6)
C7	0.3121 (3)	0.8251 (2)	1.0192 (2)	0.0460 (6)
C8	0.0856 (2)	0.7755 (2)	0.9839 (2)	0.0457 (7)
C9	0.0242 (2)	0.8431 (2)	1.0363 (2)	0.0451 (6)
C10	−0.1134 (3)	0.8549 (2)	0.9951 (2)	0.0522 (7)
H10	−0.1563	0.8996	1.0295	0.063*
C11	−0.1880 (3)	0.8015 (2)	0.9038 (2)	0.0563 (7)
C12	−0.1244 (3)	0.7362 (2)	0.8531 (2)	0.0619 (8)
H12	−0.1735	0.7007	0.7911	0.074*
C13	0.0134 (3)	0.7226 (2)	0.8936 (2)	0.0563 (8)
H13	0.0560	0.6776	0.8593	0.068*
C14	0.3662 (3)	0.5694 (2)	1.0060 (3)	0.0523 (7)
C15	0.2725 (3)	0.4813 (3)	1.0018 (3)	0.0784 (10)
H15A	0.2235	0.4650	0.9286	0.118*
H15B	0.2114	0.5019	1.0382	0.118*
H15C	0.3216	0.4197	1.0359	0.118*
C16	0.0468 (3)	0.9576 (2)	1.1848 (2)	0.0664 (8)
H16A	−0.0043	1.0124	1.1391	0.100*
H16B	0.1152	0.9894	1.2436	0.100*
H16C	−0.0109	0.9156	1.2118	0.100*
C17	−0.3382 (3)	0.8185 (3)	0.8597 (3)	0.0794 (10)
H17A	−0.3803	0.7545	0.8231	0.095*

H17B	−0.3719	0.8312	0.9186	0.095*
C18	−0.3746 (4)	0.9083 (4)	0.7853 (5)	0.1113 (15)
H18	−0.3360	0.9723	0.8151	0.134*
C19	−0.4456 (4)	0.9161 (4)	0.6917 (4)	0.1173 (16)
H19A	−0.4886	0.8564	0.6548	0.141*
H19B	−0.4565	0.9817	0.6575	0.141*
O1	0.41425 (17)	0.60944 (14)	1.10660 (15)	0.0517 (5)
O4	0.22313 (15)	0.75382 (14)	1.03094 (14)	0.0512 (5)
O2	0.3986 (2)	0.60219 (17)	0.93385 (16)	0.0684 (6)
O3	0.28083 (19)	0.90274 (16)	0.96467 (17)	0.0681 (6)
O5	0.10653 (18)	0.89150 (15)	1.12579 (15)	0.0582 (5)

Atomic displacement parameters (Å^2)

	U^{11}	U^{22}	U^{33}	U^{12}	U^{13}	U^{23}
C1	0.0542 (17)	0.0533 (16)	0.0632 (19)	−0.0034 (13)	0.0263 (16)	−0.0015 (15)
C2	0.0491 (18)	0.076 (2)	0.073 (2)	−0.0145 (16)	0.0245 (17)	−0.0128 (18)
C3	0.0390 (16)	0.095 (3)	0.063 (2)	0.0017 (17)	0.0112 (15)	−0.0091 (19)
C4	0.0483 (17)	0.0676 (19)	0.0568 (19)	0.0101 (15)	0.0115 (15)	0.0017 (16)
C5	0.0426 (15)	0.0516 (16)	0.0420 (15)	0.0016 (12)	0.0139 (13)	−0.0015 (13)
C6	0.0431 (14)	0.0494 (15)	0.0435 (15)	0.0003 (12)	0.0177 (13)	−0.0015 (13)
C7	0.0468 (15)	0.0433 (15)	0.0506 (16)	−0.0001 (12)	0.0197 (13)	0.0004 (14)
C8	0.0382 (14)	0.0477 (15)	0.0518 (17)	0.0038 (12)	0.0153 (13)	0.0055 (13)
C9	0.0424 (15)	0.0473 (14)	0.0446 (16)	0.0017 (12)	0.0127 (13)	0.0014 (13)
C10	0.0426 (15)	0.0616 (17)	0.0552 (18)	0.0108 (13)	0.0196 (14)	0.0042 (15)
C11	0.0407 (15)	0.076 (2)	0.0493 (17)	−0.0026 (14)	0.0110 (14)	0.0006 (17)
C12	0.0541 (18)	0.075 (2)	0.0530 (18)	−0.0113 (15)	0.0126 (15)	−0.0093 (16)
C13	0.0573 (18)	0.0576 (17)	0.0596 (19)	−0.0026 (14)	0.0268 (16)	−0.0067 (15)
C14	0.0411 (15)	0.0520 (16)	0.0586 (19)	0.0105 (12)	0.0090 (15)	−0.0025 (16)
C15	0.064 (2)	0.072 (2)	0.093 (3)	−0.0104 (17)	0.0180 (19)	−0.015 (2)
C16	0.073 (2)	0.071 (2)	0.0570 (19)	0.0082 (17)	0.0236 (17)	−0.0100 (16)
C17	0.0428 (17)	0.115 (3)	0.074 (2)	−0.0002 (19)	0.0107 (17)	0.005 (2)
C18	0.048 (2)	0.131 (4)	0.138 (4)	0.020 (2)	0.009 (3)	0.003 (4)
C19	0.080 (3)	0.128 (4)	0.121 (4)	0.026 (3)	0.001 (3)	0.012 (3)
O1	0.0531 (11)	0.0474 (11)	0.0527 (12)	0.0001 (9)	0.0146 (10)	0.0027 (9)
O4	0.0372 (9)	0.0490 (10)	0.0684 (13)	0.0070 (8)	0.0188 (9)	0.0113 (9)
O2	0.0699 (14)	0.0784 (15)	0.0533 (13)	0.0034 (11)	0.0150 (12)	−0.0029 (11)
O3	0.0577 (12)	0.0600 (13)	0.0808 (15)	0.0010 (10)	0.0144 (11)	0.0244 (12)
O5	0.0482 (11)	0.0671 (13)	0.0543 (12)	0.0069 (10)	0.0097 (10)	−0.0124 (10)

Geometric parameters (Å, °)

C1—C6	1.390 (4)	C11—C17	1.521 (5)

C1—C2	1.387 (4)	C12—C13	1.392 (5)
C1—H1	0.9300	C12—H12	0.9300
C2—C3	1.364 (5)	C13—H13	0.9300
C2—H2	0.9300	C14—O2	1.189 (4)
C3—C4	1.376 (5)	C14—O1	1.357 (4)
C3—H3	0.9300	C14—C15	1.477 (4)
C4—C5	1.383 (4)	C15—H15A	0.9600
C4—H4	0.9300	C15—H15B	0.9600
C5—C6	1.390 (4)	C15—H15C	0.9600
C5—O1	1.394 (4)	C16—O5	1.423 (3)
C6—C7	1.492 (4)	C16—H16A	0.9600
C7—O3	1.195 (3)	C16—H16B	0.9600
C7—O4	1.346 (3)	C16—H16C	0.9600
C8—C13	1.363 (4)	C17—C18	1.465 (6)
C8—C9	1.386 (4)	C17—H17A	0.9700
C8—O4	1.412 (4)	C17—H17B	0.9700
C9—O5	1.364 (3)	C18—C19	1.226 (6)
C9—C10	1.389 (4)	C18—H18	0.9300
C10—C11	1.382 (4)	C19—H19A	0.9300
C10—H10	0.9300	C19—H19B	0.9300
C11—C12	1.368 (4)		
C6—C1—C2	121.5 (3)	C13—C12—H12	119.7
C6—C1—H1	119.3	C8—C13—C12	119.7 (3)
C2—C1—H1	119.3	C8—C13—H13	120.1
C3—C2—C1	119.8 (3)	C12—C13—H13	120.1
C3—C2—H2	120.1	O2—C14—O1	123.0 (3)
C1—C2—H2	120.1	O2—C14—C15	126.7 (3)
C2—C3—C4	120.1 (3)	O1—C14—C15	110.4 (3)
C2—C3—H3	119.9	C14—C15—H15A	109.5
C4—C3—H3	119.9	C14—C15—H15B	109.5
C3—C4—C5	120.2 (3)	H15A—C15—H15B	109.5
C3—C4—H4	119.9	C14—C15—H15C	109.5
C5—C4—H4	119.9	H15A—C15—H15C	109.5
C4—C5—C6	121.0 (3)	H15B—C15—H15C	109.5
C4—C5—O1	116.6 (3)	O5—C16—H16A	109.5
C6—C5—O1	122.3 (3)	O5—C16—H16B	109.5
C1—C6—C5	117.5 (3)	H16A—C16—H16B	109.5
C1—C6—C7	116.7 (3)	O5—C16—H16C	109.5
C5—C6—C7	125.8 (2)	H16A—C16—H16C	109.5
O3—C7—O4	123.2 (3)	H16B—C16—H16C	109.5
O3—C7—C6	124.4 (2)	C18—C17—C11	112.3 (3)
O4—C7—C6	112.4 (2)	C18—C17—H17A	109.1

C13—C8—C9	121.1 (3)	C11—C17—H17A	109.1
C13—C8—O4	119.7 (2)	C18—C17—H17B	109.1
C9—C8—O4	118.9 (3)	C11—C17—H17B	109.1
O5—C9—C8	115.9 (3)	H17A—C17—H17B	107.9
O5—C9—C10	125.9 (2)	C19—C18—C17	132.9 (5)
C8—C9—C10	118.3 (3)	C19—C18—H18	113.5
C11—C10—C9	121.3 (2)	C17—C18—H18	113.5
C11—C10—H10	119.4	C18—C19—H19A	120.0
C9—C10—H10	119.4	C18—C19—H19B	120.0
C12—C11—C10	119.1 (3)	H19A—C19—H19B	120.0
C12—C11—C17	121.2 (3)	C14—O1—C5	118.3 (2)
C10—C11—C17	119.7 (3)	C7—O4—C8	118.8 (2)
C11—C12—C13	120.5 (3)	C9—O5—C16	117.7 (2)
C11—C12—H12	119.7		
C6—C1—C2—C3	1.2 (4)	C9—C10—C11—C12	−0.4 (4)
C1—C2—C3—C4	0.1 (4)	C9—C10—C11—C17	−178.6 (3)
C2—C3—C4—C5	−0.7 (4)	C10—C11—C12—C13	0.9 (4)
C3—C4—C5—C6	0.1 (4)	C17—C11—C12—C13	179.1 (3)
C3—C4—C5—O1	−176.5 (2)	C9—C8—C13—C12	−0.1 (4)
C2—C1—C6—C5	−1.7 (4)	O4—C8—C13—C12	173.8 (2)
C2—C1—C6—C7	176.9 (2)	C11—C12—C13—C8	−0.6 (4)
C4—C5—C6—C1	1.0 (4)	C12—C11—C17—C18	−90.0 (5)
O1—C5—C6—C1	177.5 (2)	C10—C11—C17—C18	88.1 (4)
C4—C5—C6—C7	−177.5 (2)	C11—C17—C18—C19	122.2 (5)
O1—C5—C6—C7	−1.0 (4)	O2—C14—O1—C5	4.8 (4)
C1—C6—C7—O3	21.1 (4)	C15—C14—O1—C5	−176.0 (2)
C5—C6—C7—O3	−160.4 (3)	C4—C5—O1—C14	−113.6 (3)
C1—C6—C7—O4	−156.7 (2)	C6—C5—O1—C14	69.8 (3)
C5—C6—C7—O4	21.8 (3)	O3—C7—O4—C8	−5.3 (4)
C13—C8—C9—O5	179.9 (2)	C6—C7—O4—C8	172.4 (2)
O4—C8—C9—O5	6.0 (3)	C13—C8—O4—C7	105.2 (3)
C13—C8—C9—C10	0.6 (4)	C9—C8—O4—C7	−80.8 (3)
O4—C8—C9—C10	−173.3 (2)	C8—C9—O5—C16	−177.1 (2)
O5—C9—C10—C11	−179.6 (2)	C10—C9—O5—C16	2.2 (4)
C8—C9—C10—C11	−0.4 (4)		

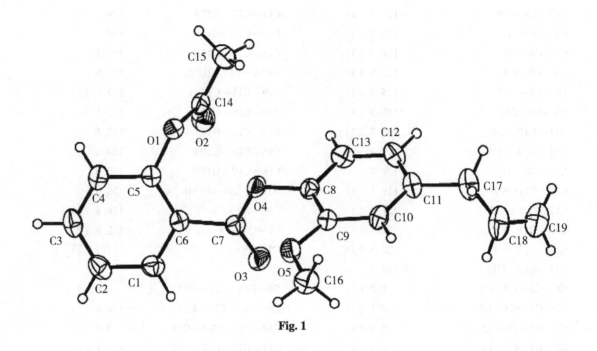

Fig. 1

(发表于《Acta Crystallographica Section E》,院选 SCI, IF: 0.413)

Simultaneous Identification of *FecB* and *FecX*^{G} Mutations in Chinese Sheep Using High Resolution Melting Analysis

YUE Yao-Jing, YANG Bo-Hui*, LIANG Xia, LIU Jian-Bin,
JIAO Shuo, GUO Jian, SUN Xiao-Ping,
NIU Chun-E, FENG Rui-Lin

(Lanzhou Institute of Animal and Veterinary Pharmaceutics Sciences, Chinese Academy of Agricultural Sciences, Lanzhou 730050, China)

Abstract: The aim of this research was to investigate the genetic structure at *FecB* (*BMPR 1B*), and *FecX*^{G} (BMP15) prolificacy genes in five sheep breeds (strains) reared in china: Small Tail Han, Gansu Alpine Fine Wool Sheep, Poll Dorset Small Tail Han F_2 (DH F_2), Borderdale×Small Tail Han F_2 (BH F_2) and Borderdale×Mongolian Sheep F_2 (BM F_2). Using an high-resolution melting (HRM) method for the simultaneous screening of *FecB* and *FecX*^{G} mutations, *FecB* mutation was discovered in Small Tail Han, DH F_2 and BH F. Out of 98 individuals of Small Tail Han, 40 were homozygous (BB), 47 heterozygous (B+) and 11 non-carriers (++) for the *FecB* mutation. The frequency of the *FecB* allele in this sample was about 0.65. Results indicated that the frequency of the *FecB* mutation is high in Small Tail Han, but heterozygous genotype (B+) was only found in DH F_2 and BH F_2. None of the sheep breeds carried the *FecX*^{G} mutation. These findings verify that HRM is a reliable, rapid, and sensitive method for *FecB* and *FecX*^{G} mutations screening. Furthermore, our study will facilitate the use of *FecB* allele and *FecX*^{G} in improving the prolificacy of non-prolific sheep breeds of China.

Key words: Sheep breeds; *FecB*; *FecX*^{G}; High-resolution melting

INTRODUCTION

In recent years, many studies on the genetics of prolificacy in sheep (*Ovis aries*) lead to highlight the importance of two major genes: *FecB* (*BMPR 1B*) and *FecX*^{G} (BMP15), which have been shown to affect ovulation rate and litter size through different mechanisms (Davis 2005).

The bone morphogenetic protein receptor type 1B (*BMPR 1B*) gene has been mapped to sheep chromo-some 6 and its coding sequence contains 10 exons (Mulsant et al. 2001). In 2001 a

---

* Corresponding author. Email: yangbh2004@163.com

G/A transition was detected at nucleotide position 746 of *BMPR 1B* cDNA, which has been associated with the 'hyper prolific phenotype' of the Booroola sheep, indicated as *FecB* (Souza et al. 2001; Wilson et al.2001).The *FecB* allele displays an additive effect for ovulation rate and leads to an increase in litter size (Davis 2004; Kumar et al. 2006). Bone morphogenetic protein 15 (*BMP*15) is a growth factor and a member of the transforming growth factor beta (TGF-β) super family that is specifically expressed in oocytes.The sheep *BMP*15 gene maps to the X chromosome (Galloway et al.2000).The *FecX^G* mutation (Q239Ter) in the *BMP*15 gene was associated with increased ovulation rate and sterility in Cambridge and Belclare sheep (Hanrahan et al.2004).

The Small Tail Han is a prolific local sheep breed in China.The Small Tail Han ewes carrying in both *FecB* and *FecX^G* mutations had greater litter size than those with either mutation alone (Chu et al.2007).Assay like single strand conformation polymorphism (SSCP) (Hanrahan et al.2004), polymerase chain reaction-restriction fragment length polymorphism (PCR) -RFLP (Davis, 2004; Kumar et al.2006) have been utilised to detect *FecB* and *FecX^G* mutations as well as genotyping the animals.However, these methods are expensive, time consuming, or have a reduced sensitivity.

The recent development of platforms and soft-ware for high-resolution melting (HRM) analysis (Wittwer et al.2003) has provided the introduction of this technology in the clinical laboratory.The great advantages of HRM (sensitivity, rapidity, simplicity, low cost and high throughput) have justified the widespread use of these methods for detection of many small genetic variations (Wittwer et al.2003; Vaughn and Elenitoba-Johnson 2004; Jones et al.2008).We report the results obtained in 683 sheep by using a fast, sensitive, and reliable HRM method that allows the simultaneous screening of *FecB* and *FecX^G* mutations.

MATERIALS AND METHODS

Blood samples were taken from Gansu province of China, using Flinders Technology Associates (FTA®) paper (Whatman Inc.), from 683 sheep along with data on litter size at the last lambing (Table 1).The ewes of the Small Tail Han were chosen at random from flocks of 300 ewes. Gansu Alpine Fine Wool Sheep were chosen as litter size for single ($n=50$) and twin ($n=48$).DH F_2, BH F_2 and BM F_2 ewes were chosen at three months old from a flock of 300 ewes.

Table 1 Mean litter size, sampling site and breed mean litter size of various sheep beads

Breed	Sample size	Mean litter size	Sampling site
STHS	98	2.13	Ding XI Gansu Province
GASFWS	98	1.49	Zhang Ye Gansu Province
DH F_2	121	None	Yong Chang Gansu Province
BH F_2	234	None	Yong Chang Gansu Province
BM F_2	132	None	Yong Chang Gansu Province

For the blood samples on FTA® paper, DNA was extracted from each 2.0 mm punch.The punch was transferred into a 0.2 mL PCR tube.An aliquot of 80 mL of 50 mM NaOH solution was added to the tube containing the blood disk and it was incubated for 10 minutes at 988C.The tube was inverted occasionally during incubation.The solution was then discarded, and the disk was washed in 100 μL TE^{-1} buffer twice. After the removal of the TE^{-1}, the disk was air-dried.

Oligos were synthesised at Takra Biotechnology (DaLian) Co., Ltd. Primers were designed using Primer 5.0 (Rozen and Skaletsky 2000). Primers and probes were chosen to detect *FecB* and *FecX^G* mutations (Souza et al. 2001; Wilson et al. 2001; Hanrahan et al. 2004). The primer pairs and the genotyping methods utilised are described in Table 2. The unlabeled probes had either a 3'-phosphate or the 3'-amino modifier (Takra Biotechnology (DaLian) Co., Ltd) incorporated to prevent extension during PCR. Probes were designed to analyze the mutations. For showing the six different melting curve plots that are produced by the two unlabeled probes of the *FecB* and *FecX^G* samples, six different genotypes of *FecB* and *FecX^G* plasmids were constructed as *FecB* (++), *FecB* (B+), *FecB* (BB), *FecX^G* (++), *FecX^G* (+G) and *FecX^G* (GG), which were all sequenced on the ABI PRISM 3730 genetic analyzer (Takra Biotech-nology (DaLian) Co., Ltd).

Table 2 Oligo nucleotides used for detection and sequencing

Name	Sequence 5'-3'	Length (bp)	3' Modification	Reference
BMPR 1BF	AGATTGGAAAAGGTCGCTATG		None	Chu et al. (2007)
BMPR 1BR	ACCCTGAACATCGCTAATACA	187	None	Chu et al. (2007)
BMPR 1BT	GACAGAAATATATCGGACGGTGTTGATGAGGCATG		C6 amino	None
BMP15F	GGCACTTCATCATTGGACACTG		None	None
BMP15R	GCCTTTAGGGAGAGGTTTGGT	87	None	None
BMP15T	TTCAATGACACTCAGAGTGTTCAGAAG		C6 amino	None

Note: Unmodified oligo nucleotides were used as primers for PCR amplification and sequencing, and oligo nucleotides with 3' modifications were used as probes.

All sheep of DNA were amplified in a 15 mL final volume containing 1×TIANGEN BIOTECH *Taq* PCR Master Mix, 0.1 mmol/L forward primer (*BMPR 1BF* & *BMP15F*), 0.5 mmol/L reverse primer (*BMPR 1BR* & *BMP15R*), 0.5 mmol/L probe (*BMPR 1BT* & *BMP15T*) and 1× LCGreenPlus+. Two DNA normal controls and one positive mutated DNA sample (*FecB* (*BMPR 1B*) (++) and *FecX^G* (*BMP15*) (++) unpublished) were included in each run.

The PCR program consists of an initial preheating at 958℃ for 5 minutes to activate the *Taq* DNA polymerase, followed by 50 amplification cycles. Each cycle comprises an annealing step at 60.58℃ for 15 seconds, an elongation step at 728℃ for 25 seconds, and denaturation at 948℃ for 15 seconds. The final melting program consists of three main steps begin-ning with a denaturation at 958℃ for 1 minute, renaturation at 258℃ for 1 minute.

The HR-1 (Idaho Technology) is an instrument that measures high-resolution DNA melting curves from samples in a single LightCycler capillary. This is achieved by monitoring the fluorescence change of the fluorescent DNA intercalating dye, LCGreenPlus, as the sample is melted. Turnaround time per sample is approximately 1 – 2 minutes, depending on how broad the temperature range is required to be. LightCycler capillaries were trans-ferred to the HR-1 instrument and heated at 0.38℃/second. *FecB* and *FecX^G* mutations were simulta-neous analyzed between 60 and 908℃ with a turn-around time of approximately 2 minutes/sample. The light-emitting diode (LED) power was auto adjusted to 90% fluorescence.

Genotyping for the HybProbe assay was based on the melting temperature (T_m) calculated from the derivative melting temperature plots of the Light Cycler data. A/a homozygous sample has a single peak of a certain T_m, a b/b homozygous sample has a single peak of a different T_m, and a/b heterozygous sample has both peaks (Wittwer et al.2003).

The data for each sample was analyzed by a melting curve analysis program written by R.Palais (University of Utah Department of Mathematics) using Lab VIEW (National Instruments). The *FecB* and *FecX^G* genotypes of each sample were then scored according to the number of peaks and the melting temperatures of that peaks. Based on the genotypes of *FecB* and *FecX^G* locus in five Chinese sheep breeds, the genotypic frequencies and the allelic frequencies were directly calculated.

RESULTS AND DISCUSSION

Study of the high-resolution melting method

Six different genotypes of *FecB* and *FecX^G* plasmids were screened for the presence of *FecB* and *FecX^G* mutations. The data from each high-resolution melt was displayed using dF/dT versus temperature plots or melting peak plots. Example data of HybP-robe analysis of *FecB* and *FecX^G* are shown in Figure 1. Homozygous samples have clearly defined single peaks, and heterozygous samples contain both peaks. The two melting peaks for the *FecB* unlabeled probe showed up at 70.718℃ for the mismatched base A and 72.618℃ for the perfectly matched base G. The two melting peaks for the *FecX^G* unlabeled probe showed up at 59.718℃ for the mismatched base T and 66.618℃ for the perfectly matched base C. Six different genotypes of *FecB* and *FecX^G* plasmids were se-quenced and analyzed showing that amplified fragments had a 100% of identity with the six different melting curve plots that are produced by the two unlabeled probes of the *FecB* and *FecX^G* samples, respectively. Our data support the finding that the HRM method here established allows the distinction of *FecB* and *FecX^G* mutations. Further-more, other relevant advantages of the method are rapidity, low cost and high-throughput capability, allowing the simultaneous detection of 48 or 192 duplicate samples for the respective 96 or 384 multi-wall plates (Jones et al.2008). These characteristics give the HRM set up clear advantages over the methods reported to date (Davis 2004; Hanrahan et al.2004; Kumar et al.2006). Previously, the screening of *FecB* and *FecX^G* mutations in sheep used techniques such as single-stranded conformation polymorphism analysis (Hanrahan et al.2004), re-striction enzyme digestion of PCR products (Davis 2004; Kumar et al.2006) one by one. In our report, we describe an HRM method for the simultaneous screening of both the *FecB* and *FecX^G* mutations.

In addition to successful extraction of genomic DNA from blood by kits (Wittwer et al.2003; Vaughn and Elenitoba-Johnson et al.2004; Jones et al.2008), melting analysis could be obtained using sheep DNA on FTA cards combined with whole genome amplification. This is ideal for screening, as samples from remote locations could be collected and sent at ambient temperature through the normal postal system to the laboratory for analysis.

Allelic and genotypic frequencies in sheep

In the present study, the *FecB* and *FecX^G* mutations were simultaneous screened by HRM analysis using the HR-1 (Idaho Technology) in five Chinese sheep breeds. *FecB* mutation was

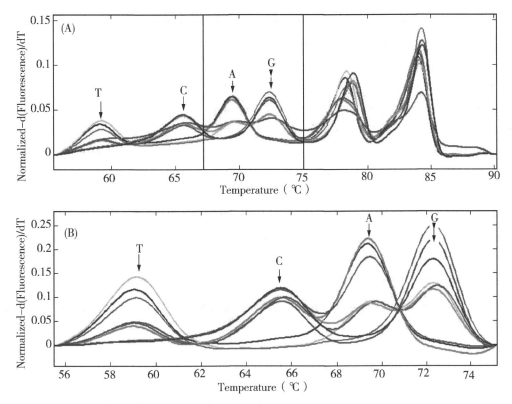

Fig. 1 Graph showing the six different melting curve plots that are produced by the two unlabeled probes of the *FecB* and *FecX^G* samples

Note: (A) The sample data are temperature shifted or corrected by selecting the PCR product melting portion of each sample and forcing them to overlay the selected sample graph. (B) The blue curve genotype is T/T at -single nucleotide polymorphism (SNP) position *FecB*, and A/A at SNP position *FecX^G*. The light grey curve genotype is T/T at SNP position *FecB*, and G/G at SNP position *FecX^G*. The orange curve genotype is T/T at SNP position *FecB*, and A/G at SNP position *FecX^G*. The pink curve genotype is C/T at SNP position *FecB*, and A/G at SNP position *FecX^G*. The dark red curve genotype is C/T at SNP position *FecB*, and G/G at SNP position *FecX^G*. The green curve genotype is C/T at SNP position *FecB*, and A/A at SNP position *FecX^G*. The red curve genotype is C/C at SNP position *FecB*, and A/A at SNP position *FecX^G*. The purple curve genotype is C/C at SNP position *FecB*, and G/G at SNP position *FecX^G*. The light blue curve genotype is C/C at SNP position *FecB*, and A/G at SNP position *FecX^G*.

discovered in Small Tail Han, DH F_2, BH F_2, while it is not found in Gansu Alpine Fine Wool Sheep and BM F_2. A total of 98 individuals of Small Tail Han sheep were analyzed for the *FecB* mutation, out of which 40 were homozygous (BB), 47 heterozygous (B+) and 11 non-carriers (++) (Table 3). Therefore, 88.7% of Small Tail Han sheep were found to be carriers for the *FecB* mutation and the frequency of B allele was about 0.648. Results indicated that the frequency of the *FecB* mutation is high in Small Tail Han. The *FecX^G* mutation was analyzed in Small Tail Han. All sheep were found non-carriers for the *FecX^G* mutation (Table 3). Recently, Chu et al. (2007) reported that the small Tail Han sheep carried both the *FecB* and *FecX^G* mutations. This study also showed that the *FecB* mutation is present in Small Tail Han sheep, while none of sheep were found to be carriers of the $FecX^G$ mutation. The variability is possibly caused by the potential differences in genetic background, since our samples were from Gansu, not Shandong (Chu et al. 2007).

The $FecX^G$ mutation was also analyzed in DH F_2 and BH F_2. None of sheep breeds carried the $FecX^G$ mutation in the BMP15 gene (Table 3). The result indicated that the twinning/multiple lambing in these breeds is not linked with the $FecX^G$ mutation (Kumara et al. 2008). Only heterozygote $FecB$ (B+) was found in DH F_2 and BH F_2. The frequency of heterozygous (B+) in DH F_2 and BH F_2 was 0.339, 0.355%, respectively. The results suggest breeding strategies that will allow adoption of MAS (marker-assisted selection, MAS) for improving the prolificacy of non-prolific sheep breeds (Hua and Yang 2009).

$FecB$ and $FecX^G$ mutations were not carried in Gansu Alpine Fine Wool Sheep and BM F_2 (Table 3). This indicates the possibility that other major genes influencing fecundity exist in Gansu Alpine Fine Wool Sheep. That is to say the prolificacy of ewes cannot be totally and accurately predicted only by the $FecB$ genotypes. Other major genes controlling fe-cundity remain to be explored.

Table 3 Distribution of the $FecB$ and $FecX^G$ genotype and allele frequency in five Chinese sheep ewes

Breeds	No. of ewes	FecB					$FecX^G$				
		Allelic frequency		BB	Genotypic frequency		Allelic frequency		GG	Genotypic frequency	
		B	+		B+	++	G	+		G+	++
STHS	98	0.648	0.352	0.408 (40)	0.480 (47)	0.112 (11)	0	1	0 (0)	0 (0)	1 (98)
G AWS	98	0	1	0 (0)	0% (0)	1 (98)	0	1	0 (0)	0 (0)	1 (98)
DH F_2	121	0.170	0.830	0 (0)	0.339 (41)	0.661 (80)	0	1	0 (0)	0 (0)	1 (121)
BH F_2	234	0.178	0.822	0 (0)	0.355 (83)	0.645 (151)	0	1	0 (0)	0 (0)	1 (234)
BM F_2	132	0	1	0 (0)	0 (0)	132 (132)	0	1	0 (0)	0 (0)	1 (132)

Note: B, $FecB$ mutation; G, $FecX^G$ mutation; +, wild-type. Numbers in parentheses are numbers of individuals that belong to the respective genotypes.

In conclusion, we have set up a sensitive, specific, low-cost, high-throughput HRM method that allows for the simultaneous screening of $FecB$ and $FecX^G$ mutations.

ACKNOWLEDGEMENTS

This work was supported by the Central Level, Scientific Research Institutes for Basic R & D Special Fund Business (No. BRF100102), by the Earmarked Fund for Modern China Wool & Cashmere Technology Research System (No. nycytx-40-2), by the National High Technology Research and Development Program of China (863 Pro-gram) (No. 2008AA101011-2).

References

[1] Chu MX, Liu ZH, Jiao CL, He YQ, Fang L, Ye SC, Chen GH, Wang JY. 2007. Mutations in BMPR 1B and BMP 15 genes are associated with litter size in Small Tail Han sheep. Journal of Animal Science 85: 598 603.

[2] Davis GH. 2004. Fecundity genes in sheep. Animal Reproduction Science 80: 247-253.

[3] Davis GH. 2005. Major genes affecting ovulation rate in sheep. Genetics Selection Evolution 37 (Suppl. 1):

S11-S23.

[4] Galloway SM, McNatty KP, Cambridge LM, Laitinen MPE, Juengel JL, Jokiranta TS, McLaren RJ, Luiro K, Dodds KG, Montgomery GW, et al.2000. Mutations in an oocyte-derived growth factor gene (BMP15) cause increased ovulation rate and infertility in a dosage-sensitive manner.Nature Genetics 25: 279-283.

[5] Hanrahan JP, Gregan SM, Mulsant P, Mullen M, Davis GH, Powell R, Galloway SM.2004. Mutation in the genes for oocyte-derived growth factors GDF9 and BMP15 are associated with both increased ovulation rate and sterility in Cambridge and Belclare sheep (*Ovis aries*). Biology of Reproduction 70: 900-909.

[6] Hua GH, Yang LG.2009. A review of research progress of *FecB* gene in Chinese breeds of sheep.Journal of Animal Science 116: 1-9.

[7] Jones AV, Cross NC, White HE, Green AR, Scot LM.2008. Rapid identification of JAK2 exon 12 mutations using high resolution melting analysis.Haematologica 93: 1 560-1 564.

[8] Kumar S, Kolte AP, Mishra AK, Arora AL, Singh VK.2006. Identification of the *FecB* mutation in Garole Malpura sheep and its effect on litter size.Small Ruminant Research 64: 305-310.

[9] Kumar S, Mishrab AK, Koltea AP, Dashd SK, Karimc SA.2008. Screening for Booroola (*FecB*) and Galway ($FecX^G$) mutations in Indian sheep.Small Ruminant Research 80: 57-61.

[10] Mulsant P, Lecerf F, Fabre S, Schibler L, Monget P, Lanneluc I, Pisselet C, Riquet J, Monniaux D, Call-ebaut I, et al.2001. Mutation in bone morphogenetic protein receptor-1B is associated with increased ovula-tion rate in Booroola Merino ewes.Proceedings of the National Academy of Sciences 98: 5 104-5 109.

[11] Rozen S, Skaletsky H.2000. Primer 5 on the WWW for general users and for biologist programmers. Methods in Molecular Biology 132: 365-386.

[12] Souza CJH, MacDougall C, Campbell BK, McNeilly AS, Baird DT.2001. The Booroola (*FecB*) phenotype is associated with a mutation in the bone morphogenetic receptor type 1 B (BMPR1B) gene. Journal of Endo-crinology 169: R1-R6.

[13] Vaughn CP, Elenitoba-Johnson KSJ.2004. High-resolution melting analysis for detection of internal tandem duplications.Journal of Molecular Diagnostics 6: 211-216.

[14] Wilson T, Wu XY, Juengel JL, Ross IK, Lumsden JM, Lord EA, Dodds KG, Walling GA, McEwan JC, O'Connell AR, McNatty KP, et al.2001. Highly prolific Booroola sheep have a mutation in the intracellular kinase domain of bone morphogenetic protein 1B receptor (ALK-6) that is expressed in both oocytes and granulose cells.Biology of Reproduction 64: 1 225-1 235.

[15] Wittwer CT, Reed GH, Gundry CN, Vandersteen JG, Pryor RJ.2003. High-resolution genotyping by ampli-con melting analysis using LCGreen.Clinical Chemis-try 49: 853-860.

（发表于《Journal of Applied Animal Research》，院选 SCI IF: 0.2179）

Clinical Study on the Treatment of Piroline against Bovine Mastitis

LIANG Jian-Ping[1,2], HAO Bao-Cheng[1]*, WANG Xue-Hong[1],
GUO Zhi-Ting[1], GUO Wen-Zhu[1], SHANG Ruo-Feng[1], TAO Lei[1],
LIU Yu[1], LI Zhao-Zhou[1], HUA Lan-Ying[1], WANG Shu-Yang[3]

(1. Key Laboratory of New Animal Drug Project of Gansu Province/Key Laboratory of New Animal Drug Project of CAAS, Lanzhou Institute of Animal Science and Veterinary Pharmaceutics, Chinese Academy of Agricultural Sciences, Lanzhou 730050, China; 2. Institute of Modern Physics, Chinese Academy of Sciences, Lanzhou 730050, China; 3. Lanzhou University, Lanzhou 730000, China)

Abstract: The study aims to investigate the efficacy of piroline and antibiotics in the treatment of bovine mastitis caused by *Streptococcus uberis* (*S. uberis*) and *Escherichia coli* (*E. coli*) during dry-milk period. 1880 cows in dry-milk period were divided into 4 groups and treated with penicillin G, ammonia benzyl penicillin, ceftiofur and piroline, respectively. The efficacy of each medicine in treating mastitis caused by *E. coli* intramammary infection (*E. coli* IMI) was followed: 31.2% for penicillin G, 36.9% for ammonia benzyl penicillin, 61.3% for ceftiofur, and 64.4% for Piroline. For those caused by *S. uberis* intramammary infection (*S. uberis* IMI), the efficacy of ceftiofur was 90% and piroline was 94.4%. The results indicated that piroline was more effective than the other three in treating the disease. The following analysis on milk samples demonstrated that there was no piroline residue in those treated cows' milk. Based on these data, it can be predicted that piroline will have a bright future in treating cow intramammary mastitis.

Key words: Bovine mastitis; Efficacy; Piroline

INTRODUCTION

Bovine mastitis is not only a major disease affecting the dairy industry, but also one of the major influencing factors in milk production (Yuan et al., 1992). It causes great economic losses and decreases animal health. Although much progress has been made in control of cow mastitis, producers still cannot prevent and cure it effectively. As reported in a literature (Yuan et al,

---

* Corresponding author. E-mail: haobaocheng@sina.cn

1992), 3650 strains of 24 species of bacteria and fungi were isolated and identified from 3006 milk samples, in which 2060 strains of 12 species were closely related to cow mastitis. The isolation rate of pathogenic bacteria was 62.5%. Bacteria related to mastitis were mainly *S.uberis* (38.11%), *E. coli* (7.14%), etc. Buddle and Cooper reported (Buddle and Cooper, 1980) that about 32% of found cure spontaneously over the dry period.

Dry cow treatment (DCT) is an important step of a mastitis control program, the advantages of DCT include reducing incidence of intramammary infections (IMI) at parturition and increasing cure rate of IMI (Buddle and Cooper, 1980). The aims of DCT are to cure existing intramammary infection (IMI), and to prevent new infection during the dry period.

A previous study that bacteriological cure rates for various intramammary dry cow treatments ranged from 25% to 75%. Systemic DCT was previously studied in an attempt to improve the cure rates. The systemic administration of antibiotics was evaluated in different studies. Owing to antibacterial drug resistance, only a few of antimicrobial agents demonstrated an improved cure rate over conventional intramammary DCT.

Piroline is a mainly active component isolated from the *Rubia Cordifolia*, which has been used in China to treat bovine mastitis for a long time (Liang et al, 1993; Liang et al, 2000). The pharmacological and antibacterial properties suggested that there was much less or no drug resistance to piroline, and thus it might be valuable for the treatment of bovine mastitis. The purpose of the present study was to compare the efficacy of piroline with that of other antibiotics commonly used in the systemic DCT. The results would be helpful for better prevention and cure of bovine mastitis.

MATERIALS AND METHODS

Herds: The DCT field trials were conducted in three factory-supply dairy herds under seasonal-calving conditions in Lanzhou, Gansu of China. Owners agreed to the following conditions: 1) provide calculated calving dates; 2) schedule dates for drying off, sampling, and treatment; and 3) permit some cows to serve as untreated controls. General information on several herds is presented in table 1.

Udder selection and milk sample collection: Udders and milk samples were collected through recommended procedures. Composite milk samples were collected from 1880 lactating cows from three herds within 4 weeks prior to drying off. Duplicate quarter milk samples were collected at drying off. And within 21 days, single sample was collected from all quarters of cows in treatment 1 and 3 at the prepartum period prior to infection of the lactating cow product (LCP).

Drugs: Penicillin G (batch number, h0605071) and Ammonium benzyl penicillin (batch number h06050714) was purchased from North China Pharmaceutical Group Corporation, Shijiazhuang city, China. Ceftiofur (batch number y06030914) was from Heibei Yuan Zheng Pharmaceutical Co., Ltd. Shijiazhuang, China. And Piroline was made by Lanzhou Institute of Animal & Veterinary Pharmaceutics Sciences of Chinese Academy of Agricultural Sciences, Lanzhou city, China.

Assignment of cows to treatment groups: Cows with *E.coli* IMI, based on cultural results of composite milk samples, were assigned randomly to the four treatment groups, a minimum of 80 *E.coli*

IMI was included in each of the four treatment groups. More than 90 SU IMI were included initially in treatment 1 and 3 in anticipation of missed infusions due to early calving or incorrect calculated calving dates. Cow missed scheduled prepartum treatment were included in treatments 2 and 4, respectively. All four treatment groups were represented within each herd in table 2.

Treatment regimens: Each group of treatments was listed in table 2. Treatment 1 received an infusion at drying off with penicillin G 2 000 000 IU/50 mL (water), treatment 2 with ammonium benzyl penicillin 0.5 g/50 mL (water), treatment 3 with ceftiofur 0.5 g/50 mL (water), and treatment 4 with piroline 0.4 g/50 mL (water) (Brander, 1969).

Table 1 Information on cooperation dairy farms

Herd	Management Practices				
	Number of cows	Wash udders	Spray treats	Treat dry-cows	Overall management
A	620	+	+	+	Good
B	708	+	+	+	Good
C	552	+		+	Fail

Table 2 Experimental design of dry-cow therapy field trial

Treatment	Time of treatment	
	Drying off	Prepartum
1	+	+
2	+	+
3	+	+
4	+	+

Note: +: Intramammary infections

Microbiological procedures: Method recommended by the National Mastitis Council, U.S.A., was followed. Presumptive identification of the following microorganisms was made; *Escherichia coli* (*E.coli*);

Staphylococcus epidermidis (SE); *Streptococcus galactiae* (SAg); *Streptococcus uberis* (*S. uberis*); other streptococci (OS); *Corynebacterium bovis* (CB) and Coliforms (CO).

Definition of terms

Infection: the number of somatic cells in the milk samples exceeds regular range and determination of pathogenic bacteria in the milk shows positive.

Cure: clinical symptoms ease off or disappear, the number of somatic cells in the milk returns to regular range, and determination of pathogenic bacteria in the milk shows negative.

Fail: clinical symptoms do not ease off, the number of somatic cells in the milk does not return to regular range, and determination of pathogenic bacteria in the milk shows positive.

Statistical analysis: An analysis of variance was conducted only on the *E.coli* and *S.uberis* data. The number of IMI with other pathogens was insufficient to conduct valid analysis. Developed (%) = (developed quarters/treated quarters) × 100%, and Cured% = (cured quarters/treated quarters) x 100%.

RESULTS

The efficacy was 61.3% against *S.uberis* IMI in treatment 3 and 64.4% in treatment 4 (Table 3), and the difference was not significant. The efficacy of treatment 4 was significantly greater ($p < 0.01$) than the 36.9% in treatment 2, and treatment 4 was also significantly different from treatment 1 (31.2%, ($p < 0.01$). Treatment 4 and 3 were significantly different from treatment 1 and 2.

Approximately 10.5% of quarters developed from new IMI with *E.coli* or *S.uberis* during the dry period (Table 4). These resulting pathogens were accounted for 93% of new IMI. Other new IMI were caused by SAg and OS, about 3% for each one. Though differences were observed between the treatment groups, the incidence of the new IMI was similar to *E.coli* and *S.uberis*. In treatment 1 and 2, the incidence of IMI with S.uberis almost doubled than that in treatments 3 and 4 (Liang et al., 1993; Liang et al., 2000). Efficacy of prepartum treatment with the ceftiofur and piroline against new SAg IMI was low, but it was 61.3% and 64.4% against the new IMI with *S.uberis* respectively (Table 3).

The rate of new dry period IMI with SU ranged from 3.6% to 4.9% of quarters for treatment 1, 2 and 3. 10.5% of quarters become infected with SU in treatment 4 (Table 4), prior prepartum samples were analyzed from cows in treatment 3 and 4, so efficacy of ceftiofur and piroline were excellent, 90% and 94.4%, respectively for treatment 3 and 4. No prepartum samples were analyzed from cows in treatment 1 and 2 and spontaneous recoveries were determined for the entire dry period using samples collected postpartum.

Table 3 Efficacy of dry-cow therapy against *S.uberis*

Treatment	Cows	Quarters	Quarter Number		Efficacy (%)
			Intramammary infections		
			Drying off	Postpartum	
1	91	243	99	61	31.2
2	102	288	102	64	36.9
3	83	211	89	66	61.3
4	98	270	91	70	64.4
Totals	374	1012	381	261	

Table 4 New dry-period intramammary infections

Group	Number.		Intramammary infections					
			E.coli			*S.uberis*		
	Cows	Quarters	Quarter Number	Developed[a]	Cured[b]	Quarter Number	Developed[a]	Cured[b]
1	98	270	21	7.8	0	10	3.9	0
2	102	288	10	3.6	0	10	3.6	0

(continued)

Group	Number.		Intramammary infections					
			E.coli			S.uberis		
	Cows	Quarters	Quarter Number	Developed[a]	Cured[b]	Quarter Number	Developed[a]	Cured[b]
3	91	243	8	3.4	45	12	4.9	90
4	83	211	15	7	60	22	10.5	94.4
Totals	374	1012	54	5.3	54	5.3		

Note: [a] Developed after prepartum treatment, [b] Expressed as percentage cure of IMI that developed prepartum sampling.

DISCUSSION

Treatment 4, treated with piroline, did not significantly reduce the number of E.coli infections postpartum when compared with treatment 3. On the contrary, ceftiofur showed good effect in reducing the number of E.coli IMI significantly when compared with treatment 1 and 2. Treatment 4, treated with piroline, did not significantly reduce the level of E.coli IMI only with ammonium benzyl penicillin for prepartum treatment as compared with penicillin. The rate of spontaneous recovery for E.coli during dry period was consistent with the figures previously reported. Efficacy of piroline was 91.3% when compared with other reports on piroline against E. coli. The range in efficacy among herds was from 83% to 95%, which corresponded with earlier studies (Liang et al., 1993).

The rate of new E.coli dry period IMI could be reduced 50% by DCT and supported earlier work. The incidence of new IMI with S.uberis was similar to that observed with E.coli and was reduced about 50% by DCT (Brander, 1969; Brown et al, 1969; Christie et al, 1974; Buddle and Cooper, 1980). However, the incidence of new S.uberis IMI was similar to that in treatment 3 and 4. Further studies are required to explain these results since no prepartum samples were collected and detected from cows in treatment 1 and 2. Additionally, the rate of spontaneous recovery was high for new S.uberis IMI during the late dry or early postpartum period. Philpot (Philpot, 1969; Philpot, 1979) reported that spontaneous recovery over the dry period was 70% for Streptococci, compared with 27% for E.coli.

Prepartum therapy with piroline was effective in eliminating over 90% of new S.uberis IMI, but it was not effective in eliminating new S.uberis IMI (Liang, 2000). The average efficacy against new E. coli IMI was less than 60%, but the number of new E.coli was largely reduced. A wide variation in the efficacy was observed when compared with the earlier reports, but these data may not be conclusive. The results validated many earlier reports on piroline reducing the incidence of new dry period IMI with E. coli and S.uberis. Prepartum therapy with piroline appeared to be of marginal benefit and probably would be of practical value in dairy herds experiencing significant clinical mastitis cows. Philpot reported a 68% reduction of clinical cases during the first week of lactation when cows received penicillin G at parturition and the end of lactation. Results from this field trial provided supportive evidence of the effectiveness of piroline, eliminating many new IMI by educing levels of infected quarters before and after parturition, especially for the IMI infected with S. uberis.

ACKNOWLEDGEMENT

The present study was supported by The Institute of Traditional Chinese Veterinary, Chinese Academy of Agricultural Science and National Nature Science Foundation of China (grant No. 10275084). The authors are very grateful to professor Li Zhang and Yanbin Si for assistance in experimental design and review of manuscript.

References

[1] Wilson, C.D.1964. The use of antibiotics in the control of mastitis. Proceedings of the Royal Society of Medicine 12 (57): 1 088-1 090.

[2] Brown, R.W, Morse, G.E and Newbould, F.H.S, 1969. Microbiological Procedures for the Diagnosis of Bovine Mastitis. National Mastitis Council, Washington DC.

[3] Budddie, G.E.and Cooper.1980. Dry-cow therapy for *Staphylococcus aureus* mastitis. New Zealand Vet J. 28: 51-53.

[4] Christie, G.J, Keefe, T.J and Strom, P.W.1974. Cloxacillin and the dry cow. Vet Med Small Anim Clin.69: 1 403-1 408.

[5] Dodd, F.H.and Griff, T.K.1975. The role of antibiotic treatment at drying off in the control of mastitis. In Proceedings of the Int Dairy Fed Seminar on Mastitis Control. Reading, UK: 282-302.

[6] Heald, C.W., Jones, G.M., Nickerson, S.and Bibb, T.L.1977. Mastitis control by penicillin and novobiocin at drying off. Can Vet J.18: 171-179.

[7] Kingwill, R.G., Neave, F.K., Dodd, F.H., Griffin, T.K., Westgarth, D.R.and Wilson, C.D. 1970. The effect of a mastitis control system on levels of subclinical and clinical mastitis in two years. Vet Rec.87: 94-99.

[8] Liang, J.P., Wei, Z.Q., Zhang, L., and Yu, J., 2000. Use of Piroline for treatment of bovine mastitis. Indian Vet J.77: 553.

[9] Liang, J.P., Zhang, J.Y., Zhao, R.C., Xu, Z.Z., Li, J.S.and Yu, J.1993. Preparation and application of piroline in veterinary clinic. China J Vet Sci Tech.12 (23): 14-16.

[10] Natzke, R.P. 1971. Therapy: One component in a mastitis control system. J Dairy Sci. 54: 1 895-1 901.

[11] Philpot, W.N.1979. Control of mastitis by hygiene and therapy. J Dairy Sci.62: 168-176.

[12] Philpot, W.N.1969. Role of therapy in mastitis control. J Dairy Sci.52: 708-713.

[13] Smith, A., Rautenbach, H.F.P., Dodd, F.H.and Bander, G.C.1975. The effect of udder infection of varying the levels and persistency of antibiotic in the dry period. In Proceedings I Int Dairy Fed Seminar on Mastitis Control, Reading. UK: 345-348.

[14] Swenson, G.H. 1979. Posology and field efficacy study with novobiocin for intramammary infusion in nonlactating dairy cows. Canadienne De Médecine Comparée.43: 440-447.

[15] Yuan, Y.L, Zhang, L.H.and Liu, C.C.1992. A survey of the pathogens of bovine mastitis in China. Scientia Agriculturea Sinica 25 (4): 70-76.

（发表于《The Thai Journal of Veterinary Medicine》，院选 SCI，IF: 0.118）

甘肃地区牛源金黄色葡萄球菌分子鉴定及 RAPD 分型

邓海平，蒲万霞*，梁剑平，倪春霞，孟晓琴

（中国农业科学院兰州畜牧与兽药研究所 中国农业科学院新兽药重点开放实验室甘肃省新兽药工程重点实验室，兰州 730050）

摘 要：本研究目的是分离鉴定引起甘肃地区奶牛乳房炎的金黄色葡萄球菌，掌握其基因型情况。利用 16S、23SrRNA 保守序列 PCR 扩增对乳房炎奶样中的金黄色葡萄球菌进行鉴定，并进行 RAPD 基因分型。结果表明，310 份奶样中共分离出金黄色葡萄球菌 100 株，RAPD 结果显示这 100 株金黄色葡萄球菌均可得到清晰的 RAPD 指纹图谱，扩增产物在 2~7 条带之间，具有多种带型组成。通过聚类分析 100 株菌产生 11 个基因型，其中 I 型 4 株，II 型 4 株，III 型 10 株，IV 型 13 株，V 型 7 株，VI 型 24 株，VII 型 16 株，VIII 型 6 株，IX 型 4 株，X 型 10 株，XI 型 2 株。VI 型为该地区的流行优势菌群，不同牛场各基因型菌株分布有明显差异。本研究说明牛场的环境与养殖条件对病原菌流行传播有明显的影响，这一结果对地区性奶牛乳房炎的防治提供了可靠的理论依据。

关键词：奶牛乳房炎；金黄色葡萄球菌；16S rRNA；23S rRNA；RAPD；基因型

奶牛乳房炎是奶牛的常见病、多发病。不仅影响奶牛健康，降低乳制品质量，给奶牛养殖业造成巨大损失，还在一定程度上威胁到了公共安全。及时准确地掌握病原菌的种类和分子流行病学趋势对于该病的临床治疗和疫苗的研制都非常重要[1-2]。DNA 随机多态性扩增（Random amplified polymor-phic DNA，RAPD）技术是一种简便、灵敏、可行的遗传标记技术。该技术问世后迅速在国内外被广泛应用于微生物的基因分型与鉴定[3]。近年来有关金黄色葡萄球菌基因分型的研究主要集中于人医来源的菌株[4]，畜源菌株的基因分型研究相对较少，特别是奶牛乳房炎病原菌基因分型的相关研究在国内还罕有报道。本研究利用保守序列对引起甘肃地区奶牛乳房炎的金黄色葡萄球菌进行鉴定，并进行基因分型研究，探讨金黄色葡萄球菌性奶牛乳房炎分子流行病学特点，为地区性奶牛乳房炎的防治和基因工程疫苗的研究提供理论依据。

---

* 收稿日期：2011-01-08

基金项目：国家科技支持计划（2006BAD04A05）；中央级公益性科研院所基本科研业务费专项资金项目（1610322011010）

作者简介：邓海平（1983— ），男，甘肃兰州人，助理研究员，硕士，主要从事兽医微生物研究，E-mail：denghaiping_001@163.com

* 通讯作者：蒲万霞，E-mail：wanxiapu@yahoo.com.cn

1 材料与方法

1.1 试剂及仪器

电泳分子量标准 DNA Ladder Marker（相对分子质量依次为 4 500、3 000、2 250、1 000、750、500、250bp）、DL2000 Marker、Premix EX Taq 酶均购自宝生物工程有限公司；DNA 提取试剂盒（离心柱型），购自 TIANGEN 生物制品公司；金黄色葡萄球菌标准菌 CVCC2246（购自中国兽医药品监察所）；基因扩增仪（Biometra, GER）。

1.2 金黄色葡萄球菌分离鉴定

1.2.1 细菌纯化培养

310 份临床型乳房炎患牛奶样分别来自甘肃地区 3 个奶牛场。奶样接种于绵羊血平板培养基和却甫曼琼脂培养基纯化培养。

1.2.2 金黄色葡萄球菌生化鉴定

将菌落形态、镜检结果符合葡萄球菌特征的菌株参考相关生化特性进行金黄色葡萄球菌生化鉴定。

1.2.3 金黄色葡萄球菌分子生物学鉴定

根据已发表的金黄色葡萄球菌的 16S rRNA 和 23S rRNA 基因保守序列（GenBank Accession No.X68417, No.X68425），并参照相关文献[5-6]，应用 Oligo6.67 软件设计 2 对引物（表1）。提取菌落形态、镜检结果符合葡萄球菌特征菌株基因组 DNA 作为模板进行 PCR 扩增。采用 25μL 反应体系，扩增反应条件：94℃ 5min，94℃ 60s，56.5℃ 40s，72℃ 40s，30 个循环后 72℃延伸 10min。扩增产物-20℃冻存。

表1 扩增引物序列

目标序列	引物
16S rRNA	P1：5-GCGGTCGCCTCCTAAAAG-3′ P2：5-TCCCGGTCCTCTCGTACTA-3′
23S rRNA	P3：5-ACGGAGTTACAAAGGACGAC-3′ P4：5-AGCTCAGCCTTAACGAGTAC-3′

1.3 金黄色葡萄球菌分离株 RAPD 基因分型

1.3.1 引物设计与筛选

根据文献报道[7-8]，选择序列 OLP13：5′-ACCGCCTGCT-3′，作为扩增引物。

1.3.2 RAPD 反应条件

采用 25μL 反应体系，体系参数：Premix 混合液 12.5μL，DNA 模板 2μL，引物 2μL，去离子水 8.5μL。PCR 循环参数：94℃ 5min，37℃ 5min，72℃ 5min，循环 4 次；94℃ 1min，37℃ 1min，72℃ 2min，循环 30 次；72℃延伸 10min。扩增产物-20℃冻存。

1.4 扩增产物凝胶电泳分析

取分子鉴定及 RAPD 扩增产物，1.0%琼脂糖凝胶上电泳，凝胶由 Quantity One 1-D 成像系统紫外成像。

1.5 数据分析

利用 Bio Numerics 软件对 RAPD 扩增电泳结果进行数据处理，图像采用 Dice 相关系数和非加权组算术平均法 UPGMA（unweighted pair group method with arithmetic mean）来进行聚类分析。

2 结果

2.1 细菌分离鉴定结果

菌株经生化鉴定，符合金黄色葡萄球菌生化指标菌株为 95 株；经金黄色葡萄球菌 16S rRNA 和 23S rRNA 保守序列 PCR 检测，有 100 株菌分别产生了 450 和 1 250bp 左右的目的片段（图1、2）。结果表明生化鉴定结果与分子鉴定结果基本一致。分子鉴定为金黄色葡萄球菌的菌株情况见表2。

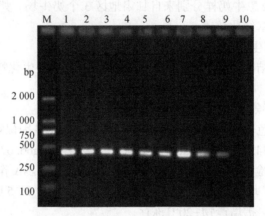

M. DL2000 相对分子质量标准；1~8. 随机选择的阳性菌株；9. CVCC2246；10. 表皮葡萄球菌阴性对照

图 1　金黄色葡萄球菌 16S rRNA 鉴定结果

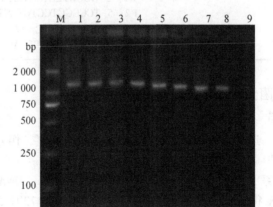

M. DL2000 相对分子质量标准；1~7. 随机选择的阳性菌株；8. CVCC2246；9. 表皮葡萄球菌阴性对照

图 2　金黄色葡萄球菌 23S rRNA 鉴定结果

表2 金黄色葡萄球菌菌株鉴定结果

牛场	菌株数	编号
QY	26	QY01~QY26
HG	39	HG01~HG39
KY	35	KY01~KY35

2.2 RAPD结果

对分离鉴定出的100株金黄色葡萄球菌基因组DNA模板进行随机多态性扩增,扩增产物经凝胶电泳后结果显示所有菌株基因组DNA经扩增后均产生了清晰、可分辨的带谱,扩增产物在2~7条带,具有多种带型组成。利用Bio-Numerics分析软件对100株金黄色葡萄球菌RAPD产物电泳图谱进行聚类分析,建立了亲缘关系聚类图(图3)。

可以看出,以相似性0.4为标准(竖线所示),100株金黄色葡萄球菌根据其相似性可分为11个聚类群。其中Ⅰ型4株,Ⅱ型4株,Ⅲ型10株,Ⅳ型13株,Ⅴ型7株,Ⅵ型24株,Ⅶ型16株,Ⅷ型6株,Ⅸ型4株,Ⅹ型10株,Ⅺ型2株。

3 讨论

3.1 奶样中金黄色葡萄球菌分离鉴定

奶牛乳房炎是影响奶牛养殖业的主要疾病之一,是发病率最高、造成损失最大的奶牛疾病。据世界奶牛协会统计,全世界奶牛中有50%左右患各种类型的乳房炎[9-10]。快速准确地掌握引起奶牛乳房炎病原菌的种类和分布是防治奶牛乳房炎和疫苗研制的首要条件。近年来,分子生物学技术在家畜疫病诊断中广泛应用。16S和23S rRNA序列既具有高度保守性又具高变性,既能体现不同菌属之间的差异,又能利用测序技术快速得到核酸序列,已成为最理想的基因鉴定靶序列[11]。作者利用金黄色葡萄球菌16S和23S rRNA保守序列设计扩增引物对奶样中分离出的金黄色葡萄球菌进行了PCR鉴定,并与传统鉴定方法进行了比较。结果表明16S和23S rRNA保守序列鉴定法和传统鉴定法两种方法的鉴定结果基本吻合。有2株菌传统方法未鉴定出,可能是因为受抗生素影响,菌体已经破裂死亡,所以无法检出。PCR检测是以细菌基因组DNA保守序列为目标进行检测,病原菌的存活状态对其没有影响。而且分子鉴定的方法只需1~2d即可得出结果,缩短了细菌性奶牛乳房炎的诊断时间,为及时的临床治疗提供了保障。

3.2 金黄色葡萄球菌RAPD分型

目前有多种方法用于金黄色葡萄球菌的分型,可归为两大类,一是表型分型,二是基因分型。随着分子生物学时代的到来,依赖于细菌DNA的分型方法更加受到关注。常用的金黄色葡萄球菌基因分型方法主要有高压脉冲场凝胶电泳分型(PFGE)、多位点测序分型(MLST)、多位点数目可变串联重复序列指纹图谱分型(MLVF)等[12]。其中PFGE是目前公认的金葡菌分型的"金标准"。但是PFGE也存在有2个缺点,首先费时,不适于实验室大量样本的分析。另外,还需特殊设备,从而限制了其适用范围[13-14]。RAPD是近年来兴起的一种揭示基因多态性的基因分型方法,据国外研究报道,利用RAPD方法对医院分离的金黄色葡萄球菌进行分型并与PFGE方法分型结果进行比较后发现两种方法分辨力相当且分辨结果高度一致[15]。说明RAPD与PFGE在金黄色葡萄球菌分型方面具有相同的高效性。

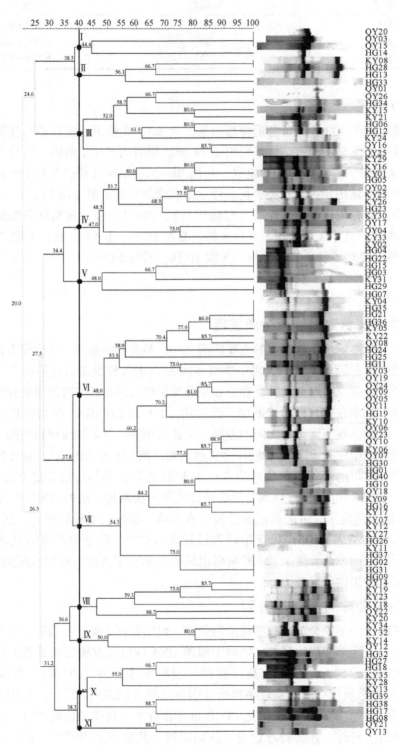

图 3 100 株金黄色葡萄球菌相似性关系树

早在 1997 年 Fitzgerald 等[16]就利用 RAPD 技术对牛源金黄色葡萄球菌进行了基因分型研究，但是近年来 RAPD 技术还是更多的应用于人医临床病原菌的分型，在兽医病原菌方面的应用较少。乌兰巴特尔等[17]对内蒙古地区分离到的 63 株引起奶牛乳房炎的金黄色葡萄球菌进行了 RAPD 分型，共分为了 12 个基因型；王琳等[18]将新疆地区两个奶牛场分离到的 16 株金黄色葡萄球菌进行 RAPD 分型并探索了最适反应条件。由于菌株数量较少，所以只分出了 2 个基因型，且其中 14 株为同一基因型。本研究对采自甘肃地区乳房炎奶样中的 100 株金黄色葡萄球菌进行了 RAPD 分后，所有菌株染色体 DNA 均扩增得到了清晰可分辨的条带，清楚地反映了菌株间的基因多态性特征。根据聚类分析树状图结果，100 株菌显示出 11 种基因型，Ⅵ型为甘肃地区主要流行的优势菌群。这与乌兰巴特尔报道的结果相似，但是与 Fitzgerald 的结果不同，这可能与地理气候环境差异对病原菌交叉传播的影响不同有关。不同基因型菌株在各牛场的分布有明显差异，牛场 QY 的菌株主要为Ⅵ型（10/26），且Ⅰ型和Ⅲ型菌株只在该牛场出现。其他 2 个牛场的菌株在各基因型均有部分分布，且没有明显占优势的基因型菌群。这在一定程度上说明了牛场 QY 中菌株与其他牛场流行传播相对较少，所以基因型也会较集中地分布。而另外 2 个牛场中菌株交叉传播频繁，产生了多种显著的变异，菌株基因型分布广泛。同一省份 3 个牛场乳房炎金黄色葡萄球菌分离株基因型分布的差异一定程度上反映了不同地理环境对牛场病原菌流行传播的影响，但更多的反映出的是养殖、管理水平差异对牛场病原菌流行的影响。

4 结论

环境差异和养殖条件对乳房炎病原菌的种类和流行传播趋势起至关重要的作用，及时准确掌握病原菌种类及基因型情况可为乳房炎的治疗提供依据。应用保守序列分子鉴定和 RAPD 技术对畜源病原菌进行快速流行病学调查，为兽医传染病分子流行病学研究提供了切实可行的研究方法，特别是对一些未建立标准分型和缺乏血清学分型等方法的菌属鉴定和分型尤为实用。其在兽医分子流行病学和传染病学等领域具有广阔的应用前景。

参考文献

［1］ 路桂环，陈敏，尹秀生，等．奶牛乳房炎发病原因及危害［J］．畜牧兽医科技信息，2009，12：41．

［2］ 易明梅，冯华，朱建国，等．生物制剂在奶牛乳房炎防治中的应用［J］．中国奶牛，2007，10：45-47．

［3］ ELINA R, SUSANA B, CECILIA F, et al. RAPDPCR analysis of *Staphylococcus aureus* strains isolated from bovine and human hosts［J］．*Microbiol Res*, 2004, 159: 245-255.

［4］ 朱成宾，高应东．RAPD 在金黄色葡萄球菌分型中的应用研究进展［J］．临床输血与检验，2009，11（3）：283-285．

［5］ FITZGERAIL J R, MONDAY S R, FOSTER T J, et al. Characterization of a putative pathogenicity island from bovine *Stap hylococcus aureus* encoding multiple superantigens［J］．*J Bacteriol*, 2001, 183（1）：63-70.

［6］ MARTIN M C, FUEYO J M, GONZALEZ M A, et al. Genetic procedures for identification of enterotoxigenic strains of *Staphylococcus aureus* from three food poisoning outbreaks［J］．*Int J Food Microbiol*, 2004, 94（3）：279-286.

［7］ NEELA V, MARIANA N S, RADU S, et al. Use of RAPD to investigate the epidemiology of *Staphylo-*

coccus aureus infection in Malaysian hospitals [J]. *World J Microbiol & Biotechnol*, 2005, 21: 245-251.

[8] RANDA G N, SALWA M, BDOU R HM, et al. Ente-rotoxicity and genetic variation among clinical *Staph-ylococcus aureus* isolates in Jordan [J]. *J Med Micro-biol*, 2006, 55: 183-187.

[9] 尹柏双, 李国江. 奶牛乳房炎的研究新进展 [J]. 中国畜牧兽医, 2010, 37 (2): 182-184.

[10] MARCUS V C, JANALNA S N, PATRLCIA C, et al. Activity of Staphylococcal bacteriocins against *Staphylococcus aureus* and *Streptococcus agalactiae* involved in bovine mastitis [J]. *Res Microbiol*, 2007, 158 (7): 1-6.

[11] 李鹏, 马艳娇, 赵云. 16S rRNA、23S rRNA 及 16S~23S rRNA 基因在细菌分离与鉴定中的应用 [J]. 现代畜牧兽医, 2008, 7: 49-52.

[12] 倪春霞, 蒲万霞, 胡永浩, 等. 金黄色葡萄球菌基因分型方法研究进展 [J]. 中国动物检疫, 2009, 26 (10): 64-68.

[13] BOU G, CERVERO G, DOMINGUEZ M A, et al. PCR-based DNA fingerprinting REP-PCR, AP-PCR and pulsedfield gel electrophoresis characterization of a nosocomial outbreak caused by imipenem and mero-penem resistant Acinetobacter baumannii [J]. *Clin Microbiol Infect*, 2000, 6: 635-643.

[14] SABOUR P M, GILL J J, LEPP D, et al. Molecular typing and distribution of *Stap hylococcus aureus* iso-lates in Eastern Canadian dairy herds [J]. *J Clin Mi-crobiol*, 2004, 42 (8): 3 449-3 455.

[15] CHI O C, WEI H, YANG L. Comparison of pulsed-field gelelect rophoresis and coagulase gene restriction profile analysis techniques in the molecular typing of *Staphylococcus aureus* [J]. *J Clin Microbiol*, 2000, 38: 2 186-2 190.

[16] FITZGERALD J R, MEANEY W J, HARTIGANIP J. Fine-structure molecular epidemiological analysis of *Stap hy lococcus aureus* recovered from cows [J]. *Epidemiol Infect*, 1997, 119 (5): 261-269.

[17] 乌兰巴特尔, 郝永清, 海岩, 等. 奶牛乳房炎金黄色葡萄球菌的分子流行病学调查 [J]. 中国兽药杂志, 2007, 41 (2): 21-23.

[18] 王琳, 杨学云, 王治才, 等. 奶牛乳房炎病例中金黄色葡萄球菌 RAPD-PCR 基因分型的研究 [J]. 中国畜牧兽医, 2007, 34 (11): 76-78.

(发表于《畜牧兽医学报》)

犬血浆中塞拉菌素含量的高效液相色谱-荧光检测方法的建立

汪 芳，李 冰，周绪正，张继瑜*，李剑勇，李金善，
牛建荣，魏小娟，杨亚军

(中国农业科学院兰州畜牧与兽药研究所 农业部兽用药物创制重点实验室甘肃省新兽药工程重点实验室，兰州 730050)

摘 要：本研究以多拉菌素为内标，建立了犬血浆中塞拉菌素的高效液相色谱-荧光检测方法。用乙腈作为血浆蛋白沉淀剂，用 Sam_{pli1} C18 型 30E 固相萃取柱进行净化，用 1-甲基咪唑和三氟乙酸酐衍生化处理。色谱条件为 OD3-1 色谱柱（250mm×4.6mm），流动相为甲醇：乙腈：1%1-庚烷磺酸：0.4%乙酸=40.0：57.6：0.9：1.5（v/v/v/v），流速 1mL·min^{-1}，进样量 20μL，激发波长 355nm，发射波长 465nm，柱温 30℃。方法的检测限为 0.25ng·mL^{-1}，线性范围 0.5~50.0$ng·mL^{-1}$，变异系数为 1.47%~3.31%，方法的样品平均回收率为 94.0%。结果表明，所建方法准确、灵敏度高和重复性好，可用于犬体内塞拉菌素血药含量的测定。

关键词：塞拉菌素；犬；血浆；多拉菌素；高效液相色谱-荧光检测法

阿维菌素类和米尔贝霉素类药物是兼具抗体内和体外寄生虫双重作用的抗虫药，塞拉菌素（图1）是阿维菌素类新型体内外抗寄生虫药[1]，其对蜱、螨、蚤、虱、恶丝虫和部分线虫都有很好的驱杀效果[2-7]，而且对利犬具有很好的安全性[8]。塞拉菌素目前已被开发为透皮制剂，广泛应用于宠物寄生虫病的治疗。

塞拉菌素血药浓度的检测方法国内未见报道。本研究以多拉菌素（图2）为内标，建立了犬血浆中塞拉菌素含量的高效液相色谱-荧光检测法，该方法具有较高的灵敏度和准确性，为研究塞拉菌素在犬体内的药代动力学和生物等效性奠定了方法学基础。

1 材料与方法

1.1 材料

1.1.1 仪器

Waters2695 高效液相色谱系统，Waters2475 荧光检测器，Empower 工作站，Scien-home Kromasil ODS-1 250 mm×4.6 mm 色谱柱，SampliQ C18 型 SPE 固相萃取柱（安捷伦公司产品，规格 100 mg·1 mL）。美国 Milipore 水纯化系统，KL512 型氮吹仪（北京康林科技有限责任公司产品），Thermo 离心机，涡旋振荡仪（Heros BIO），针筒式微孔滤膜过滤器（上海

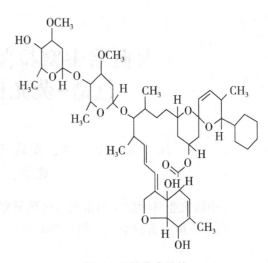

图1 塞拉菌素结构　　　　　　　　图2 多拉菌素结构

Fig. 1　Structural formula of selamectin　　　Fig. 2　Structural formula of doramectin

兴亚净化材料厂产品，孔径 0.45 μm）。

1.1.2　药品与试剂

塞拉菌素原料药（含量 97.3%）、多拉菌素标准品（含量 96.0%，批号 0901）以及塞拉菌素制剂（透明液体，规格 0.75 mL，45.0 mg）均由浙江海正药业股份有限公司提供；.N-甲基咪唑（分析纯，A Johnson MaltheyCompany）；三氟乙酸酐和冰乙酸（分析纯，上海中美化学试剂有限公司）；乙腈（色谱纯，天津市光复精细化工研究所）；甲醇和无水乙醇（分析纯，天津市百世化工有限公司）；1-庚烷磺酸钠（离子对试剂，国药集团化药试剂有限公司）；乙二胺四乙酸二钠（分析纯，天津市福晨化学试剂厂）。

1.1.3　色谱条件

ODS-1 色谱柱（250 mm×4.6 mm，5 μm，大连依利特公司）；流动相，甲醇∶乙腈∶1% 1-庚烷磺酸钠∶0.4%乙酸 = 40.0∶57.6∶0.9∶1.5（v/v/v/v）；流速 1.0 mL·min^{-1}；进样量 20 μL；激发波长 355 nm，发射波长 465 nm；柱温 30℃。

1.2　方法

1.2.1　溶液的配制

塞拉菌素和多拉菌素储备液：精密称取塞拉菌素和多拉菌素各 200mg，分别置于 100.0mL 棕色容量瓶中，用乙腈配制成浓度为 2.0mg·mL^{-1} 的储备液，置于 4℃ 冰箱保存，待需要时用乙腈逐级稀释配成所需浓度。

4% EDTA 溶液：称取 4.0mg 乙二酸四乙酸二钠，加蒸馏水定容至 100.0mL，60℃ 水浴加热至融解。

衍生化试剂：A 溶液，1-甲基咪唑（1-MIZ）∶无水乙腈 = 1∶1（v/v）；B 溶液，三氟乙酸酐（TFAA）∶无水乙腈 = 1∶2（v/v）。

1.2.2　血样采集及处理

将 0.25 mL 4%ED-TA 抗凝剂加入 5.0 mL 离心管中，置于 60℃ 烘箱干燥，将采集的空白血浆置于 5.0 mL 抗凝管中，4 000 r·min^{-1} 离心 15 min，吸取上清液于另一管中保存于 −20℃ 冰箱中。

1.2.3 血浆样品溶液的制备

先后用 5.0 mL 甲醇和 5.0 mL 水平衡 SPE 固相萃取柱，后精密量取 0.5 mL 空白血浆，加入 0.25 mL 塞拉菌素溶液和 0.25 mL 多拉菌素标准品溶液，涡旋振荡 1 min，4 000 r·min^{-1}离心 10 min，取上清液过 SPE 固相萃取柱，用注射器抽空柱内的液体，随后用 2.0 mL 水和 1.0 mL 乙醇溶液（水：乙醇=75：25，v/v）淋洗，流速 3.0 mL·min^{-1}。萃取柱用氮气流干燥 20 s，再用 2 mL 乙腈洗脱，流速 3 mL·min^{-1}，收集洗脱液，25℃用氮气流吹干。衍生化之前保证提取物完全干燥，在干燥的提取物中加入 100.0 μL A 溶液，混合 5 min，然后加入 150.0 μL B 溶液并立即涡旋约 20 s，避光静置 15 min，用 0.45 μm 微孔滤膜滤过后上样测定。

1.2.4 空白溶液的制备

精密量取 0.5 mL 空白血浆，加入 0.5 mL 乙腈溶液，涡旋振荡 1 min，4 000 r·min^{-1}离心 10 min，后续步骤同"1.2.3"。

1.2.5 对照品溶液的制备

精密量取 0.5 mL 空白血浆，加入 0.25 mL 塞拉菌素标准品溶液和 0.25 mL 多拉菌素标准品溶液，涡旋振荡 1 min，4 000 r·min^{-1}离心 10 min，后续步骤同"1.2.3"。

1.2.6 内标溶液的制备

精密量取 0.5 mL 空白血浆，加入 0.25 mL 多拉菌素标准品溶液和 0.25 mL 乙腈溶液，涡旋振荡 1 min，4 000 r·min^{-1}离心 10 min，后续步骤同"1.2.3"。

1.2.7 标准曲线与最低检测限的测定

精密称取 0.5 mL 空白血浆，加入 0.25 mL 塞拉菌素溶液和 0.25 mL 多拉菌素标准溶液，使血浆中多拉菌素的终浓度为 100.0 ng·mL^{-1}，塞拉菌素的终浓度分别是 0.5、1.0、2.0、10.0、20.0、50.0 ng·mL^{-1}，按步骤"1.2.3"操作进行处理，每个样品重复 5 次，将塞拉菌素和多拉菌素的峰面积比值作为纵坐标，塞拉菌素的血浆浓度作为横坐标绘制标准曲线，得回归方程。

1.2.8 精密度及回收率的测定

分别配制塞拉菌素低、中、高 3 个浓度（0.5、5.0、50.0 ng·mL^{-1}）的血浆样品 5 份，内含多拉菌素 100.0 ng·mL^{-1}，按步骤"1.2.3"操作，于日内及连续不同日测定，计算日内变异系数（Intra-day CV）和日间变异系数（Inter-day CV）。

配制塞拉菌素低、中、高 3 个浓度（0.5、5.0、50.0 ng·mL^{-1}）的血浆样品溶液，内含多拉菌素 100.0 ng·mL^{-1}，按步骤"1.2.3"处理后进行 HPLC-FLD 检测，每个样品重复 5 次，计算峰面积比值并求出在标准曲线上对应的塞拉菌素浓度 C_1。

将空白血浆样品按步骤"1.2.3"进行过柱处理，分别在 0.5 mL 血浆样品中加入 0.25 mL 塞拉菌素溶液和 0.25 mL 多拉菌素标准品溶液，使塞拉菌素血浆浓度分别为 0.5、5.0、50.0 ng·mL^{-1}，多拉菌素血浆浓度为 100.0 ng·mL^{-1}，按步骤"1.2.3"进行后续处理，每个样品重复 5 次，计算峰面积比值并求出在标准曲线上对应的塞拉菌素浓度 C_2。C_1/C_2 即为该方法对塞拉菌素的回收率。

1.2.9 数据处理与统计分析

采用 MicrosoftExcel 进行标准曲线的绘制。

2 结果

2.1 方法专属性考察

按上述色谱条件将塞拉菌素对照品溶液和多拉菌素内标溶液、血浆样品溶液、空白溶液进行测定，结果空白无干扰，样品与内标分离度良好，塞拉菌素和多拉菌素的保留时间分别为 4.0 和 20.2 min，色谱图见图 3~6。

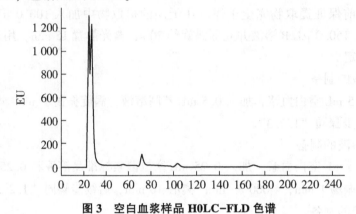

图 3 空白血浆样品 HOLC-FLD 色谱
Fig. 3 Chromatograms of blank plasma by HPLC-FLD

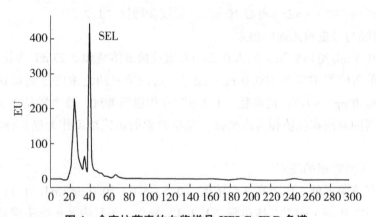

图 4 含塞拉菌素的血浆样品 HPLC-FLD 色谱
Fi.4 Chromatograms of selamectin in plasma by HPLC-FLD

2.2 线性关系考察

经对不同浓度的血浆样品进行测定，得回归方程：$y = 0.0146x + 0.0451$，$R^2 = 0.9993$，线性范围为 $0.5 \sim 50.0$ ng·mL^{-1}，检测限为 0.25 ng·mL^{-1}，满足测定要求。标准曲线见图 7。

2.3 精密度的测定

本方法测得日内、日间变异系数见表 1。结果表明，本试验条件下该方法的精密度符合要求，测定方法可靠。

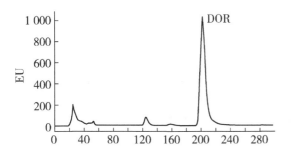

图 5 含多拉菌素的血浆样品 HPLC-FLD 色谱

Fig. 5 Chromatograms of doramectin in plasma by HPLC-FLD

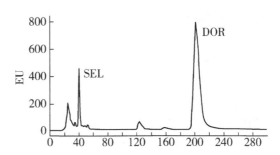

图 6 含塞拉菌素和多拉菌素的血浆样品 HPLC-FLD 色谱

Fig. 6 Chromatograms of selamectin and doramectin in plasmab HPL-FLD

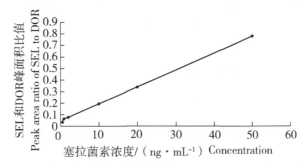

图 7 血浆塞拉菌素检测标准曲线

Fig. 7 The standard curve of selametin in plasma by HPLC-FLD

2.4 回收率的测定

本方法测得的回收率见表 2。结果表明，本试验条件下该方法的回收率可满足测定要求。

表 1 塞拉菌素在血浆样品中测定的精密度 ($n=5$)

Table 1 The precision of selamectin in plasma ($n=5$)

塞拉菌素浓度（ng·mL^{-1}） Selamectin concentration	日内变异 Intra-day CV		日间变异 Inter-day CV	
	$\bar{x}$—σ	RSD/%	$\bar{x}$—σ	RSD/%
0.5	0.50±0.02	2.13	0.50±0.01	2.08
5.0	5.08±0.10	2.89	5.14±0.15	3.31
50.0	50.32±0.69	1.47	49.85±1.03	3.12

表 2 塞拉菌素在血浆样品中提取方法回收率 ($n=5$)

Table 2 The recovery of selamectin with the methodin plasma ($n=5$)

塞拉菌素浓度（ng·mL^{-1}） Selamectin concentration	回收率 Recovery（%, $\bar{x}$—σ）	RSD/%
0.5	95.03±2.30	2.51
5.0	94.42±2.68	2.99
50.0	92.56±2.77	3.03

3 讨论和结论

3.1 内标物的选择

设立内标对血液样品中药物的检测非常必要，尤其是阿维菌素类药物，由于血药浓度较低，检测中还涉及固相萃取和衍生化处理等步骤，样品回收会受到一定影响，因此设定适合的内标可以大大降低检测误差。方法中选择的内标物应与待测物结构和性质相似[9]，多拉菌素和塞拉菌素均有一个大环内酯环，只有C5位的基团不同，两者理化性质相似，有相似的激发波长和发射波长[10-11]，故多拉菌素是本试验理想的内标物。

3.2 血浆蛋白沉淀剂的选择

本研究中血浆蛋白沉淀剂的选择比较关键，要求既能与水混溶，又不使待测物和内标变性，同时还能完全去除蛋白。本研究用乙腈作为蛋白沉淀剂，不仅因为其能降低水的介电常数，导致具有表面水层的生物大分子脱水，相互聚集，使蛋白能很好的析出，还因为其是塞拉菌素和多拉菌素很好的溶剂，且与水互溶。

3.3 样品处理过程和色谱条件的优化

固相萃取是血浆样品处理过程中关键的一步，SampliQ C18型SPE固相萃取柱能帮助笔者得到较纯的萃取物，提高了回收率。塞拉菌素在色谱分析时，很容易电离被硅烷醇或二氧化硅滞留。庚烷磺酸钠是离子对试剂，通过烷基与色谱柱吸附结合，而磺酸基与样品中的碱性基团吸附，与被测组分离子结合形成中性的离子对化合物，使得被测物在非极性固定相中溶解度增大，从而改变样品的保留，并解决了峰形拖尾的现象。

本研究选择多拉菌素为内标，建立的犬血浆中塞拉菌素含量测定的高效液相色谱-荧光检测方法灵敏度高、准确、重复性好，适用于塞拉菌素血药浓度的检测，为该药物在动物体内的检测研究提供了有效可行的途径。

参考文献

[1] BISHOP B F, BRUCE C I, EVANS N A, et al. Selamectin: a novel broad-spectrum endectocide for dogs and cats [J]. *Vet Parasitol*, 2000, 91 (3-4): 163-176.

[2] CLEMENCE R G, SARASOLA P, GENCHI C, et al. Efficacy of selamectin in the prevention of adultheartworm (Dirofilariaimmitis) infection in dogs in northern Italy [J]. *Vet Parasitol*, 2000, 91 (3-4): 251-258.

[3] ENDRIS R G, COOKE D, AMODIE D, et al. Repellency and efficacy of 65% permethrin and selamectin spot-on formulations against ixodes ricinus ticks on dogs [J]. *Vet Ther*, 2002, 3 (1): 64-71.

[4] GUNNARSSON L, CHRISTENSSON D, PALMER E. Clinical efficacy of selamectin in the treatment of naturally acquired infection of sucking lice (Linognathus setosus) in dogs [J]. *J Am Anim Hosp Assoc*, 2005, 41 (6): 388-394.

[5] KURTDEDE A, KARAER Z, ACAR A, et al. Use of selamectin for the treatment of psoroptic and sarcoptic mite infestation in rabbits [J]. *Vet Dermatol*, 2007, 18 (1): 18-22.

[6] MCCOY C, BROCE A B, DRYDEN M W. Flea blood feeding patterns in cats treated with oral nitenpyram and the topical insecticides imidacloprid, fipronil and selamectin [J]. *Vet Parasitol*, 2008, 156 (3-4): 293-301.

[7] WANG T, YANG G Y, YAN H J, et al. Comparison of efficacy of selamectin, ivermectin and mebendazole for the control of gastrointestinal nematodes in rhesus macaques, China [J]. *Vet Parasitol*, 2008, 153 (1-2): 121-125.

[8] NOVOTNY M J, KRAUTMANN M J, EHRHART J C, et al. Safety of selamectin in dogs [J]. *Vet Parasitol*, 2000, 91 (3-4): 377-391.

[9] 斯奈德 L R, 柯克兰 J J. 现代液相色谱法导论 [M]. 高潮, 陈新民, 高虹, 译. 第2版. 北京: 化学工业出版社, 1988: 527-533.

[10] SUTRA J F, CADIERGUES M C, DUPUY J, et al. Determination of selamectin in dog plasma by high performance liquid chromatography with automated solid phase extraction and fluorescence detection [J]. *Vet Res*, 2001, 32 (5): 455-461.

[11] 张萍. 多拉菌素在犬体内的药动学及其组织残留研究 [D]. 南京: 南京农业大学, 2006.

(发表于《Appl Microbiol Biotechnol》, 院选 SCI, IF: 3.689)

伊维菌素纳米乳注射液的研制与质量安全性评价

刘根新[1,3]，张继瑜[2]*，吴培星[2]，李剑勇[2]，刘英[1]，
周绪正[2]，魏小娟[2]，牛建荣[2]，胡宏伟[1]

(1. 甘肃农业大学，兰州　730060；2. 中国农业科学院兰州畜牧与兽药研究所，兰州　730050；3. 甘肃畜牧工程职业技术学院，武威　733006)

摘　要：本研究旨在研制伊维菌素纳米乳注射液（5%）并对其理化性能、载药量、稳定性和急性毒性进行评价。采用伪三元相图法进行处方筛选，在常温下，以黏度、电导率、折光率、Z电位、平均粒径为指标考察纳米乳的理化性能；在不同的温度、光照和湿度条件下以药物含量为指标考察纳米乳的稳定性；采用HPLC法测定载药量，用小鼠灌胃的方法进行急性毒性研究。结果表明，该纳米乳在透射电镜下观察其乳滴呈球状，平均粒径为70nm；不同温度、光照和湿度条件下含量无明显变化，稳定性良好；小鼠急性毒性试验表明该纳米乳注射液（0.1%）属于实际无毒。结果提示，本研究研制的伊维菌素纳米乳注射液（5%）制备简单，是一种质量稳定、安全性高的新制剂。

关键词：伊维菌素；纳米乳；伪三元相图；性能评价

伊维菌素（Ivermectin，IVM）是阿维菌素的衍生物，是目前世界上较优秀的广谱抗寄生虫药，其主要成分为22，23二氢阿维菌素Bla。IVM具有广谱、高效、用量小和安全等优点，对体内外寄生虫特别是线虫和节肢动物均具有高效驱杀作用[1]。但是IVM几乎不溶于水，严重制约着该类药物的使用。目前，虽然伊维菌素已制成预混剂、片剂、软膏剂、溶液剂、注射剂、透皮剂等常规剂型及缓释丸剂、埋植剂、微球剂等新剂型[2-7]，但未见有伊维菌素纳米乳注射液的研究报道。阿维菌素作为一类重要的抗生素，已经成为一种农用和兽用的高效生物源杀虫剂[8]。研究伊维菌素纳米乳剂对于广泛研究阿维菌素类药物以及其他水难溶性药物的纳米乳剂具有重大的意义。

纳米乳（N anoemulsion）是由表面活性剂、助表面活性剂、油相、水相组成的一种稳定透明的胶体分散系统，其粒径在10~100nm。纳米乳制备不需要特殊的设备，操作简单，且易于保存。由于粒径小，可采用过滤灭菌，作为一种新型药物载体，纳米乳制剂具有低黏度、稳定性好、吸收迅速、靶向释药、能显著提高药物的生物利用度、能极大提高难溶性药物在水中的溶解性、可以降低药物的毒副作用等特点[9-12]。近年来，国内外用新型纳米乳载体对传统的、难溶性、药效高的药物进行改造，制备出了良好的药物传递系统[13-21]。纳米乳可以广泛应用于农药、食品保鲜剂、防腐剂、兽药研制等，应用前景广阔[22]。随着研究的深入及应用进一步引起了人们对药用纳米乳剂研究的重视。

本研究通过制备、评价伊维菌素的纳米乳剂新剂型，为水难溶性药物纳米乳剂的制备提供可靠的研究方法和思路。

1 材料与方法

1.1 试验材料

JEM1200EX 投射分析电镜（日本电子光学公司）；Zetasizer Nano ZS ZEN 3600 激光动态散射仪（英国 Malvern 公司）；NDJ 1 型旋转式黏度计（上海天平仪器厂）；Agilent 1100 series 高效液相色谱仪（美国安捷伦公司）；RH KT/C 型恒温磁力搅拌器（IKA®，德国）；TGC 16C 台式高速离心机（上海安亭科学仪器厂）；BS214S 电子天平（德国 Satorius 公司）；HZ8802S 恒温水浴震荡床（化利达实验设备有限公司）。

IVM 原药（由中国药品生物制品检定所提供）；泊洛沙姆 188、聚氧乙烯氢化蓖麻油 Cremophor RH 40（德国 BASF 公司）；大豆卵磷脂（天津市光复精细化工研究所）、Tween 80、OP 乳化剂（天津福晨化学试剂厂）；蓖麻油、油酸乙酯（天津光复化学试剂厂）；注射用大豆油、肉豆蔻酸异丙酯（IPM，上海高维实业有限公司）；异丙醇、乙醇、丙三醇、1,2 丙二醇、正丁醇等均为分析纯。

1.2 试验方法

1.2.1 表面活性剂的筛选

将非离子型表面活性剂（SF）泊洛沙姆 188、Cremophor RH 40（聚氧乙烯氢化蓖麻油）、OP 乳化剂（聚氧乙烯辛基苯基醚）、大豆卵磷脂、Tween 80 配成 18 mg·mL^{-1}的胶束水溶液，考察 IVM 在各胶束溶液中的溶解度。用 HPLC 法测定 IVM 在各胶束溶液中的溶解度。以胶束溶液中 IVM 溶解度最大的作为试验用表面活性剂。

1.2.2 油相的筛选

选择大豆油、蓖麻油、肉豆蔻酸异丙酯、油酸乙酯不同碳链长短的油，黏度大小依次为蓖麻油>大豆油>肉豆蔻酸异丙酯>油酸乙酯。分别取上述四种油适量，置于具塞锥形瓶中，加入过量的 IVM，60 水浴振摇 24h 达到平衡。用 HPLC 法测定 IVM 在各油相中的溶解度。以 IVM 溶解度最大的油为试验用油。

1.2.3 助表面活性剂的选择

将 1.2.1 所选的表面活性剂和 5 种短链的助表面活性剂（CoSF），即乙醇、异丙醇、1,2 丙二醇、正丁醇、丙三醇按质量比（K_m 为 1:1、2:1、3:1）混合作为混合表面活性剂，再将"1.2.2"选择的油相分别与各种混合表面活性剂按质量比 1:9、2:8、3:7、4:6、5:5、6:4、7:3、8:2、9:1 混匀，在磁力搅拌下用超纯水滴定，观察各体系的变化，记录体系由浑浊变澄清时的临界加水量。滴定过程可借助丁达尔现象和偏光显微镜来判断是否形成了纳米乳。评价标准：液体外观透明，流动性良好且有丁达尔现象，偏光显微镜观察无双折射现象，投射电镜观察其粒径在 10~100 nm 则为纳米乳；液体透明，黏度大有双折射现象则为液晶态；液体不透亮成白色乳样则为普通乳。通过体系能否形成纳米乳从而确定助表面活性剂。

1.2.4 IVM 纳米乳的制备

采用滴定法绘制伪三元相图[23]，将 Tween 80 与正丁醇按 K_m 为 1:1、2:1、3:1 混匀后，再与油酸乙酯按不同比例混合，称取一定量的 IVM 加入其中混匀，磁力搅拌下用超纯水滴定到体系形成透亮的纳米乳（载药量达到 5%），记录临界加水量。分别以 SF-CoSF、Oil、Water 作为相图的 3 个顶点绘制相图确定纳米乳的界限。

1.2.5 纳米乳类型的鉴别

采用离心法[24]、染色法[25]鉴别。取 1.2.4 制备的 IVM 纳米乳适量,离心(15 000r·min^{-1})15min。结果未见分层,仍为澄清,淡黄色纳米乳。染色法是利用水溶性染料亚甲蓝(蓝色)和油溶性染料苏丹红(红色)在纳米乳中扩散的快慢来判断纳米乳的类型,若蓝色扩散快于红色则为 O/W 型纳米乳,反之为 W/O 型纳米乳。

1.2.6 纳米乳形态观察

取试验制备的 IVM 纳米乳(5%)适量,用 2% 的磷钨酸负染,于透射电镜下观察。

1.2.7 纳米乳粒度分析

取试验制备的 IVM 纳米乳(5%)适量,用适量超纯水稀释后用激光粒度测定仪进行测定。

1.2.8 IVM 纳米乳含量的测定

用 HPLC 法进行测定,色谱条件:甲醇:水 = 85:15;流速 1.0mL·min^{-1};检测波长 245nm;柱温 40℃;进样量 10μL。纳米乳加流动相适量,超声 30min,放冷至室温,摇匀,用微孔滤膜(0.45μm)滤过,即可进样检测。

1.2.9 IVM 纳米乳的急性毒性试验

健康小白鼠观察喂养 3d 后,剔除不合格小鼠,称体质量编号随机分组,每组 10 只,雌雄各半。采用灌胃法,按 5、6.58、8.66、11.4、15g·kg^{-1} 经口一次给稀释后的 IVM 纳米乳(0.1%),给药前禁食 8h,不限制饮水,观察记录各组动物给药后的症状、体征、死亡数目和时间。

2 结果

2.1 处方设计与筛选

2.1.1 表面活性剂的筛选

IVM 的溶解度大小顺序:Cremophor RH 40>泊洛沙姆 188>OP 乳化剂>Tween 80>大豆卵磷脂。前三者对 IVM 的溶解性差别不是很大。故初步选前三者为表面活性剂。

2.1.2 油相的筛选

IVM 在油酸乙酯中的溶解度最大,为 26mg·kg^{-1},IPM 次之,为 24mg·kg^{-1}。这可能与油酸乙酯的碳链及黏度有关,有利于 IVM 在油中的分布和溶解。故选油酸乙酯为油相。

2.1.3 助表面活性剂的选择

纳米乳能形成的体系:OP 乳化剂 1,2-丙二醇(丙三醇、PEG400、正丁醇)油酸乙酯-水;Cremophor RH40-1,2-丙二醇(PEG400)油酸乙酯-水;Tween 80 正丁醇(1,2-丙二醇、PEG400)-油酸乙酯体系形成纳米乳的难易程度、载药量、稳定性及毒性试验综合确定最佳体系为 Cremophor RH40-1,2-丙二醇(PEG400)-油酸乙酯-水。

2.2 IVM 纳米乳的制备

IVM 纳米乳的伪三元相图见图 1,结果表明,当 Km = 2:1 时纳米乳区面积最大。

2.3 纳米乳类型的鉴别

纳米乳类型的鉴别结果表明,本试验研制的 IVM 纳米乳为 O/W 型。

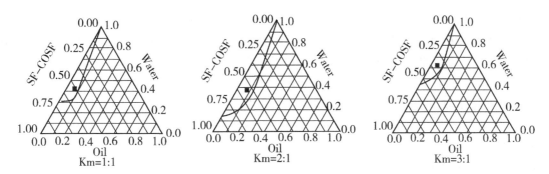

图1 Tween-80+正丁醇+油酸乙酯体系的伪三元相

Fig.1 Pseudotertiary phase diagram composed of Tween-80 n butanol ethyloleate

2.4 IVM 纳米乳的含量测定

2.4.1 方法专属性考察

分别取伊维菌素对照品、伊维菌素微乳，分别进样 10μL，分别测定，记录色谱图。结果见图2和图3。由图3可知，辅料和试剂不干扰药物的测定。

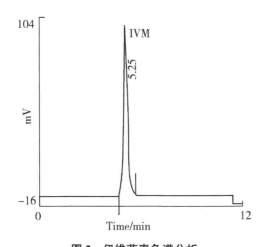

图2 伊维菌素色谱分析

Fig.2 Chromatogram of ivermectin

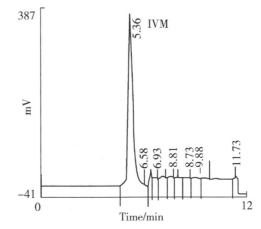

图3 伊维菌素微乳色谱分析

Fig.3 Chromatogram of ivermectin nanoemulsion

2.4.2 对照品溶液制备

精密称取伊维菌素对照品 0.011 1g，以流动相溶解配置成浓度 111.0μmL^{-1} 作为储备液。精密量取上述储备液 0.2、56、1.0、3.0、6.0、10.0mL，分别置 10mL 的容量瓶中，加流动相定容、摇匀，作为对照品溶液。

2.4.3 供试品溶液制备

精密称取样品适量，置于100mL容量瓶中，加流动相适量，超声30min，放冷至室温，摇匀，用微孔滤膜（0.45μm）滤过，即得。

2.4.4 标准曲线与线性范围

分别取上述对照溶液 10L 进样测定。以对照品溶液浓度（c）为纵坐标，峰面积的平均值（A）为横坐标作图，绘制标准曲线，并计算得回归方程：$y = 0.032x + 0.280\ 3$。$R^2 = 0.999\ 9$。结果表明，伊维菌素浓度为 2.22~111.0μg·mL^{-1} 时有良好的线性关系。

2.4.5 精密度试验

取伊维菌素标准曲线中 60μg·mL⁻¹ 的对照品溶液，重复进样 5 次，每次进样 10μL，依法测定峰面积，计算得出 RSD 为 0.31%，结果证实方法精密度良好。

2.4.6 回收率试验

分别精密量取伊维菌素对照品储备液 3.0、6.0、7.0mL，置 10mL 容量瓶中，加入处方量的空白微乳，按外标法进行测定，并计算回收率。其平均回收率为 99.6%，RSD 为 0.60%。

2.4.7 样品含量测定

精密称取伊维菌素微乳液约 0.2g，置于 100mL 的容量瓶中，按"2.4.3"方法制备供试品溶液，按外标法计算伊维菌素含量。按上述色谱条件测定峰面积，计算样品中伊维菌素含量。

2.5 纳米乳的性能评价

2.5.1 纳米乳的形态

纳米乳呈球型，大小均匀，微乳滴之间界限清晰，表面平滑，乳滴平均粒径为 53nm±2.3nm（图4）。

图4 IVM 纳米乳的透射电镜照片

Fig. 4 The electron mlcmscope of ivermectin nanoemlsion

2.5.2 粒径和粒度分布研究

纳米乳的粒径与粒度分布见图5。结果表明：稀释后的 IVM 纳米乳粒径主要分布在 10~100nm，平均粒径 70nm（图5）。

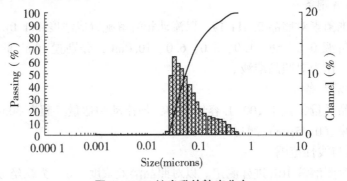

图5 IVM 纳米乳的粒度分布

Fig. 5 The diameter distributes of ivermectin nanoemlsion

2.5.3 纳米乳的基本理化性质考察

分别考察了空白纳米乳和"2.2"制备的伊维菌素纳米乳（5%）的黏度、电导率、折光率、Z电位和平均粒径（表1）。

表1 纳米乳的理化性质测定（$n=5$）
Table 1 The physicochemical property measurement of nanoemulsion（$n=5$）

样品 Sample	黏度/（$mm^2 \cdot s^{-1}$） Idex of viscosity	电导率/ms^{-1} Conductivity	折光率 Refractive property	Z电位/mV Z potentiometer	平均粒径/nm Average diameter
空白纳米乳 Blank nano emulsion	7.02±0.01	139.00±1.43	1.26±0.04	-1.35	62.0±1.2
IVM纳米乳 Nano emulsion of IVM	7.13±0.06	144.00±2.31	1.37±0.01	-1.41	70.0±2.3

空白纳米乳中加入药物以后，电导率显著增加，粒度增加，Z电位、黏度、折光率无明显变化。

2.5.4 稳定性考察

IVM纳米乳（5%）于室温放置150d仍保持澄清透明，无分层、药物析出现象。将IVM纳米乳放置于高温（40、60℃）、低温（4℃）及光照（4 500±500）lx下进行试验，于5、10d取样测定，含量考察结果见表2。

IVM纳米乳的质量稳定，在上述条件下粒度和含量无明显变化。将"2.2"制备的IVM纳米乳分别加水稀释至浓度分别为4%、3%、2%、1%、0.5%，采用上述同样的方法进行试验均质量稳定。

表2 稳定性试验结果
Table 2 The results of stable tests （%）

时间/d Time	4℃	40℃	60℃	光照 Illumination	相对湿度92.5% Relative humidity
0	100.0	99.7	99.6	100.1	99.7
5	99.7	99.8	100.0	99.9	100.2
10	99.3	100.0	99.4	99.7	99.8

2.6 IVM纳米乳的急性毒性试验

IVM纳米乳（0.1%）给药剂量达到15g·kg^{-1}时，小鼠的死亡率为0%，属于实际无毒范围，该试验也说明IVM纳米乳在这个浓度条件下其所采用的辅料也属实际无毒。

3 讨论

3.1 O/W型纳米乳剂可以有效提高药物的溶解度[10]。IVM几乎不溶于水，作为良好的抗寄生虫药物，研究广泛[27]。目前还没有见到该药物的纳米乳剂剂型的研究报道。本研究成功制备了IVM的O/W型纳米乳剂，且通过一系列试验证明其性质稳定，可以加水大量稀释。同时，该纳米乳剂可以有效提高IVM的药物含量，载药量高达5%。

3.2 纳米乳剂中的不同组分对药物的性质和结构没有直接影响。本研究采用HPLC法测定

IVM 的含量，直接用调好 pH 值的流动相破乳、稀释、定容，用 0.45m 微孔滤膜过滤器过滤后进样，结果表明：纳米乳中其他辅料对 IVM 的测定没干扰，该方法简便易行灵敏度高，重现性好。

3.3 纳米乳是含有一定量表面活性剂、助表面活性剂、油和水的一个有机体，其对药物的增溶作用取决于油相、表面活性剂和助表面活性剂[26]。本研究通过试验比较了不同油相、表面活性剂、及助表面活性剂对 IVM 的增溶作用，充分证明了这一点。但是，纳米乳剂因含有一定量的表面活性剂和助表面活性剂，这些成分对动物体都有不同程度的毒性或者副作用，而且这些成分和载药量的大小也对纳米乳剂的稳定性有一定的影响[27-28]。临床应用时需要对其进行安全性和稳定性等方面的综合评价。该研究对由 Tween 80、正丁醇和油酸乙酯组成的纳米乳剂进行了综合评价，认为该制剂为稳定、安全、成本低廉的良好药物剂型。

3.4 采用拟三元相图结合纳米乳的稳定性考察能够在有效降低表面活性剂和助表面活性剂的情况下制备出适宜的药用纳米乳，而且制备条件简单、易行。本试验采用拟三元相图成功制备了 IVM 纳米乳剂。

通过对伊维菌素纳米乳的制备及检测方法的研究为阿维菌素类药物的广泛使用开创了新途径，对开发利用纳米乳作为药物剂型并解决水难溶性药物的使用提供了新的研究途径。

4 结论

本研究采用拟三元相图用 Tween-80、正丁醇、油酸乙酯和水成功制备了伊维菌素 O/W 型纳米乳剂（5%）。该纳米乳呈乳滴呈球状，平均粒径为 70nm；不同温度、光照和湿度条件下含量无明显变化，稳定性良好；小鼠急性毒性试验表明该纳米乳注射液（0.1%）属于实际无毒。

参考文献

[1] D AV ID W F.A naly tical profiles o f drug substances [J].lvarmectin, 1988, 17: 155.

[2] 王敏儒，陈杖榴，冯淇辉.伊维菌素聚乳酸微球制备条件的优化及药物含量测定 [J].华南农业大学学报, 1998, 19 (4): 5-9.

[3] 蒲文兵，杨帆，宋刚华，等.伊维菌素（维力加）注射液对羊体内外寄生虫的驱虫试验 [J].中国动物传染病学报, 2009, 17 (1): 54-57.

[4] 赵永星，张雪晓，孙倩，等.注射用伊维菌素亚微乳的制备及其稳定性研究 [J].中国兽药杂志, 2010, 44 (6): 27-29.

[5] 朱晓娟，李引乾，侯勃，等.伊维菌素脂质体的制备及其质量评价 [J].西北农林科技大学学报（自然科学版）, 2010, 38 (4): 24-30.

[6] SHOOPWL, MROZIKH, FISHERMH .St ructur e and activ ity o f av ermectina and milber mycins in animal health [J].*Vet Parasitol*, 1995, 59: 139-156.

[7] 夏晓静，周建平，王翔，等.伊维菌素聚乳酸微球的制备 [J].中国药科大学学报, 2004, 35 (5): 429-432.

[8] BLOOMRA, MATHESONIII.Envir onmental assessmentof aver mectin by the U S food and drug administrat ion [J].*Vet Parasitol*, 1993, 48 (14): 281-294.

[9] UCHEGBUIF, VYASSP. Non io nic surfactant based vesicles (niosomes) indr ug deliver y [J].*I nt J P har m*, 1998, 172 (1): 33-70.

[10] 王晓黎，蒋雪涛，刘皋林，等.O/W 型微乳对水难溶性药物增溶作用的研究 [J].第二军医大学学报, 2002, 23 (1): 84 86.

[11] YILMAZE, BORCHERTHH. Design of a phyto sphingosine containing, positively charged nanoemulsion as a colloidal carrier system for dermal application of ceramides [J]. *Eur J Pharm Biopharm*, 2005, 60: 91-98.

[12] NORNOOAO, CHOWDS. Cremophor free intravenous microemulsions for paclitaxel II. Stability, *in vitro release and pharmacokinetics* [J]. *Int J Pharm*, 2008, 349: 117-123.

[13] TRIMMLKET, CHAIXC, DELAIRT. Interfacial depositon of functionalized polymers onto nanoemulsions produced by the solvent displacement method [J]. *Colloid Polym Sci*, 2001, 279 (1): 784-792.

[14] SHAFIQ S, SHAKEEL F, TALEGAONKAR S, et al. Development and bioavailability assessment of ramipril nanoemulsion formulation [J]. Eur J Pharm Biopharm, 2007, 66 (2): 227-243.

[15] 毛世瑞, 张磊, 李茗, 等.硫酸沙丁胺醇油包水型口服纳米乳的制备及小肠吸收考察 [J].沈阳药科大学学报, 2005, 22 (6): 401-404.

[16] 吴旭锦, 欧阳五庆, 朱小甫, 等.黄芩甙纳米乳的制备 [J].精细化工, 2007, 24 (5): 470-472.

[17] 龚明涛, 张钧寿, 戴晓鸣.羟喜树碱纳米乳在大鼠体内的药代动力学研究 [J].中国药科大学学报, 2004, 35 (4): 324-327.

[18] 何蕾, 王桂玲, 张强.紫杉醇纳米乳剂的体内外考察 [J].药学学报, 2003, 38 (3): 227-230.

[19] 孙红武, 欧阳五庆.黄连素口服纳米乳的研制、质量及安全性评价 [J].上海交通大学学报（农业科学版）, 2007, 25 (1): 60-65.

[20] 李先红, 刘榛榛.醋酸泼尼松龙纳米乳的制备 [J].海峡药学, 2006, 18 (1): 31-33.

[21] 杨青艳, 吴道澄, 吴红, 等.氟尿嘧啶纳米乳剂的制备与性质 [J].解放军药学学报, 2005, 21 (3): 170-173.

[22] LAWRENCE MJ, REESGD. Microemulsion based media as novel drug delivery systems [J]. *Adv Drug Deliv Rev*, 2005, 45 (1): 89-121.

[23] PONSR, CARRERAI, CAELLESJ, et al. Formation and properties of miniemulsions formed by microemulsions dilution [J]. *Adv Colloid Interface Sci*, 2003, 106: 129-146.

[24] SCHECHTER RS. Microemulsions and related systems [M]. New York: Marcel Dekker, 1998: 1-200.

[25] HOHO, HSIAOCC, SHEUMT. Preparation of microemulsions using polyglycerol fatty acid esters as surfactant for the delivery of protein drugs [J]. *J Pharm Sci*, 1996, 85 (2): 138-143.

[26] 姚静, 周建平, 杨宇欣, 等.微乳对难溶性药物增溶机理的研究 [J].中国药科大学学报, 2004, 35 (6): 495-498.

[27] 沈熊, 吴伟.自乳化和自微乳化释药系统 [J].复旦大学学报（医学版）, 2003, 30 (2): 180 183.

[28] 陈婧, 刘彩霞, 朱晓薇.影响微乳形成因素的研究 [J].天津中医药大学学报, 2007, 26 (1): 31-33.

（发表于《畜牧兽医学报》）

奶牛乳房炎金黄色葡萄球菌凝固酶基因型研究

倪春霞[1,2]，蒲万霞[1]，胡永浩[2]，邓海平[1]

(1. 中国农业科学院兰州畜牧与兽药研究所/新兽药重点开放实验室/甘肃省新兽药工程重点实验室，兰州 730050；2. 甘肃农业大学动物医学院，兰州 730070)

摘 要：【目的】了解上海和贵州地区引起奶牛乳房炎的金黄色葡萄球菌凝固酶基因型情况，为奶牛乳房炎的防治提供理论依据。【方法】利用16S rRNA保守序列设计引物，PCR扩增鉴定金黄色葡萄球菌，并利用凝固酶基因及其酶切产物多态性对分离鉴定出的金黄色葡萄球菌进行了凝固酶基因分型。【结果】共鉴定出78株金黄色葡萄球菌，有74株金黄色葡萄球菌扩增凝固酶基因片段，分为5个基因型和两个亚型。PCR1型为贵州地区优势基因型，PCR 3型为上海地区优势基因型。【结论】两地区各基因型菌株分布比例有显著的地域性差异，这与两地区地理环境和养殖水平差异对病原流行传播的影响有关。

关键词：奶牛乳房炎；金黄色葡萄球菌；分子鉴定；凝固酶基因分型

引言

【研究意义】奶牛乳房炎是一种主要由病原微生物引起的奶牛常见病，是造成奶牛养殖业生产经济损失的重要因素之一，不仅影响奶牛养殖业、乳品工业的发展，也给公共卫生以及食品安全带来一定隐患。

金黄色葡萄球菌是引起奶牛乳房炎最主要的病原菌之一[1]。对其进行及时准确的基因分型研究，掌握的流行趋势，可为奶牛乳房炎的防治和疫苗的研制提供重要的依据。【前人研究进展】近年来有关奶牛乳房炎病原菌的研究报道多为不同地区菌株分离鉴定、耐药性监测和血清学分型等方面[2-8]，而有关菌株基因分型方面的研究在国内还少有报道。金黄色葡萄球菌感染是近年来引起奶牛乳房炎的主要病因[9-10]，血浆凝固酶（coagulase）是金黄色葡萄球菌重要的致病因子，具有保护病原菌不被吞噬或免受抗体结合等作用。由于凝固酶是致病性特有的致病因子，所以临床上经常将其作为鉴定金黄色葡萄球菌的重要指标[11]。随着分子生物学技术在病原菌鉴定方面的广泛应用，凝固酶基因也越来越多的被应用于致病性金黄色葡萄球菌的鉴定，朱战波、任宪刚等率先对牛源金黄色葡萄球菌进行了凝固酶基因分型并分析了不同基因型菌株致病因子的特征[12,13]。凝固酶基因分型具有操作简单，重复性好，费用低等优点，但是该基因保守性较高，进行PCR分型后区分力较差，无法对菌株进行细致的分型，一定程度上制约了该方法在细菌分型研究上的应用。【本研究切入点】本研究先根据菌株凝固酶基因保守序列设计引物，扩增凝固酶基因，进行凝固酶基因分型，再用内切酶消化凝固酶基因PCR产物，由于凝固酶基因3'末端重复序列可变区具有多态性，故酶切位点也各有差异，所以可根据酶切后片段长度多态性进行对凝固酶基因进行二次分型，以弥补其区分力不足的缺点。【拟解决的关键问题】本试验通过对169份临床型乳房炎奶样中的金黄色葡萄球菌进行分子鉴定及凝固酶基因分型，了解上海、贵州两地区奶牛乳房炎金黄色葡萄球菌凝固酶基因型分布情况及地区性差异特征，为地区性奶牛乳房炎的防治提供可靠的病原分子流行病学依据。

1 材料与方法

1.1 菌株及奶样

标准菌株CVCC2246购自中国兽医药品监察所，经血浆凝固酶试验验证为阳性后，保存、使用；阴性菌株为本实验室分离鉴定的表皮葡萄球菌（凝固酶阴性）。169份临床型乳房炎患牛奶样分别来自上海地区某大型奶牛场和贵州山区奶牛场。

1.2 培养基及主要试剂

普通营养琼脂培养基，绵羊鲜血培养基，却甫曼琼脂培养基（内含7.5%NaCl，用于选择纯培养葡萄球菌），营养肉汤按常规方法配制[14]，生化鉴定所用的微量发酵管均购自兰州荣昌微生物试剂公司。dNTPs、TaqDNA聚合酶、Alu I、Premix Ex Taq、溶菌酶及DL2000 Marker均购于大连宝生物公司。细菌基因组DNA提取试剂盒购于TIANGEN生化科技（北京）有限公司。

1.3 金黄色葡萄球菌分离鉴定

1.3.1 细菌纯化培养

将169份临床型乳房炎患牛奶样分别接种于绵羊血平板培养基和却甫曼琼脂培养基纯化培养。

1.3.2 金黄色葡萄球菌生化鉴定

将菌落形态、镜检结果符合葡萄球菌特征的菌株参考相关生化特性，选择凝固酶、甘露醇发酵、O/F等实验进行金黄色葡萄球菌生化鉴定。

1.3.3 金黄色葡萄球菌分子生物学鉴定

提取菌落形态、镜检结果符合葡萄球菌特征菌株基因组DNA（试剂盒法）。根据已发表的金黄色葡萄球菌的16SrRNA基因序列（GenBank Accession No.X68417），并且参照文献[15-17]，应用Oligo6.67及DNAStar软件设计一对针对金黄色葡萄球菌的16S rRNA基因的引物P1和P2，由大连宝生物工程有限公司（大连TaKaRa公司）合成。

P1：5′-GCGGTCGCCTCCTAAAAG-3′
P2：5′-TCCCGGTCCTCTCGTACTA-3′

纯化后的反应体系为25μL：10×PCR buffer（Mg^{2+} Plus）2.5μL，dNTP Mixture（2.5mmol·L^{-1}）2μL，Taq酶（5U）0.125μL，引物各1μL，模板2μL，加灭菌超纯水补足25μL。

PCR扩增反应条件为：94℃预热5 min，94℃ 60 s，57.5℃ 40 s，72℃ 40 s，进行30个循环，然后72℃延伸10 min。PCR反应结束后，取3μL PCR产物用1%的琼脂糖凝胶电泳检测结果。

1.4 凝固酶基因扩增

1.4.1 引物合成

根据GenBank已发表的金黄色葡萄球菌凝固酶基因序列（GenBank Accession No.

D00184），并且参照文献[12,18-19]，设计一段针对凝固酶基因的特异性引物。

P1：5'-ACCACAAGGTACTGAATCAACG-3'；

P2：5'-TGC TTTCGATTGTTCGATGC-3'（大连宝生物工程有限公司合成）。

1.4.2 凝固酶基因扩增

经纯化后的PCR反应体系为25μL：Premix Ex Taq 12.5μL，引物各1μL，模板2μL，灭菌超纯水补足25μL。PCR扩增反应条件为：

94℃ 5min，94℃ 30s，62℃ 45s，72℃ 45s，进行30个循环，然后72℃延伸10min。PCR反应结束后，取3μL PCR产物用1%的琼脂糖凝胶电泳检测结果。

1.5 产物酶切反应

取经电泳检测的凝固酶基因扩增产物10μL，加入酶切反应缓冲液2μL，*Alu*I酶1μL（10U），无菌双蒸水7μL，37℃水浴3h后产物经2%琼脂糖凝胶电泳检测结果。

1.6 凝固酶基因多态性分型的区分系数计算

通过区分系数公式（the numerical index of discrimination）计算凝固酶基因多态性分型的区分系数[20]，此公式可以用来对相同菌株的不同分型方法进行比较，通过比较量化后的数值来判定何种分型方法的区分力好。公式如下：

$$D = 1 - \frac{1}{N(N-1)} \sum_{j=1}^{s} n_j(n_j - 1)$$

D = 区分系数；s = 总分型数；n_j = 第j个型中含有多少个菌株；N = 分型的菌株总数。

2 结果

2.1 菌株的分离及鉴定

菌株经生化鉴定，符合金黄色葡萄球菌生化指标菌株为78株。经16S rRNA基因PCR扩增后，有78株菌产生了450bp左右的目的片段（但由于菌株过多，只列出其中一部分菌株的电泳图，图1）。结果表明生化鉴定结果与分子鉴定结果完全一致。这些菌株分别为贵州地区38株，上海地区40株。

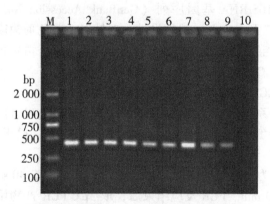

M：DL2000 Marker；1-8：随机选择菌株；9：CVCC2246阳性对照；10：表皮葡萄球菌阴性对照

图1 金黄色葡萄球菌16S rRNA鉴定

2.2 凝固酶基因扩增

共有74株金黄色葡萄球菌分离株经凝固酶基因PCR扩增后获得目的片段，根据其产物片段大小，可分为5种PCR型（图2）。各基因型菌株分布情况见表1。

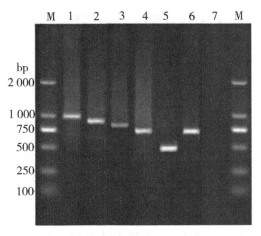

M：DL2000 Marker；1-5：1-5型菌凝固酶基因片段；6：CVCC2246阳性对照；7. 阴性对照

图2 凝固酶基因分型

表1 金黄色葡萄球菌凝固酶基因扩增

Table 1 Results of coagulase gene PCR amplification on S.aureus

PCR型（片段长度） PCR type (length of fragment)	菌株数 Number			所占比例 Proportion（%）		
	上海 Shanghai	贵州 Guizhou	总数 Total	上海 Shanghai	贵州 Guizhou	总数比 Total proportion
1（1000bp）	9	18	27	22.5	47.4	34.18
2（900bp）	11	9	20	27.5	23.7	25.32
3（800bp）	14	6	22	35	15.8	27.85
4（750bp）	0	3	3	0	7.9	5.06
5（500bp）	1	1	2	2.5	2.6	2.53
未扩增菌 Without amplification	3	1	4	7.5	2.6	5.06

2.3 PCR产物酶切

凝固酶基因产物用限制性内切酶 $AluⅠ$ 消化后，5个基因型显示了6种酶切结果（图3）。其中PCR 2型显示两种酶切结果，进一步将其分为两个不同的亚型，即亚型Ⅰ和亚型Ⅱ，两亚型菌株分布情况见表2。

表2 PCR 2型酶切后亚型菌株分布

Table 2 The result of subtype of PCR 2 type after digested by AluI

酶切亚型 Subtype of enzyme digestion	菌株数 Number		
	上海 Shanghai	贵州 Guizhou	总计 Total
Ⅰ	9	3	12
Ⅱ	2	6	8

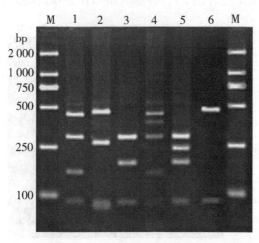

1：PCR 1 型片段；2：PCR 2 型片段 I；3：PCR 2 型片段 II；4：PCR 3 型片段；5：PCR 4 型片段；6：PCR 5 型片段

图 3　凝固酶基因 Alu I 酶切

2.4　凝固酶基因多态性分型的区分系数

本试验分型中 $s=6$、$N=162$，带入计算公式可以得出凝固酶基因多态性分型的区分系数为 0.717。

3　讨论

致病性金黄色葡萄球菌能多产生血浆凝固酶，常作为鉴别葡萄球菌有无致病性的重要标志。Jonsson 等[21]发现 α-HL 和凝固酶能协同作用引发小鼠乳腺炎。王英杰[22]等研究发现凝固酶在金黄色葡萄球菌性奶牛乳房炎致病机理中具有重要意义。因此，对金黄色葡萄球菌进行凝固酶基因的检测和分型对于奶牛乳房炎的防治具有重要意义。凝固酶基因分型方法操作简单，重复性好，费用低，因此较多的应用于动物源分离株的分型[18]。谭磊等[23]对分离自黑龙江、内蒙古、河北、山东及湖北的 55 株牛源金黄色葡萄球菌进行了凝固酶基因扩增，分为 8 型；Güler 等[24]对土耳其国内某地区 7 年内所分离到的 125 株金黄色葡萄球菌进行了凝固酶基因扩增，分为 4 型；Montesinos 等[25]对加拿大的 124 株耐甲氧西林金黄色葡萄球菌进行了凝固酶基因扩增，分为 4 型；Raimundo[26]等报道了澳大利亚不同地区 7 个农场的 151 株金黄色葡萄球菌凝固酶基因分型情况。Güler 和 Raimundo 的分型结果显示产物大小约 1 000bp 左右的 PCR1 型最多，为主要的基因型。本研究对分离自贵州和上海地区的 169 份奶样中的 79 株金黄色葡萄球菌进行了分子鉴定和凝固酶基因分型。有 74 株金黄色葡萄球菌扩增出了目的片段，并根据片段大小分为 5 个 PCR 型。其中片段大小为 1 000bp 左右的 PCR1 型为贵州地区的菌株流行优势基因型，这与 Güler 和 Raimundo 的分型结果相同；片段大小为 800bp 左右的 PCR3 型为上海地区的菌株流行优势基因型，PCR1、PCR2 和 PCR3 型菌株在上海地区分布比例较为平均。对扩增产物进行酶切后，20 株 PCR2 型菌株又分为两个不同的亚型。78 株菌中有 4 株未扩增出凝固酶基因片段，原因可能是由于基因突变致使基因保守性发生变化，引物无法识别目的片段所致，具体原因还需进一步研究[22]。本试验中凝固酶基因多态性分型的分型率为 94.93%，区分系数为 0.717，这与 Aarestrup 和

Elizabete 的报道[18,19]稍有不同，可能与金黄色葡萄球的地方流行性差异有关。

优势型金葡菌株会对牛中性粒细胞的吞噬和杀灭作用有更强的抵御能力，并且在牛的乳腺中有着更强的致病力，同时这些菌株也在农场中有着更强的传播能力[27]。上海地区菌株中这3个基因型菌株分布较平均，反应了该地区菌株流行广泛、频繁的特征，出现了多个较为优势的流行菌株基因型。这可能与该地区的金黄色葡萄球菌克隆株较多有关，并呈地方流行性，也可能与气候条件有关。两个地区的PCR2型呈不同的亚型，说明不同地区的分离株虽呈现同一基因型，但还是有一定的差别的。贵州地区奶样主要来自山区，奶牛养殖户较少，饲养的奶牛数量也较小，没有大型的奶牛养殖场，所以病原菌只会在较小的范围内传播，地区间流行传播的机会较少，所以基因型较集中；而上海地区的分离株来自规模较大的牛场，饲养的奶牛种类和数量较多，因此菌株的流行传播也较频繁，基因型分布也较广。也进一步表明奶牛乳房炎病原菌的感染率因地区、气候不同而异，且还和牛场环境及现代化水平相关。

4 结论

本研究首次对来自贵州和上海两个不同地区的临床型奶牛乳房炎金黄色葡萄球菌分离株进行了分子鉴定、凝固酶基因检测及多态性分型，比较了平原和山区牛源金黄色葡萄球菌凝固酶基因分型特征，结果显示菌株共分为5个PCR型及两个亚型，两地区菌株基因型组成具有明显的地域性差异。研究结果为地区性奶牛乳房炎的防治提供可靠的病原分子流行病学依据。

参考文献

[1] Jürgen S H, Bärbel K, Wilfried W, Michael Z, Axel S, Klaus F.The epidemiology of *Staphylococcus aureus* infections from subclinical mastitis in dairy cows during a control programme.*Veterinary Microbiology*, 2003, 96: 91-102.

[2] 刘文进, 陈创夫, 滕文军, 等.新疆垦区奶牛乳房炎致病菌的调查.家畜生态学报, 2005, 26 (4): 45-47.
Liu W J, Chen C F, Teng W J, Wu J, Ren X Y.Investigation on mastitis pathogens for diary cattle in Xinjiang.*Ecology of Domestic Animal*, 2005, 26 (4): 45-47. (in Chinese)

[3] Li J P, Zhou H J, Yuan L, He T, Hu S H.Prevalence, genetic diversity, and antimicrobial susceptibility profiles of *Staphylococcus aureus* isolated from bovine mastitis in Zhejiang Province, China .*Journal of Zhejiang University (Science B)*, 2009, 10 (10): 753-760.

[4] 邓海平, 俞诗源, 王玲, 等.内蒙古地区奶牛乳房炎病原菌的分离鉴定及药敏试验研究.安徽农业科学, 2009, 37 (2): 595-596.
Deng H P, Yu S Y, Wang L, Meng X Q.Separation and identification on pathogens from cow mastitis in Inner Mongolia area and their drug sensitive experiment.*Journal of Anhui Agricultural Sciences*, 2009, 37 (2): 595-596. (in Chinese)

[5] 倪春霞, 蒲万霞, 胡永浩, 等.上海地区奶牛乳房炎病原菌的分离鉴定及药敏实验.黑龙江畜牧兽医, 2009, 10: 79-80.
Ni C X, Pu W X, Hu Y H, Wang L, Meng X Q, Deng H P.Isolation, identification and drug sensitive test of pathogenic bactcria causing dairy cattle mastitis in Shanghai area.*Heilongjiang Animal Science and Veterinary Medicine*, 2009, 10: 79-80. (in Chinese)

[6] Kerro D O, Tareke F.Bovine mastitis in selected area of southern Ethiopia.*Tropical Animal Heath and*

Production, 2003, 35: 197-205.

[7] 倪春霞, 蒲万霞, 胡永浩, 等.奶牛乳房炎病原菌的分离鉴定及耐药性分析.西北农业学报, 2010, 19 (2): 20-24.

Ni C X, Pu W X, Hu Y H, Deng H P, Wang L, Meng X Q.Isolation, identification and drug sensitive test of pathogenic bactcria causing dairy cattle mastitis.*Acta Agriculturae Boreali-Occidentalis Sinica*, 2010, 19 (2): 20-24. (in Chinese)

[8] Kalmus P, Viltrop A, Aasmae B, Kask K.Occurrence of clinical mastitis in primiparous Estonian dairy cows in different housing conditions.*Acta Veterinaria Scandinavica*, 2006, 48 (1): 21.

[9] 周绪正, 张继瑜, 王有祥, 等. "乳源康" 治疗奶牛乳房炎药效学研究.中国兽药杂志, 2005, 39 (8): 45-48.

Zhou X Z, Zhang J Y, Wang Y X, Li J Y, Li J S, Pu W X, Wei X J, Zhang Z F, Yang H, Yang M C, Tian J M, Li H S, Xu Z Z.Effects of Ruyuankang on milk cow's mastitis.*Chinese Journal of Veterinary Drug*, 2005, 39 (8): 45-48. (in Chinese)

[10] 易明梅, 黄奕倩, 朱建国, 等.上海 地区奶牛乳房炎主要病原菌的分离鉴定及耐药性分析.中国兽医学报, 2009, 29 (3): 360-363.

Yi M M, Huang Y Q, Zhu J G, Sun H C, Hua X G, Yuan Y M, Wang Z J.Isolation, identification and drug sensitivity test of main pathogenic bacteria causing dairy cow mastitis in shanghai area Chinese. *Journal of Veterinary Science*, 2009, 29 (3): 360-363. (in Chinese)

[11] Swee H G, Sean K, Byrne J, Zhang L, Anthony W C.Molecular typing of *staphylococcus aureus* on the basis of coagulase gene polymorphisms.*Journal of Clinical Microbiology*, 1992, 30 (6): 1642-1645.

[12] 朱战波, 任宪刚, 崔玉东, 等.牛源金黄色葡萄球菌凝固酶基因的分型.中国兽医学报, 2007, 27 (5): 728-730.

Zhu Z B, Ren X G, Cui Y D, Piao F Z.Coagulase gene typing of *Staphylococcus aureus* isolated from bovine. *Chinese Journal of Veterinary Science*, 2007, 27 (5): 728-730. (in Chinese)

[13] 任宪刚, 朱战波, 崔莉, 等.不同凝固酶基因型牛源金黄色葡萄球菌主要致病因子的检测及分析.中国预防兽医学报, 2008, 30 (8): 608-611.

Ren X G, Zhu Z B, Cui L, Song B F, Tan L, Cui Y D, Piao F Z.Virulence factors of *Staphylococcus aureus* of different coagulase genotype. *Chinese Journal of Preventive Veterinary Medicine*, 2008, 30 (8): 608-611. (in Chinese)

[14] 姚火春.兽医微生物学实验指导.北京: 中国农业出版社, 2002.

Yao H C.*Veterinary Microbiology Experiment Guidance*.Beijing: China Agriculture Press, 2002. (in Chinese)

[15] Niniou I, Vourli S, Lebessi E, Foustoukou M, Vatopoulos A, Pasparakis D G, Kafetzis D A, Tsolia M N.Clinical and molecular epidemiology of community-acquired, methicillin-resistant *Staphylococcus aureus* infections in children in central Greece. *European Journal of Clinical Microbiology and Infectious Disease*, 2008, 27: 831-837.

[16] Stephen M, Mary E K, Ariane D, Raf D R, Marc S, Christina E Z, Vivian F, Saara S, Jaana V V, Névine E S, Christina C, Wolfgang W, Panayotis T T, Nikolas L, Willem V L, Alex V B, Anna Vi, Idoia L, Javier G, Saara H, Barbro O L, Ulrika R, Geoffrey C, Barry C.Harmonization of pulsed-field gel electrophoresis protocols for epidemiological typing of strains of methicillin-resistant *Staphylococcus aureus*: a single approach developed by consensus in 10 European laboratories and its application for tracing the spread of related strains. *Journal of Clinacal Microbiology*, 2003, 41: 1 574-1 585.

[17] SitiIsrina O S, Zaini K, Christoph L, Michael Z. Comparative studies on pheno and genotypic

properties of *Staphylococcus aureus* isolated from bovine subclinical mastitis in central Java in Indonesia and Hesse in Germany. *Journal of Veterinary Science*, 2004, 5 (2): 103-109.

[18] Aarestrup F M, Dangler C A, Sordillo L M. Prevalence of coagulase gene polymorphism in*Staphylococcus aureus* isolates causing bovine mastitis. *Canadian Journal Veterinary Research*, 1995, 59: 124-128.

[19] Elizabete R S, Nivaldo S. Coagulase gene typing of *Staphylococcus aureus* isolated from cows with mastitis in southeastern Brazil. *The Canadian Journal of Veterinary Research*, 2005, 69: 260-264.

[20] Dag H, Heike C, Wolfgang W, Jorg R N, Hermann C, Doris T, Ulrich V. Typing of methicillin resistant *Staphylococcus aureus* in a university hospital setting by using novel software for spa repeat determination and database management. *Journal of Clinacal Microbiology*, 2003, 41 (4): 5 442-5 448.

[21] Jonsson P, Lindberg M, Haraldsson I, Wadström T. Virulence of *Staphylococcus aureus* in a mouse mastitis model: studies of alpha hemolysin, coagulase, and protein A as possible virulence determinants with protoplast fusion and gene cloning. *Infection and Immunity*, 1985, 49 (3): 765-769.

[22] 王英杰, 王长法, 陈营, 等. PCR 扩增血浆凝固酶基因检测致病性金黄色葡萄球菌. 生物技术通报, 2008, 5: 185-188.
Wang Y J, Wang C F, Chen Y, An L G, Yang J W, Zhong J F, Yang S H, Yang H J. Identification of coagulase gene of pathogenic *Staphylococcus aureus* bovine mastitis by PCR. *Biotechnology Bulletin*, 2008, 5: 185-188. (in Chinese)

[23] 谭磊. 基于 PCR 的牛源金黄色葡萄球菌基因分型研究 [D]. 大庆: 黑龙江八一农垦大学, 2008.
Tan L. The Study on PCR-based genotyping of bovine source staphylococcus aureus [D]. Daqing: Heilongjiang August First Land Agricultural University, 2008. (in Chinese)

[24] Güler L, Ok U, Gündüz K, Gülcü Y, Hadimli H H. Antimicrobial susceptibility and coagulase gene typing of *Staphylococcus aureus* isolated from bovine Clinical Mastitis Cases in Turkey. *Journal of Dairy Science*, 2005, 88 (9): 3 149-3 154.

[25] Montesinos I, Salido E, Delgado T, Cuervo M, Sierra A. Epidemiologic genotyping of methicillin-resistant*Staphylococcus aureus* by pulsed-field gel electrophoresis at a university hospital and comparison with antibiotyping and protein A and coagulase gene polymorphisms. *Journal of Clinical Microbiology*, 2002, 40 (6): 2 119-2 125.

[26] Raimundo O, Deighton M, Capstick J, Gerraty N. Molecular typing of Staphylococcus aureus of bovine origin by polymorphisms of the coagulase gene. *Veterinary Microbiology*, 1999, 66: 275-284.

[27] Larsen H D, Aarestrup F M, Jensen N E. Geographical variation in the presence of genes encoding superanfigenic exotoxins and β-hemolysin among *Staphylococcus auleus* isolated from bovine mastiffs in Europe and USA. *Veterinary Microbiology*, 2002, 85: 61-67.

(发表于《中国农业科学》)

甘肃犬瘟热病毒流行株 N 蛋白和 H 蛋白基因的序列分析

王小辉[1]，王旭荣[1*]，郭庆勇[2]，李建喜[1]，李世宏[1]

(1. 中国农业科学院兰州畜牧与兽药研究所/中国农业科学院临床兽医学研究中心，兰州　730050；2. 新疆农业大学动物医学学院，乌鲁木齐　830052)

摘　要：采用 RT-PCR 方法扩增了 3 株犬瘟热病毒（CDV）分离株的 N、H 蛋白基因，将其克隆入 pGEM-T Easy 载体并进行序列分析。结果表明，获得的 N 蛋白基因序列与 GenBank 上登录的其他国家和地区的 CDV 分离株相应序列的同源性为 92.5%~99.9%，获得的 H 蛋白基因与其他国家和地区的 CDV 分离株相应序列的同源性为 81.7%~100%。由 N 蛋白基因和 H 蛋白基因推导的氨基酸序列构建的遗传进化树可知，这 3 株 CDV 流行株均处于野毒株的分支上，在 N 蛋白基因的遗传进化树上，CDV 流行株呈现明显的地域性差异，而在 H 蛋白基因的遗传进化树上，不同国家和地区的 CDV 分离株之间虽有一定的地域联系，但与 N 蛋白基因相比，其地域性联系并不明显。说明这 3 株 CDV 流行株为野毒株，H 蛋白基因的变异水平明显高于 N 蛋白基因，而 N 蛋白基因更能显示毒株与地域的关系。

关键词：犬瘟热病毒；N 基因；H 基因；序列分析

犬瘟热是由犬瘟热病毒（canine distemper vi-rus，CDV）引起的一种急性、高度接触性传染病，是当前对我国养犬业、毛皮经济动物养殖业和野生动物保护危害最大的疾病之一[1-2]。CDV 核衣壳（N）蛋白由 1572 个核苷酸组成，其开放阅读框起始于第 53~55 位的 A5G 起始密码子，从第 108~110nt 的 A4G 到第 1677~1679nt 的终止信号 4AA 为止，共编码 523 个氨基酸。N 蛋白是 CDV 较为保守的免疫原蛋白，感染时可引起强烈的抗原抗体反应，是麻疹病毒属中主要的交叉抗原。N 蛋白上有 4 细胞表位，在细胞免疫中发挥重要作用。N 蛋白也是主要的毒力相关蛋白，CDV 的毒力在中枢神经系统持续感染中具有重要的作用。CDV 可变区的 N 蛋白分为 3 个区，即可变区 N 末端（17~159aa）、可变区 C 末端（408~519aa）和高度保守区的中间区（160~407aa），中间保守区是 N 蛋白结构和功能的核心部分[3]。血凝素（H）蛋白是 II 类糖蛋白，C 末端（35~55aa）形成了 H 蛋白仅有的疏水锚定区。H 蛋白是诱导机体产生中和抗体的主要蛋白之一，它决定 CDV 宿主的特异性，并协助 F 蛋白使 CDV 以囊膜与宿主细胞膜发生融合的方式进入宿主细胞[4-5]。据国内外的文献报道，不同地区的 CDV 流行株之间以及流行株与疫苗株之间，在主要结构蛋白的基因水平上都有一定的差异，这可能是造成 CDV 毒力增强和免疫失败的原因。因此，本研究采用 RT-PCR 方法特异性扩增 CDV 流行株的 N 和 H 蛋白基因，对 2009 年从甘肃省兰州地区疑似 CDV 感染的病犬体内分离的 3 株 CDV 流行株进行研究，通过分析该地区 CDV 流行株与疫苗株在基因水平上的差异,以及分离的 CDV 流行毒株之间和 CDV 流行毒株与疫苗毒株的关系,为探索该地区 CDV 流行毒株的起源、分子流行病学特征和抗原变异规律提供依据,也为国产 CDV 诊断试剂的研制和疫苗的改进奠定一定的理论基础。

1 材料与方法

1.1 毒株

2009 年 1—3 月从甘肃省宠物医院疑似犬瘟热的病犬体内分离获得的 CDV 流行株,由中国农业科学院兰州畜牧与兽药研究所中兽医(兽医)研究室分离和保存。

1.2 细胞、菌种和载体

Vero 细胞购自解放军第四军医大学实验动物中心,受体菌 E.coli DH5α 购自大连宝生物工程有限公司,pGEM-T Easy 载体为 Promega 公司产品。

1.3 主要试剂

RNAiso™ Plus、One Step RNA RT-PCR 试剂盒(AMV)、DNA 胶回收试剂盒、质粒微量抽提试剂盒、DL2000 DNA Marker 均为大连宝生物工程有限公司产品。

1.4 引物的设计与合成

根据 GenBank 中登录的 CDV 标准株 Onder-stepoort(AF378705)、A75/15(AF164967)、Snyder Hil(G5138403)和 CDV 3(E5726268)的基因序列分别设计了扩增 N、H 蛋白基因的引物。PN1:5′-ACCTACCAATATGGCTAGCCTTC-3′,PN2:5′-GCAGGACTGGTCTTGAATATT-3′,预期扩增片段大小为 1 600bp;PH1:5′-TCAGGTAGTCCAG-CAATGCTC-3′,PH2:5′-GTGTATCATCATACT-GTCAGG-3′,预期扩增 1 800bp 的片段。引物由大连宝生物工程有限公司合成。

1.5 病毒的增殖

将 Vero 细胞置于 37℃ 50mL/L CO_2 条件下培养 3d 长成单层后,吸弃培养基,接种前用 PBS 洗细胞单层 1 次,将分离的病毒接种于单层细胞上,接种后吸附 2h,加维持液。每隔 24h 观察 1 次,当细胞出现 80%~90% 病变时,收获病毒。

1.6 病毒 RNA 的提取

用 RNAiso™ Plus 从细胞培养液中提取病毒 RNA,操作过程按照说明书进行。

1.7 N、H 蛋白基因的扩增

按 One Step RNA PCR 试剂盒(AMV)说明进行。反应体系为 50μL:Primer 1 Step Enzyme Mix 2μL,2×1 Step Bufer 25μL,RNase Free dH_2O 16μL,上、下游引物各 1μL,模板 5μL。扩增 N、H 基因的循环参数相同:50℃ 30min,94℃ 2min;94℃ 1min,52℃ 50s,72℃ 2min,35 个循环;72℃ 7min,4℃ 保存。

1.8 目的基因的克隆与鉴定

按 pGEM-T Easy 载体说明书,将目的基因的 RT-PCR 产物克隆入 pGEM-T Easy 载体中,按常规方法转化 DH5α 感受态细胞,挑选白斑,接种于含有氨苄青霉素(100μg/mL)的 LB 液体培养基中,37℃ 振荡培养过夜,提取质粒并进行酶切鉴定,筛选阳性重组质粒,送北京六合华大基因科技股份有限公司测序。

1.9 序列分析

用 DNAStar 5.0 软件分析本研究中分离的流行毒株的 N、H 蛋白基因序列,并与 GenBank 中登录的其他 CDV 分离毒株的相应序列进行比较,构建遗传进化树,分析甘肃省兰州地区 CDV 流行株的特征。

2 结果

2.1 病毒的增殖

将分离的病毒接种于培养 48~72h 形成单层的 Vero 细胞,一般于接种后第 6~7 天收获病毒。

2.2 N、H 蛋白基因的克隆

从细胞培养物中提取病毒 RNA,通过 RT-PCR 方法扩增 N、H 蛋白基因,结果(见图 1、图 2)显示,分别扩增出了预期大小的 N 蛋白基因(约 1 600bp)和 H 蛋白基因(约 1 800bp)。

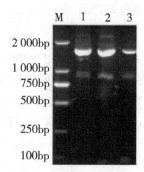

图1　CDV 分离株 N 蛋白基因的 RT-PCR 扩增

M:DNA 分子质量标准;
1~3:分别为 CDV 分离株 GS0904-7、GS0904-8、GS0904-11 的扩增产物

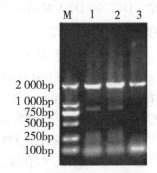

图2　CDV 分离株 H 蛋白基因的 RT-PCR 扩增

M:DNA 分子质量标准;
1~3:分别为 CDV 分离株 GS0904-7、GS0904-8、GS0904-11 的扩增产物

2.3 N 蛋白基因的序列分析

将本试验分离的 3 株 CDV 流行株的 N 蛋白基因序列登录到 GenBank 上,获得的登录号分别为 GS0904-7(HM623890)、GS0904-8(HM623892)和 GS0904-11(HM623894)。将测得的这 3 株病毒的 N 蛋白基因序列与 GenBank 上登录的 CDV 标准株 Onderstepoort(AF378705)、A75/15(AF164967)、Snyder Hil(GU138403)以及其他国家和地区分离株的相应序列用 Blast 和 DNAStar 软件进行同源性分析,并根据推导的氨基酸序列绘制遗传进化树。结果(见表1)显示,N 蛋白基因较为保守,不同地区分离株的 N 蛋白基因核苷酸序列的同源性为 92.5%~99.9%,推定的氨基酸序列的同源性为 95.7%~99.4%。从绘制的遗传进化树(见图3)可见,CDV 分离株之间具有明显的地域性。

表 1　CDV 流行毒株与其他毒株 N 蛋白基因的核苷酸和氨基酸序列的同源性比较

毒株编号 Strain No.	1	2	3	4	5	6	7	8	9	10	11	12	13
1		98.5	97.7	97.7	98.7	97.7	98.9	98.3	98.7	99.0	99.0	99.0	98.1
2	94.5		98.7	98.7	98.1	98.3	97.9	97.7	97.7	98.1	98.1	98.1	97.1
3	94.5	98.0		98.7	97.5	97.7	97.1	96.9	97.3	97.3	97.3	97.3	96.6
4	94.5	97.9	99.3		97.5	97.7	97.1	96.9	96.9	97.3	97.3	97.3	96.6
5	97.1	93.8	93.8	93.8		97.3	98.5	98.7	98.3	98.3	98.3	98.3	97.7
6	93.9	98.5	97.3	97.3	95.7		97.1	96.9	96.9	97.3	97.3	97.3	96.4
7	97.7	93.5	93.5	93.6	95.7	93.1		98.1	99.6	99.2	99.2	99.2	99.0
8	96.7	93.4	93.6	93.6	88.6	92.7	95.9		97.9	97.9	97.9	97.9	97.3
9	97.5	93.3	93.6	93.5	88.2	92.9	99.8	95.7		99.0	99.0	99.0	98.9
10	98.2	94.0	94.0	94.0	86.6	93.6	99.2	96.1	99.0		99.4	99.4	98.5
11	97.4	93.4	93.4	93.4	88.6	93.2	99.1	95.5	98.9	98.9		99.4	98.5
12	98.0	93.8	93.8	93.9	87.8	93.4	99.2	96.0	99.0	99.5	98.8		98.5
13	97.0	92.9	92.7	92.8	95.4	92.4	99.2	95.1	99.0	98.5	98.4	98.5	

注：右上部分为氨基酸序列的同源性，左下部分为核苷酸序列的同源性。下表同

Note：The upper-right is the homology of nucleotide sequences, and the lower-left is the homology of the deduced amino acid sequences in the table.

The same as folows

1：AF154957-A75/15；2：G5138403-Snyder Hil；3：AF305419-Onderstepoort；4：AF378705-Onderstepoort；5：AB474397-Japan；6：HM63009-Kazakhstan；7：DQ522030-Taiwan；8：EU489475-Harbin；9：EU589210-Harbin；10：AY390348-Xinjiang；11：GS0904-11；12：GS0904-8；13：GS0904-7

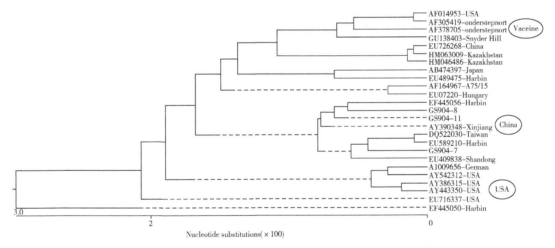

图 3　依据 CDV N 蛋白基因推导的氨基酸序列构建的遗传进化树

2.4　H 蛋白基因的同源性分析

对本试验分离的 3 株 CDV 流行株的 H 蛋白基因测序后，将其序列登录到 GenBank 上，获得的登录号分别为 GS0904-7（HM623891）、GS0904-8（HM623893）和 GS0904-11（HM623895）。将测得的这 3 株病毒的 H 蛋白基因序列与 GenBank 上登录的 CDV 标准株 Onderstepoort（AF378705）、A75/15（AF164967）、Snyder Hil（GU138403）以及分离自其他国家和地区的 CDV 毒株的相应序列进行比较，并绘制遗传进化树。结果（见表 2）显示，不同 CDV 分离株 H 蛋白基因的变异水平明显高于 N 蛋白基因，不同毒株间 H 蛋白基因核苷

酸序列及其推定的氨基酸序列的同源性分别为81.7%~100%和86.4%~99.9%。从绘制的遗传进化树（见图4）可见，不同的CDV分离株之间虽然有一定的地域联系，但与N蛋白相比，其地域性联系并不明显。

表2 CDV兰州流行毒株与其他毒株H蛋白基因的核苷酸和氨基酸序列的同源性比较

毒株编号 Strain No.	1	2	3	4	5	6	7	8	9	10	11	12	13
1		92.9	91.4	92.6	95.9	86.5	97.5	91.8	96.5	95.9	95.4	95.4	95.4
2	92.2		94.2	95.9	91.1	82.5	91.9	95.6	91.6	91.4	91.1	91.3	91.1
3	91.7	96.0		97.7	89.7	82.0	90.4	95.9	89.9	89.7	89.4	89.6	89.4
4	92.7	96.8	98.8		90.6	82.8	91.6	97.5	91.1	90.9	90.6	90.7	90.6
5	95.1	90.7	89.6	90.7		85.8	95.2	90.1	99.3	98.4	98.0	98.2	98.2
6	90.1	86.1	86.0	86.8	88.2		86.0	81.7	86.3	86.0	85.7	85.7	85.7
7	98.2	92.7	91.9	93.0	95.2	90.2		90.8	95.7	95.2	94.7	94.9	94.9
8	92.3	96.8	98.0	90.4	86.3	92.7	90.6		90.4	90.1	90.3	90.1
9	95.4	91.0	89.9	91.0	99.2	88.6	95.7	90.7		99.2	98.7	99.0	99.0
10	94.8	90.7	89.7	90.7	98.6	88.3	95.1	90.5	99.1		97.7	98.7	98.5
11	94.2	90.2	89.3	90.3	87.8	87.8	95.0	90.1	98.4	97.8		97.5	97.7
12	94.5	90.5	89.5	90.5	98.3	87.9	94.8	90.2	98.8	98.1	97.7		98.8
13	94.6	90.5	89.6	90.5	98.4	88.0	95.0	90.2	98.9	99.1	97.7	99.4	

Note: 1: AF154957-A75/15; 2: GU138403-Snyder Hill; 3: AF305419-Onderstepoort; 4: AF378705-Onderstepoort; 5: EF042818-Harbin; 6: EU743935-Harbin; 7: FJ705238-Taiwan; 8: FJ705239-Taiwan; 9: GQ332534-Jilin; 10: GQ332535-Jilin; 11: GS0904-11; 12: GS0904-8; 13: GS0904-7

3 讨论

分子流行病学在追踪CDV暴发的起源地和调查不同的病毒株在易感动物中流行的动力学方面非常有用。CDV被认为只有一种抗原型，但是发现了共流行的具有不同毒力和细胞噬性的CDV基因型，这或许可以解释CDV不同的病毒株致病力的差异及在某些情况下相同的疫苗在不同地区免疫效果的差异和免疫失败的原因。目前，针对CDV的进化研究主要集中在不同毒株的系统进化关系上。由于在研究过程中使用不同基因的全长或者部分片段，以及采用不同的系统发育统计方法，所以对各个研究的结果没有达成共识[6]。

CDV的N蛋白和H蛋白是病毒的两种主要结构蛋白，其中N蛋白是病毒结构中最多的蛋白。据Hamburger等[7]报道，N蛋白与CDV的毒力密切相关，在中枢神经系统的持续性感染中具有重要作用，并在病毒装配、转录和复制过程中起调控作用。H蛋白决定CDV宿主的特异性，并协助F蛋白使CDV的囊膜与宿主细胞膜发生融合，从而使病毒进入宿主细胞。在免疫压力的作用下，CDV抗原表位可能发生漂移，尤其是较大的H蛋白极易发生变异，从而引起CDV毒力的变化，对来自不同地理环境和不同物种的CDV毒株所做的序列分析发现，H蛋白基因根据地理位置具有基因漂移现象，大多数在不同地区的CDV毒株可被

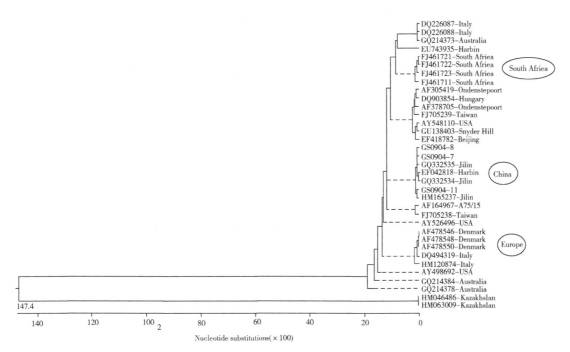

图4 依据CDV H蛋白基因推导的氨基酸序列构建的遗传进化树

归类为7个主要的基因谱系,即美洲1型(大多数为疫苗株)、美洲2型、亚洲1型、亚洲2型、欧洲型、Artic样型及欧洲野生型,尤其是H蛋白基因的第530和549位氨基酸决定了该病毒的宿主嗜性和致病性[6]。在本研究中,笔者对分离的3株病毒的N、H蛋白基因进行了RT-PCR扩增,并与GenBank上登录的标准株Onderstepoort、Snyder Hill、A75/15及其他国家和地区分离毒株的N和H蛋白基因进行了序列比较,绘制了遗传进化树,结果发现犬瘟热的流行具有明显的地域性。根据核苷酸推定的氨基酸序列可见,某一地区的氨基酸突变位点都具有一定的统一性。根据N、H蛋白基因推导的氨基酸序列绘制的遗传树,发现有的野毒株在相同的分支上相距较近,因此在自然界存在野毒株与疫苗株基因重组的可能。

另外,N、H蛋白基因核苷酸序列及其推导的氨基酸序列的同源性分析表明,N蛋白基因比H蛋白基因的保守程度高,由根据N蛋白基因推导的氨基酸绘制的遗传进化树可见,同一地区的CDV分离株都位于同一分支或相近的分支上,而不同地区的分离株则相距较远。源自中国的毒株(山东、哈尔滨、新疆及台湾地区的毒株)、欧洲地区的毒株(意大利和丹麦的毒株)和疫苗株分别处在3支不同的大分支上,而来自美国的毒株处在较远的另一分支上。由H蛋白基因推导的氨基酸序列绘制的遗传进化树可见,不同地区的CDV分离株虽然也有一定的地域性,但没有N蛋白明显,说明H蛋白基因更易发生变异,与之前的研究结果相符[8-14]。本试验结果表明,与H蛋白基因相比,分析N蛋白基因的变异对研究CDV流行的地域性差异和基因分型更具有规律性和可靠性。

参考文献

[1] 蔡宝祥.家畜传染病学[M].4版.北京:中国农业出版社,2001:347-351.
CAI Bao-xiang. *Lemology of Domestic Animals* [M].4th ed.Beijing:China Agricultural Oress,2001:347-351.(in Chinese)

［2］殷震，刘景华.动物病毒学［M］.2版.北京：科学出版社，1997：756-762.
YI.Zhen, LIU Jing-hua. Animal Virology [M].2nd ed.Bei-jing: Science Press, 1997: 756-762. (in Chinese)

［3］孟庆玲，乔军，陈创夫.犬瘟热病毒TN野毒株核蛋白基因的克隆与序列分析［J］.中国兽医科技，2004, 34 (4): 24-28.
ME.Qing-ling, QIAO Jun, CHE.Chuang-fu.Cloning and sequence analysis of nucleocapsid protein gene of canine dis-temper virus wild type strain TN [J]. *Chinese Journal of Vel-erinary Science and Technology*, 2004, 34 (4): 24-28. (in Chinese)

［4］孟庆玲，乔军，陈创夫.犬瘟热病毒TN株血凝基因的克隆与序列分析［J］.微生物学通报，2004, 31 (6): 6-10.
MENG Qing-ling, QIAO Jun, CHE.Chuang-fu.Cloning and sequence analysis of haemagglutinin protein gene of canine distemper virus strain TN [J]. *Microbiology*, 2004, 31 (6): 6-10. (in Chinese)

［5］聂福平，李作生，邱薇，等.犬瘟热病毒Yunnan株H基因的克隆与表达［J］.动物医学进展，2008, 29 (1): 9-12.
NIE Fu-ping, LI Zuo-sheng, QIU Wei, et al.Cloning and ex-pression of the H protein fragment of canine distemper virus Yunnan strain [J]. *Progress in Veterinary Medicine*, 2008, 29 (1): 9-12. (in Chinese)

［6］PRATELLI A.Canine distemper virus: The emergence of new variants [J]. Vet J, 2010Mar 15. (Epub ahead of print)

［7］HAMBURGER D, GRIOT, ZURBRIGGEN.A, et al.Loss of virulence of canine distemper virus is associated with a struc-tural change recognized by a monoclonal antibody [J]. *Expe-rientia*, 1991, 47 (8): 842-845.

［8］SEKULIN K, HAFNER M A, KOLODZIEJEK J, et al.Emer-gence of canine distemper in Bavarian wildlife associated with a specific amino acid exchange in the haemagglutinin protein [J]. *Vet J*, 2010 Jan 19. (Epub ahead of print)

［9］HARDER T C, KENTER M, VOS H, et al.Canine distemper virus from diseases large felids: biological properties and phylogenetic relationships [J]. *J Gen Virol*, 1996, 77 (Pt3): 397-405.

［10］HIRAMA K, GOTO Y, UEMA M, et al.Phylogenetic analy-sis of the hemagglutinin (H) gene of canine distemper viruses isolated from wild masked palm civets (*Paguma larvata*) [J]. *J Vet Med Sci*, 2004, 66 (12): 1 575-1 578.

［11］STETTLER M, ZURBRIGGEN A.ucleotide and deduced amino acid sequences of the nucleocapsid protein of the virulent A75/17-CDV strain of canine distemper virus [J]. *Vet Microbiol*, 1995, 44 (2/4): 211-217.

［12］YOSHIDA E, IWATSUKI K, MIYASHITA N, et al.Mo-lecular analysis of the nucleocapsid protein of recent isolates of canine distemper virus in Japan [J]. *Vet Microbiol*, 1998, 59 (2/3): 237-244.

［13］IWATSUKI K, MIYASHITA N, YOSHIDA E, et al.Molecu-lar and phylogenetic analyses of the haemagglutinin (H) proteins of field isolates of canine distemper virus from naturaly infected dogs [J]. *J Gen Virol*, 1997, 78 (0t 2): 373-380.

［14］CALDERON M G, REMORINI P, PERIOLO O, et al.Detection by RT-PCR and genetic characterization of canine distem-per virus from vaccinated and non-vaccinate dogs in Argentina [J]. *Vet Microbiol*, 2007, 125 (3/4): 341-349.

（发表于《中国兽医科学》）

奶牛乳房炎金黄色葡萄球菌基因多态性分型试验

邓海平，倪春霞，蒲万霞*

(中国农业科学院兰州畜牧与兽药研究所/中国农业科学院新兽药工程重点开放实验室/甘肃省新兽药工程重点实验室，兰州　730050)

摘　要：利用随机引物 AP-7，建立引物随机多态性扩增（RAPD）体系对 71 株引起内蒙古和贵州地区奶牛乳房炎的金黄色葡萄球菌分离株进行基因分型研究。结果表明，71 株金黄色葡萄球菌均得到清晰的 RAPD 指纹图谱，扩增产物为 2~9 条带，产物大小为 240~4 500bp。菌株共分为 6 个基因型，其中 Ⅰ 型 17 株（占 23.9%）、Ⅱ 型 3 株（占 4.2%）、Ⅲ 型 33 株（占 46.5%）、Ⅳ 型 15 株（占 21.1%）、Ⅴ 型 2 株（占 2.8%）、Ⅵ 型 2 株（占 2.8%）。Ⅰ 型为内蒙古自治区的流行优势菌群，Ⅲ 型为贵州地区的流行优势菌群。两地区各基因型菌株比例有明显差异，这可能与奶牛养殖业水平和环境差异有关。

关键词：奶牛乳房炎；金黄色葡萄球菌；RAPD；基因型

　　乳房炎是奶牛的常见多发病，是造成奶牛生产经济损失的重要因素之一，不仅影响奶牛养殖业、乳品工业的发展，也给公共卫生以及食品安全带来一定隐患。奶牛乳房炎主要是由多种非特定的病原微生物引起的，而其中以金黄色葡萄球菌感染为主。近年来有关奶牛乳房炎病原菌分离鉴定的报道中，金黄色葡萄球菌的检出率均在 30% 以上[1-4]。病原菌基因分型研究对于掌握病原菌的流行趋势、疫病的防治和疫苗的研制都非常重要。随着分子生物学的发展，越来越多的基因分型方法被应用于病原菌的分子流行病学研究。1990 年，由 Williams 等基于 PCR 技术首次推出了一种简便、灵敏可行的新的遗传标记技术-随机多态性扩增 DNA（Random amplified polymorphic DNA，RAPD）[5]。RAPD-PCR 技术不但具有简便、快速、无需预先了解被测基因组的相关分子生物学背景等优点，而且还有可用引物数量大，检测区域几乎覆盖整个基因组，多态信息含量大等特点[6]。该技术问世后迅速在国内外被广泛应用于微生物的分子分型与鉴定。近年来有关金黄色葡萄球菌基因分型的研究主要集中于人来源的菌株[7-9]，畜源菌株的基因分型研究相对较少，特别是致奶牛乳房炎金黄色葡萄球菌基因分型的相关研究在国内还罕有报道。本研究利用 RAPD-PCR 技术对我国内蒙古和贵州地区奶牛乳房炎奶样中的金黄色葡萄球菌分离株进行基因分型，探讨不同地区乳房炎金黄色葡萄球菌分子流行病学特点，为我国奶牛乳房炎的防治与基因工程疫苗的研制提供可靠的理论依据。

---

*　Corresponding author. E-mail wanxiapu@yahoo.com.cn

1 材料与方法

1.1 菌株来源

共有分离自临诊型奶牛乳房炎奶样中金黄色葡萄球菌71株，其中内蒙古自治区34株（N1~N34），贵州地区37株（G1~G37）。这些菌株经生化鉴定及分子生物学鉴定符合金黄色葡萄球菌特点。金黄色葡萄球菌标准菌CVCC2246购自中国兽药监察所。

1.2 试剂及仪器

DNA Ladder Marker（相对分子质量依次为4 500、3 000、2 250、1 000、750、500、250bp）、蛋白酶K、溶菌酶、Premix EX Taq酶均购自TaKaRa宝生物工程（大连）有限公司，细菌染色体DNA提取试剂盒（离心柱型），购自TIANGEN生物制品公司。Eppendorf高速冷冻离心机、Biometra基因扩增仪、DYY-8C型电泳仪（北京六一）。

1.3 DNA模板的制备

将菌株活化后接种于1.5mL LB培养液中，37℃培养过夜；取1mL培养物置于1.5mL Eppendorf管中，室温10 000r/min离心2min，弃上清液，沉淀滴加180μL 20g/L的溶菌酶液，吹打混匀后37℃水浴30~60min，加入20μL蛋白酶K，56℃水浴30min。其余步骤按细菌染色体DNA提取试剂盒操作。获得的DNA于-20℃保存备用。

1.4 引物设计与筛选

根据文献[10-11]方法进行引物设计与筛选，最终选择序列AP-7：5'-GTGGATGCGA-3'，作为扩增引物。引物由TaKaRa宝生物工程（大连）有限公司合成。

1.5 RAPD反应条件优化

为了能够产生分辨力较高、重复性较好的条带，在试验前先进行最适扩增条件筛选预试验。将标准菌CVCC2246 DNA按照模板浓度分为原倍、1/2倍、1/10倍3个测试组，每组5个重复；RAPD-PCR退火温度分别选择30~39℃（1℃等差）10种不同温度梯度进行比较。

1.6 RAPD反应体系

根据预试验结果确定最适的反应条件。采用25μL反应体系，体系参数为：Premix混合液12.5μL，DNA模板2μL，引物2μL，去离子水8.5μL。PCR循环参数分别为94℃5min，37℃5min，72℃5min，循环4次；94℃1min，37℃1min，72℃2min，循环30次；72℃延伸10min。

取扩增产物5μL在1.0%琼脂糖凝胶上电泳（8V/cm），电泳缓冲液为1×TAE，30~50min，DNA Ladder Marker同时进行电泳，用Quantity One 1-D凝胶成像系统紫外成像。

1.7 数据分析

用Quantity One 1-D凝胶成像系统得到RAPD-PCR结果电泳图像后，利用BioNumerics软件进行数据处理，具体参数设定标准如下：最适值为0.5%，位置公差为1.25%，阈值为80%。图像采用Dice相关系数和非加权组算术平均法UPGMA（Unweighted pair group method with arithmetic mean）来进行聚类分析。

2 结果

2.1 RAPD反应体系确定

利用标准菌株DNA模板进行预试验后表明，模板浓度为原倍和1/2倍、退火温度在35~39℃时所有菌株均可产生清晰的条带，37℃时条带效果最佳。其他温度时产生的条带不清晰或未产生条带。模板浓度为1/10倍时在设定的退火温度内只有个别菌株可产生条带。

2.2 RAPD扩增结果

通过预试验确定最适反应条件后，对72株金黄色葡萄球菌菌株（包括标准菌）的DNA模板进行随机多态性扩增，产物经凝胶电泳后利用Quantity One 1-D 凝胶成像系统得到电泳图（图1-3）。由图可见，所有菌株染色体DNA经扩增后均产生了清晰、可分辨的带谱，扩增产物在2~9条带，片段大小为240~4 500bp，具有多种带型组成。

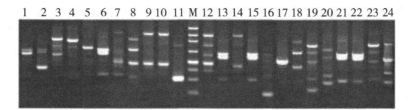

图1 内蒙地区部分金黄色葡萄球菌RAPD结果

Note：1~23. N01~N23 PCR产物；24. CVCC2246；M. DNA Ladder Marker

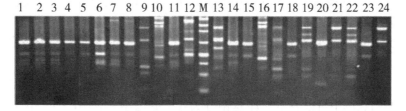

图2 贵州地区部分金黄色葡萄球菌RAPD结果

Note：1~24. G01~G24 PCR产物；M. DNA Ladder Marker

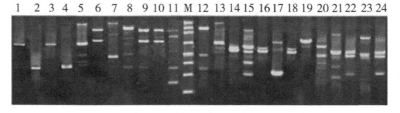

图3 其他部分金黄色葡萄球菌RAPD结果

Note：1~13. G25~G37；14~24. N24~N34；M. DNA Ladder Marker

2.3 RAPD产物的遗传相关性分析

利用Bio-Numerics分析软件对72株金黄色葡萄球菌染色体DNA的RAPD产物电泳图谱进行聚类分析，建立了亲缘关系聚类图（图4）。由图可以看出，除标准菌外，71株临床获得金黄色葡萄球菌根据其遗传差异可分为Ⅰ~Ⅵ6个主要聚类群。其中Ⅰ型17株，占总数23.9%；Ⅱ型3株，占总数4.22%；Ⅲ型33株，占总数46.5%；Ⅳ型15株，占总数21.1%；Ⅴ、Ⅵ型各2株分别占总数的2.8%。Ⅰ型和Ⅳ型内部各分为3个亚型，Ⅲ型内部分为2个亚型。分离自内蒙古地区奶样的34株菌中Ⅰ型12株，Ⅱ型3株，Ⅲ型9株，Ⅳ型8株，Ⅴ型2株。分离自贵州地区奶样的37株菌中，Ⅰ型5株，Ⅲ型24株，Ⅳ型6株，Ⅵ型2株。

3 讨论

金黄色葡萄球菌是引起奶牛乳房炎的主要病原菌之一，由其引起的炎症具有传染性强和难治愈的特点，特别是越来越多的畜源多药耐药MRSA（Methicillin-resistant *Staphylococcus aureu*）菌株的出现，使金黄色葡萄球菌性奶牛乳房炎治疗越发困难，这已经成为困扰世界奶牛养殖业的难题[12-13]。近年来在分子生物学和免疫学方面所取得的一些成果对于研制新一代有效的基因工程疫苗起了很大的推动作用。但在近几年报道的许多有关预防本病的疫苗，都由于缺乏广泛的有效性、生产成本以及某些情况下的不稳定性，而未能广泛应用。因此要研制方便、高效和安全的抗金黄色葡萄球菌感染的疫苗，首先必须要很清楚地了解各地区病原菌株的基因型和相互之间的关系[14]。

RAPD是近年来兴起的一种新的揭示基因多态性的基因分型方法，其特点是不需预先知道有关基因序列，只需采用通用的随机引物，通过DNA扩增，产生基因组指纹图谱进行细菌分型。RAPD的随机性主要取决于选择引物的随机性，RAPD指纹图条带的数目、强度与选择的引物、DNA模板的质量和PCR的反应体系及循环条件等密切相关[15-16]。因此，为了获得明显且清晰易辨的条带，并减少非特异性扩增，筛选合适的引物、优化反应条件和重复试验是很重要的。

本试验对采自内蒙古自治区和贵州地区乳房炎奶样中的71株金黄色葡萄球菌进行了RAPD分型，从RAPD产物指纹图上可以看出所有菌株染色体DNA扩增后均得到了清晰可分辨的条带，条带数在2~9条，产物大小为240~4500bp，清楚地反应了菌株间的基因多态性特征。由聚类分析树状图中可以看出，71株菌分为6个基因型，其中Ⅰ、Ⅲ和Ⅳ型菌株占了总数的90%以上，而Ⅲ型菌株占了总数的近50%，为主要流行的优势菌群。Ⅱ、Ⅴ和Ⅵ型菌株均分别来自同一地区且遗传相似性较高（N20与N27，85.7%；G26与G28，99%；N03与N04，66.7%），可能为两地区流行较少且仍未发生地区间流行的地区特有基因型菌株。

内蒙古自治区34株金黄色葡萄球菌分布在Ⅰ~Ⅴ5个基因型中，除了Ⅱ型和Ⅴ型2个地区特有基因型外，其他各型菌株比例较为平均，Ⅰ型菌为相对的流行优势菌群。12株Ⅰ型菌中有10株为I-2亚型，该型菌株间遗传相似性在70%以上，且均含有1 000bp和1 300bp左右的2条普带，所以这10株菌可能是来自同一亲本的不同克隆株。内蒙古自治区为我国最主要的奶源基地，奶牛养殖业发达，饲养的奶牛种类和数量较多，有多个万头以上大型奶牛场，所以病原微生物在牛场内、牛场之间和地区之间奶牛中的流行传播非常频繁，菌株也更容易发生基因突变。这可能就是内蒙古地区菌株基因型较多且分布较平均的原因。贵州地区37株金黄色葡萄球菌分布在Ⅰ、Ⅲ、Ⅳ和Ⅵ4个基因型，Ⅲ型菌株占总数的64.9%（24/37），为该地区的流行优势菌群。贵州地区主要为山区，奶牛养殖户较少，而且每户饲养的奶牛数量也较少，一般都在300头以下，没有大型的奶牛养殖场。所以奶牛传染病暴发几率小，即使发病，病原菌只会在较小的范围内传播，地区间流行传播的机会较少，因此多数菌株都具有较高的亲缘性。这可能

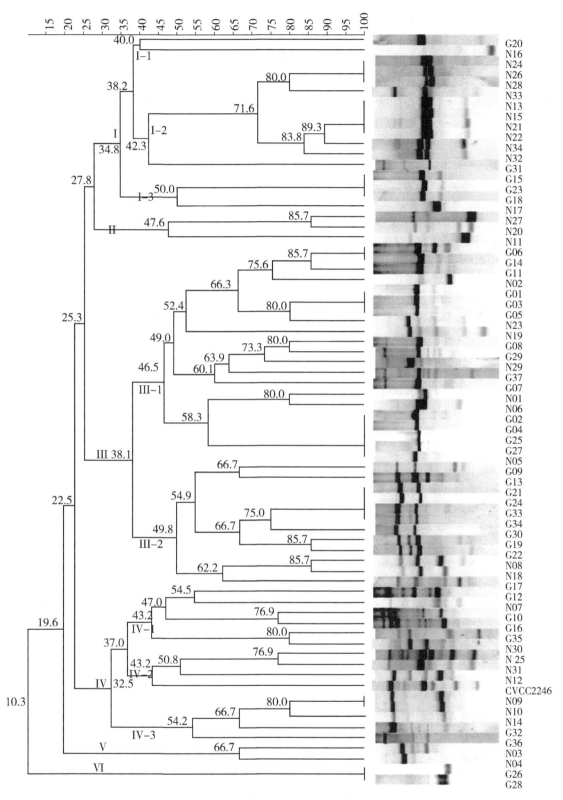

图4 72株金黄色葡萄球菌(包括标准菌CVCC2246)的RAPD扩增产物电泳图谱树状

就是贵州地区菌株多数集中在同一基因型的原因。

本试验首次对我国不同地区奶牛乳房炎金黄色葡萄球菌分离株进行了 RAPD 基因多态性研究，结果证实不同地区菌株基因型组成明显不同，分析可能与地理、气候差异，牛场环境及地区奶牛养殖业现代化水平相关。及时准确地掌握了金黄色葡萄球菌分子流行病学特征，为地区性奶牛乳房炎的防治和基因工程疫苗研制提供了可靠依据。

参考文献

[1] Jurgen S, Barbel K, Wilfried W, et al. The epidemiology of *Staphylococcus aureus* infections from subclinicalmastitis in dairy cows during a control programme [J]. Vet Microbiol, 2003, 96: 91-102.

[2] 陈智, 刘少宁, 牟凯, 等. 金黄色葡萄球菌聚集因子 A 蛋白的原核表达及抗血清活性分析 [J]. 中国兽医学报, 2010, 30 (1): 29-31.

[3] 易明梅, 黄奕倩, 朱建国, 等. 上海地区奶牛乳房炎主要病原菌的分离鉴定及耐药性分析 [J]. 中国兽医学报, 2009, 29 (3): 360-363.

[4] Kerro D O, Tareke F. Bovine mastitis in selected area of southern Ethiopia [J]. Trop Anim Heath Production, 2003, 35: 197-205.

[5] 李大刚. 随机扩增多态性 DNA 技术的原理及应用 [J]. 黑龙江畜牧兽医, 2002, 6: 34-35.

[6] 韦莉萍, 朱冰, 胡琼, 等. 金黄色葡萄球菌随机引物扩增 DNA（RAPD-DNA）多态性分型研究 [J]. 中华医学检验杂志, 2000, 23 (2): 111.

[7] 朱成宾, 高应东. RAPD 在金黄色葡萄球菌分型中的应用研究进展 [J]. 临床输血与检验, 2009, 11 (3): 283-285.

[8] Fitzgerald J R, Monday S R, Foster T J, et al. Characterization of a putative pathogenicity island from bovine *Staphylococcus aureus* encoding multiple superantigens [J]. J Bacteriol, 2001, 183 (1): 63-70.

[9] Martin M C, Fueyo J M, Gonzalez Hevia M A, et al. Genetic procedures for identification of enterotoxigenic strains of *Staphylococcus aureus* from three food poisoning outbreaks [J]. Int J Food Microbiol, 2004, 94 (3): 279-286.

[10] Elina R, Susana B, Cecilia F, et al. RAPD-PCR analysis of *Staphylococcus aureus* strains isolated from bovine and human hosts [J]. Microbiol Res, 2004, 159: 245-255.

[11] Fitzgerald J R, Meaney W J, Hartigani P J. Finestructure molecular epidemiological analysis of *Staphylococcus aureus* recovered from cows [J]. Epidemiol Infect, 1997, 119 (5): 261-269.

[12] Sabine P, Klaus O, Constanze W. Longitudinal study of the molecular epidemiology of methicillin-resistant *Staphylococcus aureus* at a university hospital [J]. J Clin Microbiol, 2006, 44 (12): 4 297-4 302.

[13] 邓海平, 蒲万霞, 俞诗源, 等. 奶牛乳房炎主要病原菌及其分离鉴定方法研究进展 [J]. 中国草食动物, 2009, 29 (2): 52-55.

[14] 乌兰巴特尔, 郝永清, 海岩, 等. 奶牛乳房炎金黄色葡萄球菌的分子流行病学调查 [J]. 中国兽药杂志, 2007, 41 (2): 21-23.

[15] 崔颖鹏, 徐鸿绪, 唐蕾, 等. 耐甲氧西林金黄色葡萄球菌基因多态性分型研究 [J]. 中国热带医学, 2007, 7 (10): 1 753-1 762.

[16] Neela V, Mariana N S, Radu S, et al. Use of RAPD to investigate the epidemiology of *Staphylococcus aureus* infection in Malaysian hospitals [J]. World J Microbiol Biotechnol, 2005, 21: 245-251.

（发表于《中国兽医学报》）

重金属 Pb^{2+} 的抗原合成与鉴定

张志强[1,2]，杨志强[2]，李建喜[2]*，王学智[2]，陈化琦[2]，

张景艳[2]，张凯[2]，孟嘉仁[2]

(1. 甘肃农业大学动物医学院，兰州　730070；2. 中国农业科学院兰州畜牧与兽药研究所，兰州　730050)

摘　要：本研究旨在为重金属铅的 ELISA 检测方法建立提供一种免疫抗原并对其进行初步评估。试验采用络合剂 CHX-A"-DTPA 双功能基团分别特异性结合重金属元素 Pb^{2+} 和蛋白载体 KLH，合成出复合络合物 KLH-CHXA"-DTPA-Pb^{2+} 作免疫抗原，最佳合成条件 pH 值、反应时间、温度分别为 9.0、(22±3) h 和 25℃；用 BCA 试剂盒、紫外扫描、三硝基苯磺酸法和原子吸收法对合成抗原做了初步评估，免疫抗原蛋白含量为 2.66g/L，络合剂与 Pb^{2+} 的偶联度为 77.84%，结构与设计相似；5 次免疫 BALB/c 小鼠后，抗血清可达 128 000 以上。结果提示，重金属 Pb^{2+} 的抗原合成成功，免疫小鼠后特异性好，可用于重金属铅的 ELISA 检测方法研究。

关键词：Pb^{2+}；抗原；小鼠

随着人们生活水平的不断提高和环保意识的增强，对重金属污染尤为关注，残留检测急需加强[1]。近年来，重金属污染导致中毒事件报道较多，铅尤为突出。机体长时间摄入过量的铅会导致免疫机能下降，诱发多种疾病，儿童铅中毒会严重影响智商发育，有导致生殖系统发育不良的报道[2]。目前，重金属铅的检测方法有原子吸收光谱、ICP/ACS、ICP/MS、FIAAS 或电化学方法等[3]，原子吸收法最为常用。但这些方法均需依靠大型设备，有费用昂贵、样品前处理繁琐、检测时间长等缺陷。ELISA 法具有检测速度快、费用低廉、灵敏度高等优点，在残留检测应用报道较多[4]。本试验利用螯合剂螯合 Pb^{2+}、偶联载体蛋白的特性，在体外合成 Pb^{2+} 免疫抗原，并对其进行鉴定和评估，旨在为重金属 Pb^{2+} 的 ELISA 检测方法建立提供一种高效抗原。

1　材料与方法

1.1　主要试剂与仪器

光谱纯硝酸铅（Pb（NO_3）$_2$），天津市光复精细化工研究所；螯合剂 CHX-A-DTPA，美国 Macrocyclics 公司；钥孔戚蛋白（KLH）、胎牛血清白蛋白（BSA）、弗氏完全佐剂、弗氏不完全佐剂、三乙胺、三硝基苯磺酸，均购自 Sigma 公司；山羊抗小鼠 IgG-HRP，中山金

---

*　Corresponding author. E-mail：lzjianxil@163.com

桥公司；其他化学试剂均为分析纯。

Molecular Devices 公司 SpectraMax M2 多功能酶标仪；Beckman 水平离心机；赛多利斯 ME235S 微量分析天平；BS224S 分析天平；岛津 UV-2501 紫外-可见分光光度计；意大利 HANNA 酸度计；上海精宏 HZP-250 型全温振荡培养箱；Eppendorf 十二道可调移液器（30~300μL）等。

1.2 实验动物

6 周龄 BALB/c 雌性小鼠，体质量 18~20g，购自兰州大学医学院实验动物中心。

1.3 抗原合成

免疫抗原和检测抗原按参考文献 [5-6] 制备，并略加改进。Pb^{2+} 来源于 $Pb(NO_3)_2$，CHX-A″-DTPA（简称 DTPA）为双位点偶联剂，载体蛋白为 KLH、BSA。按比例和顺序加入并充分混匀，25℃恒温反应 22h。将其取出放至超滤管中，用 pH 9.0 的 HEPES 缓冲液离心洗去未反应完全的小分子物质，再用反应缓冲液将体积稀释至 3mL。

1.4 抗原鉴定

1.4.1 抗原蛋白水平检测

采用 BCA 蛋白浓度测定试剂盒法，将样品蛋白的浓度调至试剂盒的蛋白浓度检测范围内，每个样品做 3 个平行。根据标准曲线计算出样品蛋白浓度。

1.4.2 完全抗原紫外检测

采用紫外分光光度扫描法[7-8]，在 200~400nm 范围内分别扫描 BSA，KLH，DTPA，免疫抗原，检测抗原。

1.4.3 游离氨基酸测定

采用三硝基苯磺酸法[9]，在波长 335nm 处检测吸光度，绘制标准曲线，按公式计算抗原的游离氨基酸百分比，并判断抗原蛋白交联度。游离氨基酸百分比 = （△A÷A1）×100%，其中 A1 为标准蛋白与对照的差值，A2 为待测蛋白与对照的差值，△A 为 A1 与 A2 的差值。

1.4.4 抗原 Pb^{2+}

含量测定 采用石墨炉原子吸收法[10]测定抗原中 Pb^{2+} 的含量。由于 Pb 属于小分子物质，未结合的 Pb^{2+} 经过过滤被除去，而采用石墨炉原子吸收法测定的 Pb^{2+} 含量为抗原中结合上的 Pb^{2+}，可以直接反映出偶联比。

1.5 小鼠免疫

将 1.0g/L 的免疫抗原与等体积弗氏完全佐剂混合，注射器双推法充分混匀，使其形成白色油包水乳状物，得到抗原乳化剂。免疫 6 周龄雌性 BALB/c 小鼠，共免疫 5 次，每隔 2 周免疫 1 次；剂量为 200μg/只，每组 6 只；在腹腔、背部、皮下、颈部多点免疫，并用苦味酸将每组小鼠编号。第 2、3、4 次免疫时，抗原与等体积弗氏不完全佐剂混合，以同样剂量加强免疫，第 5 次免疫时，抗原不加佐剂经腹腔注射进行加强免疫[11-15]。

在第 2 次免疫 7d 后，断尾采集小鼠血液，置于灭菌管中，室温放置 30min；37℃恒温箱中放置 1h，而后取出，4℃冰箱中过夜；血液 3 000r/min 离心 30min，收集上清于灭菌管中。

1.6 抗血清的评价

1.6.1 间接 ELISA 法

根据参考文献 [16-17]，按照间接 ELISA 法的步骤，筛选最佳包被量和酶浓度，检测抗血清的效价，建立间接竞争 ELISA 法。

1.6.2 抗血清特异性的检测

抗 Pb^{2+} 的特异性：取同时偶联的 Pb-DTPA-BSA 和 DTPA-BSA，均以 5μg/孔包板，同时检测 6 只小鼠的血清，按 1.6.1 的 ELISA 间接法步骤进行。

与 BSA、KLH 的交叉反应性：BSA、KLH 以 5μg/孔包板，小鼠血清倍比稀释，按 1.6.1 的 ELISA 间接法步骤进行。

2 结果

2.1 抗原蛋白浓度

将标准蛋白 BSA 和 KLH 分别用缓冲液配制成 0、0.025、0.05、0.1、0.2、0.3、0.4、0.5g/L 的标准溶液，按照 BCA 蛋白浓度测定试剂盒所示步骤，以波长为 562nm 的 D 值为 y 轴，浓度为 x 轴，制作标准曲线。标准曲线如图 1、2 所示，经计算检测抗原的质量浓度为 2.79g/L，免疫抗原的质量浓度为 2.66g/L。

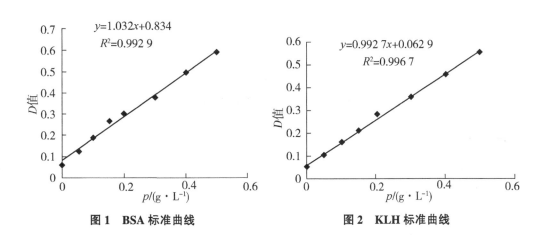

图 1 BSA 标准曲线　　　　图 2 KLH 标准曲线

2.2 抗原结构特性紫外分析

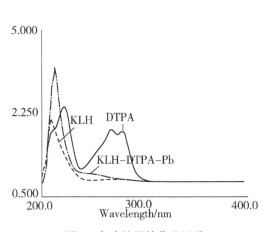

图 3 免疫抗原的紫外图谱

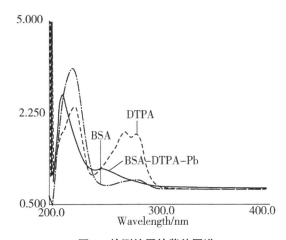

图 4 检测抗原的紫外图谱

由图 3、4 可以看出，当 DTPA 分别与 KLH、BSA 偶联并络合 Pb^{2+} 后，其紫外吸收图谱与载体蛋白 KLH、BSA 相比，吸收曲线发生了改变，在接近 280nm 处有最大吸收峰，且最大吸收峰位置吸光值有所增加，说明抗原合成成功。

2.3 抗原游离氨基酸

将标准蛋白 BSA 用反应缓冲液溶解成 0.02、0.05、0.10、0.15、0.20g/L 的标准溶液，然后按照三硝基苯磺酸法的步骤，在波长 340nm 处读取值，以 D 值为 y 轴，浓度为 x 轴，绘制标准曲线，标准曲线如图 5 所示。按照公式可计算出游离氨基酸百分比为 39.9%。

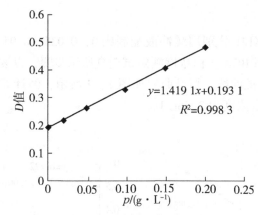

图 5　BSA 游离氨基标准曲线

2.4 抗原中 Pb^{2+}

含量的测定　通过石墨炉原子吸收法检测，检测抗原中 Pb^{2+} 的质量浓度为 115.2mg/L；通过与加入的 Pb^{2+} 相比，偶合率达到 77.84%，说明抗原中螯合上了 77.84% 的 Pb^{2+}，只有 22.16% 的 Pb^{2+} 未被螯合，通过过滤除去了；直接证明了抗原合成是成功的。

2.5 抗血清的效价检测

通过间接竞争 ELISA 法检测抗血清的效价，第 5 次免疫时效价最高，可达到 128 000（图 6）

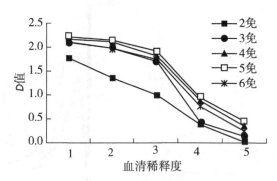

图 6　抗血清效价检测
1.1∶4 000；2.1∶8 000；3.1∶16000；4.1∶64 000；5.1∶128 000

2.6 抗原特异性检测

2.6.1 抗 Pb-DTPA 的特异性

取同时偶联的 Pb-DTPA-BSA 和 DTPA-BSA，均以 5μg/孔包板，同时检测 4 只小鼠的血清，按 ELISA 间接法步骤进行。前者的 D 值在不同稀释度都大于后者，说明产生了抗 Pb-DTPA 的抗体（图7）。

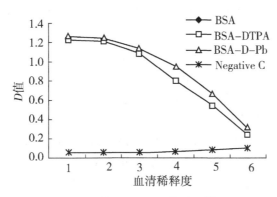

图7 抗 Pb-DTPA 的特异性

1. 1∶2 000；2. 1∶4 000；3. 1∶8 000；4. 1∶16 000；5. 1∶32 000；6. 1∶64 000

2.6.2 与 BSA、KLH 的交叉反应性

以 BSA-DTPA-Pb 为参照，由图8可以看出，当包被 BSA 时，小鼠抗血清 D_{450nm} 值与阴性血清相比无差别，而当包被 KLH、BSA-DTPA-Pb 时，小鼠抗血清为 D_{450nm} 值与阴性血清相比，有明显的变化；说明抗体与 BSA 无交叉反应。

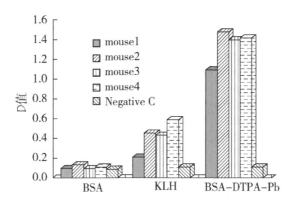

图8 抗原与 BSA 的交叉反应

3 讨论

重金属离子属小分子物质，如铅离子相对分子质量仅为 207.2，不足以形成一个抗原表位，而重金属有与络合剂结合的特性。资料报道，一些络合剂有双功能基团，既能络合金属离子又能与蛋白氨基端结合，这给重金属免疫检测方法的建立提供了可行途径。本试验先将重金属离子铅与螯合剂 DTPA 螯合后，使其形成一种螯合有铅离子，从而形成具有特异性结

构的复合物，该物质具有半抗原特征。如果再将该螯合物与载体蛋白 BSA 或 KLH 连接后，就可形成具有免疫原性的完全抗原，通过结构和其他特性的鉴定后就可免疫小鼠制备血清。鉴定结果显示，无论是免疫抗原还是检测抗原，蛋白质量浓度分别达 2.66g/L 和 2.79g/L，是加入蛋白量的 53.2% 和 55.8%，足可引进机体产生免疫反应。紫外检测结果表明，BSA、KLH、DTPA-Pb、BSA-DTPA-Pb 和 KLH-DTPA-Pb 的紫外典线和特征吸收峰既相似又有异同，这表明抗原的结构特征与设计结构的相似度较高，复合物中具有 DTPA-Pb^{2+} 基团，该方法可用于重金属免疫与检测抗原的合成。抗原免疫动物后能否产生特异性的抗血清，还与这种络合复合物中金属离子的络合比有很大关系，所以抗原合成后需对其进行偶联比测定。据资料报道，目前可用于偶联比测定的方法有游离氨基酸测定、金属离子测定和蛋白电泳等[10]。本试验中采用了前 2 种方法，即用三销基本磺酸法和原子吸收光谱法，估测了合成抗原的金属偶联比，结果合成的检测抗原和免疫抗原的偶联度分别为 39.9%、41.56%，免疫动物后有产生特异性抗血清的可能。

利用合成的免疫抗原 KLH-DTPA-Pb 免疫 BALB/c 小鼠，每隔 2 周免疫 1 次，免疫 1 周后尾静脉采血并分离血清，并用试验中建立的间接 ELISA 法检测抗血清滴度。结果显示，抗原的最佳包被量为 5mg/L，最佳酶滴度为 1∶10 000，建立的间接 ELISA 法检测抗血清滴度灵敏度最优；最佳免疫剂量为 200μg/只，最佳免疫次数为 5 次，第 1 次免疫时加完全佐剂，第 2~4 次时免疫加不完全佐剂，第 5 次免疫不加佐剂时，其免疫效果最好。用建立的 ELISA 方法检测抗血清，效价在第 5 次免疫后最高，可达 1∶128 000。特异性检测结果显示，小鼠分别针对 KLH、KLH-DTPA、KLH-DTPA-Pb 均产生了抗体，但抗 KLH-DTPA-Pb 的抗体水平相对较高，却与 KLH-DTPA 间差异不够显著。该结果提示，免疫小鼠后体内成功产生了针对 Pb-DTPA 表位的特异性抗体，虽达到了预期试验目的，可用于重金属铅的 ELISA 检测方法研究，但要制备单克隆抗体，抗原偶联比还需改善，如能对特异性抗原进行纯化，免疫效果会更理想。

参考文献

[1] 王宏镔，束文圣，蓝崇钰. 重金属污染生态学研究现状与展望 [J]. 生态学报，2005，25（3）：596-605.

[2] 覃志英，黄兆勇，陈广林，等. 食品重金属污染的研究进展 [J]. 广西预防医学，2003，9（增刊）：5-8.

[3] 姜天久，牛涛. 重金属污染物的免疫学检测技术研究进展 [J]. 生态环境，2005，14（4）：590-595.

[4] Blake D A, Jones R M, Blake Ⅱ R C. Antibody-based sensors for heavy metal ions [J]. Biosensors Bioelectronics, 2001, 16: 799-809.

[5] 刘功良，王菊芳，李志勇，等. 重金属离子的免疫检测研究进展 [J]. 生物工程学报，2006，22（6）：878-881.

[6] 袁晓科，杨志强，李建喜，等. 重金属镉离子免疫抗原初步制备与鉴定 [J]. 中兽医医药杂志，2008，27（4）：19-21.

[7] 朱曙东，赵文军，张光毅. 酪氨酸蛋白专一抗体的制备与纯化 [J]. 生物化学杂志，1997，13（5）：561-565.

[8] 石麟义，阎凤周. 动物生物化学 [M]. 长春：吉林科学技术出版社，1990.

[9] Sharpe R M. Environmental oestrogens and male infertility [J]. Pure Appl Chem, 1998, 70（9）：

1 685-1 701.

[10] 刘绮丽,农炫明.石墨炉原子吸收法测定明目地黄胶囊中重金属镉的含量[J].海峡药学,2008,20(4):24 242-24 245.

[11] Stanker L H, Muldoon M T, Burkley S A, et al. Development of a monoclonal antibody-based immunoassy to detect furosemide in cow's milk [J]. J Agric Food Chem, 1996, 44: 2 455-2 459.

[12] Abad A, Pimo J, Montoya A. Development of an enzymelinked immunosorbent assay to carbaryl. antibody production from several haptens and characterization in different immunoassy formals [J]. J AgricFood Chem, 1997, 45: 1 486-1 494.

[13] Jones R M, Yu H N, Delehanty J B, et al. Monoclonal antibodies that recognize minimal differences in the three-dimensional structures of metal-chelate complexes [J]. J Bioconjuage Chem, 2002, 13: 408-415.

[14] 蔡勤仁,曾振灵.杨桂香,等.恩诺沙星单克隆抗体的制备及鉴定[J].中国农业科学,2004,37(7):1 060-1 064.

[15] 何方洋,邱阳生,杨根海,等.克伦特罗单克隆抗体的制备与鉴定[J].中国兽医科技,2001,31(6):30 232.

[16] 赵永芳.生物化学技术原理及应用[M].3版.北京:科学出版社,2002.

[17] 石德时,周斌,覃雅丽,等.氯霉素间接竞争ELISA(ciELISA)检测方法的建立[J].中国兽医学报,2002,22(1):77-78.

(发表于《中国兽医学报》)

六茜素新制剂的急性毒性试验

王学红，梁剑平，刘宇，郭志廷，郭文柱，冯艳丽

（中国农业科学院兰州畜牧与兽药研究所/农业部兽用药物创制重点实验室/甘肃省新兽药工程重点实验室，兰州 730050）

摘 要：为了测定六茜素新制剂的急性毒性，在预试验的基础上，将60只昆明系清洁级小鼠随机分成6个组，IZl服剂量分别为400、480、576、691和829.4 mg/kg，进行了为期1周的六茜素新制剂的急性毒性试验。结果显示，六茜素的LD50为（666.45±0.02）mg/kg。95%置信区间为602.韶—736.84 me/kg。据兽药规范中毒理学评价标准可视为低毒药物。

关键词：六茜素；新制剂；急性毒性试验

（发表于《畜牧与兽医》）

塞拉菌素口服与透皮给药对小鼠和大鼠的急性毒性试验

汪芳，周绪正，李冰，李金善，李剑勇，
牛建荣，魏小娟，杨亚军，张继瑜

（中国农业科学院兰州畜牧与兽药研究所/农业部兽用药物创制重点实验室/甘肃省新兽药工程重点实验室，兰州 730050）

摘 要：为阐明塞拉菌素的毒性，指导临床用药，对其原料药和制剂进行了口服和经皮2个途径给药对小鼠和大鼠的急性毒性试验。预试验中小鼠和大鼠一次性口服不同剂量的塞拉菌素原料药和制剂，测定LD_0和LD_{100}，再根据简化寇氏法分组正式试验，测定LD_{50}和95%可信限。结果测得小鼠口服塞拉菌素原料药的LD_0、LD_{100}、LD_{50}和95%可信限分别为4g/kg、12.6 g/kg、8.72g/kg和7.73~9.82g/kg；大鼠口服塞拉菌素原料药的LD_0、LD_{100}、LD_{50}和95%可信限分别为1.6g/kg、5g/kg、2.51g/kg和2.24~2.82 g/kg；小鼠口服塞拉菌素制剂的LD_0、LD_{100}、LD_{50}和95%可信限分别为4.2g/kg、8.4 g/kg、7.71g/kg和7.26~8.19g/kg；大鼠口服塞拉菌素制剂的LD_0、LD_{100}、LD_{50}和95%可信限分别为3.4g/kg、8.4 g/kg、6.62g/kg和6.10~7.53g/kg。由于LD_{50}值均在501~15 000mg/kg，故塞拉菌素原

料及制剂判断为低毒或实际无毒物质。小鼠和大鼠分别透皮给予高剂量塞拉菌素制剂 10 g/kg，未见毒副反应，表明临床用药是安全的。

关键词：塞拉菌素；原料药；制剂；急性毒性；LD_{50}

（发表于《畜牧与兽医》）

特种经济动物纤维超微结构的研究

李维红[1,2]，郭天芬[1,2]，席　斌[1,2]，胡正艳[3]

（1. 中国农业科学院兰州畜牧与兽药研究所，兰州　730050；2. 农业部动物毛皮及制品质量监督检验测试中心，兰州　730050；3. 兰州市畜牧兽医研究所，兰州　730020）

摘　要：以银狐、蓝狐、南貉、美洲貉、紫貂、水貂为试验材料，观察特种经济动物狐、貉、貂毛绒纤维的超微结构，利用扫描电镜法比较其鳞片层结构特征。结果显示，银狐针毛翘角平均值为 34.6°；鳞片高度平均值为 10.88μm，鳞片厚度平均值为 0.48μm，银狐绒毛翘角平均值为 25.3°，鳞片高度平均值为 11.59μm，鳞片厚度平均值为 0.51μm；蓝狐针毛翘角平均值为 33.1°，鳞片高度平均值为 15.09μm，鳞片厚度平均值为 0.63μn，蓝狐绒毛翘角平均值为 25.0°，鳞片高度平均值为 9.80μm，鳞片厚度平均值为 0.55μm；南貉针毛翘角平均值为 35.1°，鳞片高度平均值为 14.54μm，鳞片厚度平均值为 0.67μm，南貉绒毛翘角平均值为 32.70°，鳞片高度平均值为 16.41μm，鳞片厚度平均值为 0.65μ脚；美洲貉针毛鳞片高度平均值为 6.05μm，鳞片厚度平均值为 0.26μm，美洲貉绒毛翘角平均值为 25.5°，鳞片高度平均值为 13.04μm，鳞片厚度平均值为 0.5μm；紫貂针毛鳞片高度平均值为 10.18μm，鳞片厚度平均值为 0.54μm，紫貂绒毛鳞片高度平均值为 9.24μm，鳞片厚度平均值为 0.52μm；水貂针毛鳞片高度平均值为 11.50μm，鳞片厚度平均值为 0.60μm，水貂绒毛鳞片高度平均值为 22.33μm，鳞片厚度平均值为 0.59μm。不同种的动物纤维具有独特的形态特征，在超微结构上存在明显的差别。不同种的动物纤维，其鳞片翘角、鳞片高度、鳞片厚度差异显著（$P<0.05$）。同一种动物纤维其针毛与绒毛的鳞片翘角、鳞片高度、鳞片厚度差异显著（$P<0.05$）。

关键词：特种经济动物；纤维；扫描电镜；超微结构

（发表于《畜牧与兽医》）

猪戊型肝炎病毒甘肃分离株衣壳蛋白基因序列分析

郝宝成[1,2]，梁剑平[1,3]，兰 喜[2]，刑小勇[3]，
项海涛[3]，温峰琴[3]，胡永浩[3]，柳纪省[2,3]

(1. 中国农业科学院兰州畜牧与兽药研究所/农业部兽用药物创制重点实验室/甘肃省新兽药工程重点实验室，兰州 730050；2. 中国农业科学院兰州兽医研究所，兰州 730046；3. 甘肃农业大学动物医学院，兰州 730070)

摘 要：为进行国内猪戊型肝炎病毒（HEV）衣壳蛋白基因特征研究，参照 GenBank 中已发表的戊型肝炎病毒核苷酸序列，设计了 1 对扩增 HEV 衣壳蛋白（Y）基因的引物，利用 RT-PCR 等方法成功克隆出 HEV 甘肃分离株 swCH189 的 Y 基因 cDNA 片段，并与国内其他 9 株典型 HEV 分离株进行序列分析，用 DNAstar 软件进行序列同源性分析，PHYLIP 软件绘制基因进化树。结果表明，swCH189 株的 Y 基因与国内 9 株典型 HEV 分离株间的核苷酸序列同源性为 79.6%～91.8%，推导的氨基酸序列同源性为 91.4%～98.85%。基因进化树显示 swCH189 与基因Ⅳ型 swCH25 株亲缘关系最近，在同一分支上，属于基因Ⅳ型。

关键词：猪戊型肝炎病毒；衣壳蛋白基因；序列分析

(发表于《畜牧与兽医》)

甘肃青海藏系绵羊微卫星遗传多样性研究

郎 侠

(中国农业科学院兰州畜牧与兽药研究所，兰州 730050)

摘 要：藏系绵羊又称西藏羊、藏绵羊、藏羊，是我国三大粗毛绵羊品种之一，是在青藏高原条件下经过长期的自然和人工选育而成的古老绵羊品种，具有耐寒、耐粗放饲养管理条件、抗病力强等优良种质特性，分布于青藏高原及其毗邻的川、滇、甘等高寒地区，是具有青藏高原特色的家畜品种资源，主要有高原型（草地型）和山谷型两大类型，总数达 3 000 多万只，是青藏高原农牧民主要的生产和生活资料。藏系绵羊是优良的地方品种，是培育绵羊新品种的重要育种素材，有着巨大的开发潜力和广阔的利用前景。但由于藏系绵羊品种资源的保护受到资金、自然条件、效益等多方面的限制，保护利用力度不够，品种退化严重。本研究旨在运用微卫星标记技术，评估甘肃、青海省的

6个藏系绵羊品种（类群）的遗传多样性，为制定藏系绵羊资源的科学保护和合理利用策略提供理论依据。

关键词：藏系绵羊；微卫星；遗传多样性

材料与方法

采用典型群随机抽样法，分别从岷县黑裘皮羊、欧拉型藏羊、乔科型藏羊、甘加型藏羊、青海贵德黑裘皮羊、青海高原型藏羊和滩羊（对照品种）的中心产区采集血样386份（每个品种随机采取不少于30个个体，公畜不少于10%，两代或三代内没有血缘关系，品种的外貌特征明显），EDTA抗凝，－20℃冻存备用。用世界粮农组织（FAO）推荐的引物，分别是BM6506、BM6526、OarFCB48、OarFCBl28、OarHH35、OarVH72、MAF70、MCM38，进行PCR扩增，银染检测。用GelQuant软件计算目的片段大小，Excel Microsatellite Toolkit软件计算等位基因频率、有效等位基因数、群体杂合度和多态信息含量。利用Dispan软件计算群体间DA遗传距离和聚类分析。

结果

7个绵羊品种/群体在8个微卫星座位中共检测出103个等位基因。平均每个位点检测到等位基因数目为12.9个，数目最多的位点是OarFCBl28为17个，片段范围为107~153bp；最少的位点是MAF70为8个，片段范围为146~172bp。多态信息含量和群体杂合度分别为0.851~0.878，0.876~0.898。7个绵羊品种的Nei氏遗传距离介于0.110 3~0.242 7，Nei氏标准遗传距离介于0.137 5~0.323 2。岷县黑裘皮羊和贵德黑裘皮羊的遗传距离最小（DA＝0.110 3，Ds＝0.145 8），青海高原型藏羊和滩羊遗传距离最远（DA＝0.242 7，Ds＝0.305 6）。7个绵羊群体聚为三大类：青海高原型藏羊、乔科型藏羊、欧拉型藏羊和甘加型藏羊聚为一类；岷县黑裘皮羊、贵德黑裘皮羊聚为一类；滩羊。

讨论与结论

遗传多样性也称基因多样性，是指动物种（群）内不同个体间遗传变异的总和，是生物多样性的重要组成部分和最本质的体现。遗传杂合度通常认为是度量群体遗传变异的最适参数，杂合度越高，群体遗传变异就越丰富，可选择的范围就广，若群体杂合度高于0.5.表明该群体没有受到高强度的选择，拥有丰富的多样性。反之，表明该群体遗传多样性较低。本研究中7个绵羊品种的群体平均杂合度在0.869 2~0.892 1，说明7个绵羊群体在所选座位具有丰富的遗传变异。微卫星的多态性水平常用多态信息含量（PIC）来表示，Botstein等提出：当PIC>0.5时，为高度多态性位点；当0.25<PLC<0.5时，为中度多态性位点；当PIC<0.25时，为低度多态性位点。本研究中8个微卫星位点的平均多态信息含量在0.851~0.878，均属高度多态位点，可作为有效的遗传标记用于绵羊品种间遗传多样性分析。遗传距离主要反映品种间亲缘关系的远近，通过精确估测品种间的遗传距离和聚类分析，研究结果符合品种的地理分布、育成史及文献记录。

（发表于《动物遗传育种研究最新进展（第十六次全国动物遗传育种学术讨论会）》）

牦牛 DRB3.2 基因 PCR-RFLP 研究

包鹏甲，阎 萍，梁春年，郭 宪，丁学智，裴 杰，褚 敏，朱新书

(中国农业科学院兰州畜牧与兽药研究所，兰州 730050)

摘 要：MHC B 基目组中多态性最丰富的基因之一，编码细胞表面糖蛋白，主要参与机体的免疫应答调控、抗原识别及诱导免疫反应，在脊椎动物的免疫系统中起着十分关键的作用。BoLA-DRB 是牛 MHC 基因家旗中的 II 类基因，编码的 MHC 抗原与免疫应激和抗病性密切相关。牦牛（Bos grunniens）是生活在青藏高原的特有牛种，能很好的适应当地高寒高海拔的恶劣环境，提供肉、乳、皮、毛等产品，是当地重要生产资料和生活资料，国内研究已证实牦牛对疾病有较高的抗性，但对该基因的研究报道较少。

关键词：牦牛；DRB3.2 基因；PCR-RFLP

(发表于《动物遗传育种研究最新进展——第十六次全国动物遗传育种学术讨论会论文集》)

骡肠阻塞时舌色舌苔的分析

严作廷[1]，巩忠福[1]，谢家声[1]，杨锐乐[1]，黄虎祖[2]，李绪权[2]，尚清彦[2]

(1. 中国农业科学院兰州畜牧与兽药研究所/中国农业科学院临床兽医学研究中心，兰州 730050；2. 甘肃农业大学动物医学院)

摘 要：通过对 265 例肠阻塞骡不同病情不同部位的口腔变化、舌色变化和舌苔变化的分析。发现：①口腔的变化在轻症时以干燥为主，占 57.17%；重症和危症时以甘臭和恶臭为主，分别占 52.75% 和 66.67%；②舌色的变化在轻症时以淡红为主，占 77.08%；重症和危症以青紫和暗红为主。分别占 32.96% 和 66.67%。其次为红舌；③舌苔的变化方面，轻症以薄白苔为主，占 77.08%；黄苔占 19.79%。重症和危症时以黄苔为主，分别占 64.84% 和 85.9%；薄白苔分别占 28.57% 和 2.56%；④经 X^2 检验口腔、舌色和舌苔的变化在轻、重、危之间差异极显著（$P<0.001$）。而在不同部位肠阻塞之间口腔、舌色和舌苔的变化无显著差异（$P>0.05$）。

关键词：舌色；舌苔；骡；肠阻塞

(发表于《甘肃畜牧兽医》)

黑曲霉高产纤维素酶菌株的高通量筛选

李兆周[1]，王学红[1]，陈化靓[2]，梁剑平[1]

（1. 中国农业科学院兰州畜牧与兽药研究所，农业部兽用药物创制重点实验室，甘肃省/中国农业科学院新兽药工程重点实验室，兰州 730050；2. 兰州理工大学生命科学与工程学院，兰州 730050）

摘　要：以黑曲霉 GSICC 60108 为出发菌株，经紫外和电子束诱变处理后，采用纤维素-刚果红平板初筛和新建立的高通量微孔板法进行复筛以获得高产纤维素酶菌株。结果表明，试验得到 1 株稳定高产的纤维素酶菌株黑曲霉 EBR-106；所建立的复筛方法的标准曲线在葡萄糖质量浓度为 0~10mg/mL 的范围内线性良好（$r=0.99975$），孔间和板间酶活力值的变异系数均在 5%以下，重现性和稳定性均能够满足高通量复筛的需要；经过对所得菌株固态发酵条件优化后发现，当碳源为稻草粉和麸皮的混合物，氮源为 2.5%硫酸铵添加量为 1.5%时的玉米浆作为生长促进剂，加水量为 35mL/（100g）的培养基产酶效果最好，在最佳培养条件下，其分泌纤维素滤纸酶活力为 378.11 FPU/mL。

关键词：黑曲霉；纤维素酶；纤维素-刚果红平板；微孔板筛选法

（发表于《甘肃农业大学学报》）

奶牛乳房炎 3 种主要病原菌血清学交叉免疫试验研究

郁　杰[1]，李宏胜[2]，罗金印[2]，李新圃[2]，徐继英[2]，张礼华

（1. 江苏畜牧兽医职业技术学院，泰州 225300；2. 中国农业科学院兰州畜牧与兽药研究所，兰州 730050）

摘　要：对我国部分地区奶牛场临床型乳房炎病乳中分离鉴定出的 24 株无乳链球菌、7 株停乳链球菌和 10 株金黄色葡萄球菌，进行 3 种菌株之间及同种异地各菌株间血清学交叉反应试验。结果表明，无乳链球菌、停乳链球菌和金黄色葡萄球菌 3 种菌株之间无免疫交叉反应，但我国各地同种异地间各菌株间存在不同程度的血清学交叉反应，而且菌株之间交叉反应程度有明显差异，表明该方法可作为制茁菌株抗原性筛选的一项重要指标。

关键词：无乳链球菌；停乳链球菌；金黄色葡萄球菌；血清学交叉反应

（发表于《甘肃农业大学学报》）

应用响应面法优化金丝桃素的纯化工艺

胡小艳[1,2]，刘宇[1]，王学红[1]，郭凯[1,2]，
华兰英[1]，石广亮[3]，梁剑平[1]

(1. 中国农业科学院兰州畜牧与兽药研究所新兽药工程重点实验室，兰州 730050；2. 中国农业科学院研究生院，北京 100081；3. 甘肃农业大学动物医学院，兰州 730070)

摘 要：在单因素试验的基础上，根据中心组合试验设计原理，选择上柱液中金丝桃素的浓度、洗脱液中乙醇的浓度和洗脱液用量为考察因素，以产物中金丝桃素含量为响应值。采用3因素5水平的响应面分析法对NKA-9型大孔树脂分离纯化贯叶连翘中金丝桃素的工艺条件及参数进行优化。结果表明，金丝桃素的最佳纯化工艺条件为上柱液中金丝桃素的浓度37.64mg/L，洗脱液乙醇的浓度69.42%，洗脱液的用量12.3BV，此时产物中金丝桃素含量理论值可达到1.42%。经验证试验表明，产物中金丝桃素含量为1.45%，与理论值相符。

关键词：贯叶连翘；金丝桃素；NKA-9型大孔吸附树脂；分离纯化；响应面优化法

（发表于《黑龙江畜牧兽医》）

3种紫花苜蓿草籽蛋白质瘤胃动态降解率的评定

乔国华，张怀山，周学辉，路远，张茜，王春梅

(中国农业科学院兰州畜牧与兽药研究所草业饲料研究室，兰州 730050)

摘 要：为了解紫花苜蓿草籽的瘤胃降解特性，试验研究了中兰1号苜蓿、甘农4号紫花苜蓿和陇中苜蓿草籽的营养成分；同时测定了用3种紫花苜蓿草籽分别作为主要蛋白质饲料来源时日粮的营养物质表观消化率。结果表明，3种苜蓿草籽的蛋白质含量较高（22.2%~26.1%）；3种苜蓿草籽蛋白质的瘤胃动态降解率参数 a 和有效降解率都比较高，分别达到了0.43~0.51和0.83~0.89；日粮粗蛋白质、脂肪、干物质、中性洗涤

纤维和酸性洗涤纤维的表观消化率没有显著变化（$P>0.05$）。

关键词：紫花苜蓿；草籽；反刍动物；瘤胃动态降解率

（发表于《黑龙江畜牧兽医》）

从生态学观点论青藏高原地区牦牛产业的可持续发展

朱新书，阎　萍，梁春年，郭　宪，裴　杰，包鹏甲，储　敏

（中国农业科学院兰州畜牧与兽药研究所，兰州　730050）

摘　要：为了研究青藏高原地区牦牛产业可持续发展，试验从生态学角度出发着重论述牦牛产业发展思路。牦牛是青藏高原高寒地区特有的畜种，是牧民不可缺少的生产生活资料。在青藏高原地区，科学地发展牦牛产业具有经济和生态双重意义。

关键词：生态学观点；青藏高原；牦牛；可持续发展

（发表于《黑龙江畜牧兽医》）

甘肃天祝牦牛乳中水解酶活力测定及分析

席　斌[1,2]，李维红[1,2]，高雅琴[1,2]，牛春娥[1,2]

（1. 中国农业科学院兰州畜牧与兽药研究所，兰州　730050；2. 农业部动物毛皮及制品质量监督检验测试中心，兰州　730050）

摘　要：为了比较甘肃天祝抓喜秀龙乡红疙瘩村、岱乾村和碳山岭镇四台沟村3种水解酶［酸性磷酸酶（ACP）、碱性磷酸酶（AKP）、淀粉酶（AMS）］活力，试验采用分光光度法对3个地区的90份牦牛乳进行检测分析。结果表明，红疙瘩村与岱乾村海拔接近、地区差异小，ACP活力、AKP活力均无明显差异，随着牦牛所处地区海拔的明显升高，ACP活力明显升高，而AKP活力明显下降，AMS活力受地区变化影响不大。胎次对牦牛乳中ACP活力、AMS活力无明显影响，随着胎次的增加，AKP活力呈现上升趋势且第5胎极显著高于其他胎次。

关键词：牦牛乳；水解酶；活力；地区；胎次

（发表于《黑龙江畜牧兽医》）

高效液相色谱法测定贯叶金丝桃粉剂中金丝桃素的含量

郭文柱[1]，梁剑平[1]，荆文魁[2]，王学红[1]，尚若峰[1]，郝宝成[1]

(1. 中国农业科学院兰州畜牧与兽药研究所农业部新兽药创制重点实验室/甘肃省新兽药工程重点实验室，兰州　730050；2. 二连浩特出入境检验检疫局技术中心，二连浩特　011100)

摘　要：为了有效控制贯叶金丝桃粉剂的质量，试验采用高效液相色谱（HPLC）法，流动相为乙腈-水-乙酸铵（三者比例为50∶40∶10；乙酸铵浓度为0.30mol/L，pH值为3.75），流速为1mL/min，检测波长为588nm，测定贯叶金丝桃粉剂中金丝桃素的含量。结果表明，金丝桃素的检测浓度在 $10\sim80\mu g/mL$ 范围内与峰面积的线性关系良好（$r=0.9996$），平均回收率为98.3%，RSD = 1.14%。说明该方法简便、准确、快速、重复性好，可用于贯叶金丝桃粉剂的质量控制。

关键词：贯叶金丝桃；金丝桃素；高效液相色谱法

(发表于《黑龙江畜牧兽医》)

贯叶连翘中金丝桃素的提取及含量测定方法

王学红，梁剑平，郭志廷，郭文柱，刘　宇，郝宝成

(中国农业科学院兰州畜牧与兽药研究所/农业部兽用药物创制重点实验室/甘肃省新兽药工程重点实验室，兰州　730050)

摘　要：金丝桃素是贯叶连翘的主要活性成分，具有抗抑郁、抗菌消炎、抗 DNA 和 RNA 病毒等药理活性，并已被作为原料制成药品在临床上广泛使用，用于治疗抑郁症、艾滋病及甲、乙型肝炎等疾病。由于金丝桃素具有多种药理作用，对其研究也越来越多。文章就近年来国内外对贯叶连翘中金丝桃素的提取及含量测定方法作一综述，为金丝桃素的进一步研究提供文献依据。

关键词：贯叶连翘；金丝桃素；提取；测定

(发表于《黑龙江畜牧兽医》)

赖氨酸对绵羊肝脏和背最长肌中 GHR 和 IGF-Ⅰ基因表达调控的影响

程胜利[1]，李建升[2]，冯瑞林[1]，岳耀敬[1]，苗小林[1]，
刘建斌[1]，郎　侠[1]，郭天芬[1]，裴　杰[1]

(1. 中国农业科学院兰州畜牧与兽药研究所，兰州　730050；2. 甘肃农业大学动物科学技术学院，兰州　730070)

摘　要：为了研究赖氨酸对绵羊组织生长的影响，试验选用 15 只年龄、体重相近的杂交母羊（陶塞特♂×蒙古羊♀）随机分为 3 组，于基础日粮中分别添加不同水平的赖氨酸（0，4，10g），采用 Real-time PCR 方法检测日粮不同赖氨酸水平对绵羊肝脏和背最长肌中 GHR 和 IGF-Ⅰ mRNA 表达的影响。结果表明，日粮赖氨酸水平影响绵羊肝脏和背最长肌中 GHR 和 IGF-Ⅰ基因的表达，随日粮赖氨酸水平的提高肝脏 GHR 和 IGF-Ⅰ mRNA 及背最长肌 IGF-Ⅰ mRNA 表达量相应增加；背最长肌中 4g 和 10g 赖氨酸组 GHR 基因相对表达量均极显著高于对照组（$P<0.01$），10g 赖氨酸组的 GHR 基因相对表达量与 4g 赖氨酸组间差异不显著（$P>0.05$）。说明添加一定量的赖氨酸可以调节绵羊组织的生长发育。

关键词：绵羊；赖氨酸；GHR 基因；IGF-Ⅰ基因；表达

（发表于《黑龙江畜牧兽医》）

两种家兔体温和部分血常规指标的比较

王东升[1]，严作廷[1]，张世栋[1]，李宏胜[1]，李世宏[1]，
龚成珍[2]，李锦宇[1]，荔　霞[1]

(1. 中国农业科学院兰州畜牧与兽药研究所中国农业科学院临床兽医学研究中心甘肃省中兽药工程技术研究中心，兰州　730050；2. 甘肃农业大学动物医学院，兰州　730070)

摘　要：为了比较大耳白兔和青紫蓝兔体温、血常规和血气指标的差异，试验将 60 只家兔按品种分为 2 组，每组 30 只。每天上午、下午分别测定体温 1 次，连续 5d，试验第 5 天每组随机抽取 10 只，心脏采血，用全自动血细胞分析仪测定白细胞数（WBC）、

红细胞数（RBC）和血小板（PLT）等16项血常规指标，用血气分析仪测定血清主要离子、血液酸碱度（pH值）和血氧饱和度（SO_2）等10项血气指标，并进行比较。结果表明，2组家兔的体温和血气指标差异不显著（$P>0.05$），血常规指标中除大耳白兔粒细胞数和粒细胞百分率显著高于青紫蓝兔（$P<0.05$）外，其余血常规指标差异不显著（$P>0.05$）。说明大耳白兔和青紫蓝兔体温、血气和血常规指标相差不大。

关键词：家兔；体温；血气；血常规

（发表于《黑龙江畜牧兽医》）

六茜素新制剂的药效学试验研究

王学红，冯艳丽，梁剑平，郭志廷，郭文柱，刘　宇

（中国农业科学院兰州畜牧与兽药研究所新兽药工程重点开放实验室，兰州　730050）

摘　要：为了测定六茜素新制剂的药效，试验以大肠杆菌、变形杆菌、无乳链球菌、停乳链球菌、金黄色葡萄球菌和绿脓杆菌6种奶牛乳房炎致病菌为受试菌株，进行体外抑菌试验和体内抗菌活性试验。结果表明，六茜素新制剂对大肠杆菌、停乳链球菌、变形杆菌、金黄色葡萄球菌和绿脓杆菌中度敏感，而对无乳链球菌高度敏感，并且对大肠杆菌感染小鼠有较强的保护作用。说明六茜素新制剂具有较强的抗菌活性且效果明显优于六茜素。

关键词：六茜素；新制剂；药效学

（发表于《黑龙江畜牧兽医》）

奶牛产奶量和质量的季节性变化规律研究

孙晓萍[1]，刘建斌[1]，杨博挥[1]，朗　侠[1]，李思敏[2]

（1. 中国农业科学院兰州畜牧与兽药研究所，兰州　730050；2. 甘肃白银市畜牧局，白银　730900）

摘　要：为了解不同季节奶牛产奶量和质量的变化规律，试验对3个奶牛场荷斯坦奶牛全年日均产奶量、泌乳期平均个体产奶量及饲养日平均个体产奶量进行了分析，得出了奶牛产奶量的季节性变化规律。结果表明，奶牛产奶量的变化规律为每年的3—6月随气温的升高逐步上升；7—10月气温上升到最高值，产奶量呈下降趋势；11—3月气温下降到0℃下，产奶量基本稳定。乳脂率一、二季度高于三、四季度；乳蛋白含量一、

四季度高于二、三季度；干物质含量一、四季度高于二、三季度。也就是说，气温高的二、三季度牛奶中的乳脂率、乳蛋白含量及干物质含量有下降趋势。试验同时测定了 Ca、Fe、Zn、Ca、Mg、Se 6 种微量元素含量，结果 Cu、Zn、Ca、Mg 含量在一、四季度最高，Fe 在二、四季度含量高，Se 含量全年没有变化。

关键词：奶牛；产奶量；质量；季节性；变化规律

(发表于《黑龙江畜牧兽医》)

陶×滩、波×滩一代杂交羔羊生长发育分析

孙晓萍[1]，李思敏[2]，朗　峡[1]，杨博辉[1]，刘建斌[1]

(1. 中国农业科学院兰州畜牧与兽药研究所，兰州　730050；2. 甘肃省白银市畜牧站，白银　730900)

摘　要：为了研究陶×滩（陶赛特×滩羊）、波×滩（波得代×滩羊）一代羊初生到3月龄的体重和生长发育情况，试验选择产羔时间相对集中的陶×滩、波×滩一代羊作为试验组，同龄滩羊为对照组分析其后代产肉性能及相关性状的变化情况。结果表明，在西北半荒漠草原生态环境条件下，陶×滩、波×滩一代羊生长发育快于滩羊，3月龄时波×滩一代公、母羔体重比滩羊公、母羔体重分别高 2.72kg（$P<0.01$）、0.51kg（$P>0.05$）；陶×滩一代公、母羔体重比滩羊公、母羔体重分别高 6.42kg（$P<0.01$）、3.04kg（$P>0.05$），陶×滩、波×滩一代羊的体高、体长、胸围极显著大于滩羊（$P<0.01$），说明陶×滩、波×滩一代羊的产肉性能好于滩羊。

关键词：陶×滩一代羊；波×滩一代羊；滩羊；生长发育；分析

(发表于《黑龙江畜牧兽医》)

天祝白牦牛被毛形态及纤维类型分析

牛春娥[1,2,3]，张利平[3]，高雅琴[1,2]，梁春年[1]

(1. 中国农业科学院兰州畜牧与兽药研究所，兰州　730050；2. 农业部动物毛皮及制品质量监督检验测试中心，兰州　730050；3. 甘肃农业大学动物科技学院，兰州　730050)

摘　要：为了研究天祝白牦牛毛绒生产性能，试验采集了 1~7 周岁天祝白牦牛体侧部、腹部、肩部、背部、股部被毛毛样进行纤维类型分析。结果表明，绒毛和两型毛含量随

年龄的增长逐渐降低,粗毛含量随年龄的增长逐渐升高。1~2岁天祝白牦牛被毛中绒毛含量高,品质好,但个体小,产量相对较低;3~6岁绒毛含量基本稳定,被毛品质较好,个体产量较高,是天祝白牦牛产绒性能最佳的阶段;7周岁以后绒毛含量明显降低,平均为18.41%,被毛品质变差,绒用性能降低。说明天祝白牦牛身体不同部位中,体侧部绒毛产量高,品质好;腹部粗毛(裙毛)产量高,品质较优。

关键词:天祝白牦牛;被毛形态;纤维类型;分析

(发表于《黑龙江畜牧兽医》)

兔毛的扫描电镜观察

李维红[1,2],席 斌[1,2],郭天芬[1,2],王宏博[1,2],牛春娥[1,2]

(1. 中国农业科学院兰州畜牧与兽药研究所,兰州 730050;2. 农业部动物毛皮及制品质量监督检验测试中心,兰州 730050)

摘 要:为了观察4种兔毛纤维的超微结构,试验以野兔、安哥拉兔、獭兔、家兔兔毛为材料,利用扫描电镜观察并比较其鳞片层的结构特征。结果表明,家兔粗毛鳞片翘角与安哥拉兔粗毛鳞片翘角之间差异不显著($P>0.05$),而与獭兔相比差异显著($P<0.05$),二者绒毛的鳞片翘角之间差异不显著($P>0.05$);安哥拉兔毛、绒鳞片高度与其他3种相比差异极显著($P<0.01$),而其他3种兔毛之间两两相比差异不显著($P>0.05$);安哥拉兔毛、绒鳞片厚度与其他3种相比差异极显著($P<0.01$),野兔兔毛与其他2种相比差异显著($P<0.05$),獭兔与家兔兔毛相比差异不显著($P>0.05$)。4种兔毛、绒之间的碳元素含量差异均不显著($P>0.05$),安哥拉兔粗毛氧元素含量与其他3种相比差异显著($P<0.05$),其他3种两两相比差异均不显著($P>0.05$);野兔硫元素含量与家兔绒毛中较低,与其他品种相比差异显著($P<0.05$),其他品种之间两两相比差异均不显著($P>0.05$);4种兔毛、绒钙元素含量均差异显著($P<0.05$)或极显著($P<0.01$)。说明不同种的动物纤维具有独特的形态特征,在超微结构上存在明显的差别。

关键词:兔毛;扫描电镜;超微结构

(发表于《黑龙江畜牧兽医》)

阿司匹林丁香酚酯栓剂的制备及质量控制

于远光[1]，李剑勇[2]，杨亚军[2]，刘希望[2]，张继瑜[2]，
牛建荣[2]，周旭正[2]，魏小娟[2]，李　冰[2]，叶得河[1]

（1. 甘肃农业大学，兰州　730070；2. 中国农业科学院兰州畜牧与兽药研究所/农业部兽用药物创制重点实验室/甘肃省新兽药工程重点实验室，兰州　730050）

摘　要：以半合成脂肪酸甘油酯为基质制备栓剂，对栓剂的外观与色泽、重量、融变时限、局部刺激性做了质量控制测定，并采用高效液相色普法测定阿司匹林丁香酚酯的含量。试验结果表明，该栓剂制备工艺可行，含量测定方法可靠；外观与色泽、重量、融变时限均符合药典规定，对兔直肠无刺激性；质量可控，稳定性好，无刺激性与不良反应，使用安全，可用于临床应用。

关键词：阿司匹林丁香酚酯；栓剂；质量控制

（发表于《湖北农业科学》）

发酵型黄芪提取物对肉仔鸡生产性能及免疫球蛋白的作用研究

张　凯，杨志强，王学智，孟嘉仁，张景艳，
秦　哲，王　龙，王　磊，李建喜

（中国农业科学院兰州畜牧与兽药研究所，兰州　730050）

摘　要：将192只21日龄肉仔鸡随机分成4组，分别饲喂正常日粮、含黄芪生药（RA）日粮、含发酵型黄芪提取物（FRAE）日粮和含中农华信黄芪多糖（AP）日粮，试验期28 d。测定肉仔鸡全程饲料利用率、日增重；50日龄心脏采血制备血清，测定其中免疫球蛋白A（IgA）、免疫球蛋白M（IgM）、免疫球蛋白G（IgG）的含量。扑杀后测定器官指数，并做组织切片镜检。结果表明，经过28 d试验后，RA组的平均日增重和日采食量明显下降（$P<0.05$），FRAE组饲料转化率明显下降（$P<0.05$）；IgA和IgM的含量没有显著性差异（$P>0.05$）。和对照组比较，AP组和FRAE组的IgG的含量无显著性差异（$P>0.05$），和对照组、AP组、FRAE组比较，RA组的IgG的含量有显著性的降低（$P<0.05$）。结果表明，在试验期内发酵型黄芪提取物对肉仔鸡的生产性

能和健康状况有明显促进作用。

关键词：发酵型黄芪提取物；免疫球蛋白；肉仔鸡

(发表于《湖北农业科学》)

宫衣净酊急性毒性研究

王东升，谢家声，李世宏，严作廷，张世栋

(中国农业科学院兰州畜牧与兽药研究所，兰州 730050)

摘 要：研究了小鼠口服宫衣净酊的急性毒性，并评价其安全性，为临床安全用药提供依据。采用最大给药量法，将40只小鼠随机分为2组，每组20只，雌、雄各半，分别为给药组和对照组，给药组按每千克体重40g宫衣净酊灌胃，对照组按每千克体重80mL生理盐水灌胃，分2次灌服，给药后连续观察7d，测定其1日最大给药量，确定宫衣净酊的急性毒性。结果表明，小鼠口服宫衣净酊的日最大给药量为40g/kg，相当于临床奶牛日用量的100倍，表明宫衣净酊较为安全。

关键词：宫衣净酊；急性毒性；小鼠；最大给药量

(发表于《湖北农业科学》)

黄河首曲——玛曲湿地沼生植物的群系分类研究

张怀山[1]，张吉宇[2]，乔国华[1]，王春梅[1]，张 茜[1]，杨 晓[1]

(1. 中国农业科学院兰州畜牧与兽药研究所，兰州 730050；2. 兰州大学草地农业科技学院，兰州 730000)

摘 要：黄河首曲——玛曲湿地是青藏高原东部湿地面积较大，植物群落特征明显，最原始、最具代表性的高寒沼泽湿地。玛曲湿地沼生植物的群落类型可分为莎草沼泽和杂类草沼泽2大类共4个群系。建群种分别为华扁穗草（*Blysmus sinocompressus*）、裸果扁穗苔草（*Blysmocarex nudicarpa*）、槽秆荸荠（*Eleocharis valleculosa*）和两栖蓼（*Polygonum amphibium*），亚建群种主要有褐鳞苔草（*Carex brunnescens*）、沿沟草（*Catabrosa aquatica*）、甘肃嵩草（*Kobresia kansuensis*）、两栖蓼和火绒草（*Leontopodium alpinum*）等多种植物。群落总盖度在5%~90%；主要分布在河道洼地、河漫滩、淘谷洼地、阶地陡坎洼地、沟源洼地等常年积水的地方，地下水位埋深大都在20~30cm。

关键词：湿地；沼生植物；群系分类；玛曲

(发表于《湖北农业科学》)

静脉灌注 LPS 对家兔的发热效应及肝蛋白的 SDS—PAGE 比较分析

张世栋[1]，李世宏[1]，王东升[1]，龚成珍[2]，李宏胜[1]，李锦宇[1]，荔 霞[1]，严作廷[1]

(1. 中国农业科学院兰州畜牧与兽药研究所/中国农业科学院临床兽医学研究中心/甘肃省中兽药工程技术研究中心，兰州 730050；2. 甘肃农业大学动物医学院，兰州 730070)

摘 要：以静脉注射内毒素（LPS）的方法建立了家兔发热模型，并以白虎汤的降温治疗作为反证，再运用 SDS-聚丙烯酰胺凝胶电泳比较和分析了 LPs 引起的发热效应对家兔肝脏总蛋白表达变化的影响以及肝组织病理变化。结果表明，以 15μg/kg 体重的 LPS 静脉注射家兔能很好地复制出动物温病模型，模型组、治疗组和对照组 3 组动物的肝蛋白表达均有各自的特异条带。所以，肝组织蛋白的表达变化可以作为动物温病研究与评价的生物学指标。

关键词：LPS；发热模型；白虎汤；SDS—PAGE；蛋白差异表达

(发表于《湖北农业科学》)

两种不同仪器分析方法对洗净山羊绒含脂率测定的比较

王宏博，常玉兰，李维红，杜天庆，梁丽娜

(中国农业科学院兰州畜牧与兽药研究所/农业部动物毛皮及制品质量监督检验测试中心（兰州），兰州 730050)

摘 要：通过索氏脂肪抽提装置与纤维油脂快速抽出器对洗净山羊绒含脂率测定结果的比较发现。两种仪器测试结果差异不显著（$P>0.05$），证实了纤维油脂快速抽出器对洗净山羊绒含脂率测定结果的可靠性。纤维油脂快速抽出器可满足现行标准测试要求。

关键词：纤维油脂快速抽出器；索氏脂肪抽提装置；比较分析

(发表于《湖北农业科学》)

绵羊毛与山羊绒的鉴别

王晓红[1]，姚 穆[2]，刘守智[3]

(1. 中国纺织大学纺织学院，上海 200051；2. 西北纺织工学院纺织科学与工程系，陕西西安 710048；3. 西安理工大学 陕西西安 710048)

摘 要：利用扫描电镜法测试分析66°的绵羊毛与一级除毛山羊绒，得到了以直径、表面鳞片高以及表面鳞片厚为指标的绵羊毛与山羊绒各自的频率分布，并计算出以这3个指标为判定依据的独立误判概率。在验证了3个指标为线性无关后，将3个独立误判概率相乘，从而大幅度降低了绵羊毛与山羊绒之间的误判概率，实验结果证实了这一结论。

关键词：绵羊毛；山羊绒；鉴别；独立误判概率；概率相乘

(发表于《湖北农业科学》)

塞拉菌素制剂的豚鼠最大化试验和 Buehler 试验

汪 芳，张继瑜，周绪正，李金善，李剑勇，李 冰，
牛建荣，魏小娟，杨亚军

(中国农业科学院兰州畜牧与兽药研究所/中国农业科学院新兽药工程重点开放实验室，兰州 730050)

摘 要：为检验塞拉菌素制剂对皮肤的过敏性，以化学药物指导原则为指导，进行了豚鼠最大化试验（GPMT）和Buehler豚鼠封闭斑贴试验（BT）。用2,4-二硝基氯苯溶液作为阳性对照物，在GPMT试验和BT试验的阳性对照组的致敏率分别为80%和70%，对豚鼠的致敏率较高，2个试验受试物组的致敏率均为0，表明塞拉菌素对豚鼠没有致敏性。

关键词：塞拉菌素；豚鼠；斑贴试验；最大化试验

(发表于《湖北农业科学》)

射干麻黄地龙颗粒治疗鸡呼吸型传染性支气管炎效果观察

谢家声，罗超应，王贵波，罗永江，辛蕊华

(中国农业科学院兰州畜牧与兽药研究所/甘肃省中兽药工程技术研究中心，兰州 730050)

摘 要：为研究射干麻黄地龙颗粒治疗产蛋鸡呼吸型鸡传染性支气管炎（IB）的效果，试验从4 260只感染鸡群中，随机抽取其中270只，分为试验组、药物时照组和空白对照3组；观察射干麻黄地龙颗粒对患病鸡群死亡率、产蛋率以及破软蛋率等指标的影响。结果表明，试验组死亡率分别比药物对照组和空白对照组降低4.45和7.78个百分点（$P<0.05$）；产蛋率分别提高6.80和6.90个百分点（$P>0.05$）；破蛋率分别降低0.58和0.45个百分点（$P>0.05$）。表明射干麻黄地龙颗粒治疗呼吸型IB的效果优于药物对照组和空白对照组。

关键词：射干麻黄地龙颗粒；鸡传染性支气管炎；死亡率；产蛋率

(发表于《湖北农业科学》)

中草药饲料添加剂"禽健宝"对蛋鸡生产性能和防病效果观察

严作廷，王东升，荔 霞，谢家声，李世宏，张世栋

(中国农业科学院兰州畜牧与兽药研究所/中国农业科学院临床兽医学研究中心，兰州 730050)

摘 要：用51周龄罗曼蛋鸡2 630羽，随机分为禽健宝组、喹乙醇组和空白对照组。试验期限为60 d。禽健宝组和喹乙醇组在基础日粮中分别按4 g/kg和35 mg/kg的量添加，空白对照组不添加任何药物。结果表明，禽健宝组产蛋率分别比喹乙醇组和对照组提高1.0%和3.9%；喹乙醇组和对照组死亡率分别为0.11%和0.91%，禽健宝组为0。禽健宝组和喹乙醇组料蛋比均比对照组降低7.72%。说明禽健宝不但能提高产蛋鸡的产蛋率，提高饲料报酬，而且具有较好的防病保健作用。

关键词：中草药；饲料添加剂；禽健宝；饲料报酬

(发表于《湖北农业科学》)

猪戊型肝炎病毒 swCH189 株衣壳蛋白基因 CP239 片段原核表达条件的优化

郝宝成[1,2]，梁剑平[1,3]，兰 喜[2]，刑小勇[3]，
项海涛[3]，温峰琴[3]，胡永浩[3]，柳纪省[2,3]

(1. 中国农业科学院兰州畜牧与兽药研究所/新兽药重点开放实验室/甘肃省新兽药工程重点实验室，兰州 730050；2. 中国农业科学院兰州兽医研究所/家畜疫病病原生物学国家重点实验室/农业部畜禽病毒学重点开放实验室/农业部草食动物疫病重点开放实验室，兰州 730046；3. 甘肃农业大学动物医学院，兰州 730070)

摘 要：为了提高猪戊型肝炎病毒 swCH 189 株衣壳蛋白基因 CP239 片段在大肠杆菌中的表达量，研究了载体、温度、转速、诱导时间以及诱导剂 IPTG 浓度等不同条件对衣壳蛋白基因 CP239 片段融合蛋白表达量的影响。结果表明，用 LB 培养基于 37℃ 培养 3.5 h 后，采用终浓度为 0.3 mmol/L 的 IPTG 在 37℃、200r/min 诱导培养 4h，pET32a-CP239 融合蛋白表达量最大；SDS-PAGE 检测结果表明 pET32a-CP239 融合蛋白的分子质量与预期大小一致，约为 45.3ku；Western blotting 结果表明，pET32a-CP239 融合蛋白可以与抗-HEV 阳性血清发生特异性反应，并具有良好的反应原性，说明衣壳蛋白基因 CP239 片段蛋白得到正确表达。

关键词：猪戊型肝炎病毒；衣壳蛋白；CP239 片段；原核表达

（发表于《华北农学报》）

牦牛 MSTN 基因内含子 2 多态性及与生长性状的相关性

梁春年，阎 萍，刑成峰，裴 杰，郭 宪，
包鹏甲，丁学智，褚 敏，朱新书

(中国农业科学院兰州畜牧与兽药研究所，兰州 730070)

摘 要：采用 PCR-SSCP 技术对大通牦牛、甘南牦牛和天祝白牦牛（共 277 头）肌肉抑制素基因（MSTN）内含子 2 的部分序列进行了多态性研究，分析该基因与牦牛生长性状的相关性。结果表明，牦牛 MSTN 基因内含子 2 存在 2 个等位基因和 3 种基因型。

在该基因座上，甘南牦牛、天祝白牦牛呈 HardyWeinberg 平衡状态，大通牦牛呈不平衡状态。3 种基因型与牦牛部分生长性状的最小二乘法分析表明，MSTN 基因内含子 2 对成年牦牛胸围、体质量、胸围指数、体长指数和肉用指数均显著相关（$P<0.05$），而不同基因型的牦牛体高、管围、体长和管围指数差异不显著。初步推断牦牛 MSTN 基因内含子 2 可作为牦牛标记辅助选择的遗传标记之一。

关键词：牦牛；MSTN 基因；多态性；遗传标记；PCR-SSCP

（发表于《江苏农业科学》）

高效液相色谱法测定美洛昔康注射液美洛昔康含量

王海为[1,2]，李剑勇[2]，杨亚军[2]，李 冰[2]，
张继瑜[2]，刘希望[2]，周旭正[2]，牛建荣[2]，魏小娟[2]

（1. 甘肃农业大学动物医学院，兰州 730070；2. 中国农业科学院兰州畜牧与兽药研究所/新兽药工程重点实验室，兰州 730050）

摘 要：建立了测定美洛昔康注射液中美洛昔康含量的高效液相色谱法。采用 C_8 色谱柱（5μm，4.6mm×150mm），流动相为 V（甲醇）：V（水，磷酸调 pH 值至 3.0）= 40：60，流速为 1mL/min，检测波长为 355mm，进样量 10μL，柱温为室温。美洛昔康浓度在 5~200μg/mL 内，峰面积与浓度的线性关系良好（$r^2=0.9998$）。平均回收率为 99.20%，RSD 为 1.30%（$n=4$）。此法简便、快速、灵敏度高、重现性好。

关键词：美洛昔康；含量测定；高效液相色谱法

（发表于《江苏农业科学》）

环形泰勒虫 suAT1 基因的克隆及序列分析

蔺红玲[1,2]，张继瑜[2]，袁莉刚[1]，魏小娟[2]，李 冰[2]

（1. 甘肃农业大学动物医学院，兰州 730070；2. 中国农业科学院兰州畜牧与兽药研究所/农业部兽用药物创制重点实验室/新兽药工程重点开放实验室，兰州 730050）

摘 要：以环形泰勒虫兰州株基因组为模板，经 PCR 扩增获得了 suAT1 的部分基因，将该基因克隆到 pMD20-T 载体，对重组质粒进行 PcR 和双酶切鉴定及序列测定。结果表明，该基因的长度为 1 377bp，编码了 459 个氨基酸。同源性分析结果显示：克隆序列

与 GenBank 收录的环形泰勒虫参考核苷酸序列同源性为 98.35%，氨基酸同源性为 97.13%，说明环形泰勒虫兰州株的 suAT1 与 GenBank 上公布的序列同源性很高，进一步说明了 suAT1 基因存在很强的保守性。

关键词：环形泰勒虫；suAT1 基因；序列分析

（发表于《江苏农业科学》）

益蒲灌注液质量标准研究

苗小棱[1]，李 芸[2]，潘 虎[1]，杨耀光[1]，苏 鹏[1]，王 瑜[1]，焦增华[1]

(1. 中国农业科学院兰州畜牧与兽医研究所/新兽药工程重点实验室，兰州 730050；2. 甘肃中医学院，兰州 730000)

摘 要：为了控制益蒲灌注液的质量，本试验采用薄层色谱法对益蒲灌注液中连翘、蒲黄进行定性鉴别，采用 HPLC 测定水苏碱的含量。结果显示，定性鉴别薄层色谱特征明显，水苏碱含量在 3.39~20.34 μg 范围内线性关系良好，回归方程 = $lgA = 1.784lgC + 3.9756$ ($r=0.9993$)，平均回收率 ($n=6$) 为 99.28%，$RSD=2.47$%。说明该方法简便、灵敏，阴性无干扰，可以有效控制益蒲灌注剂的质量。

关键词：子宫内膜炎；益蒲灌注液；薄层色谱鉴别；HPLC；水苏碱

（发表于《江苏农业科学》）

动物福利与毛皮动物处死方法

郭天芬[1,2]，李维红[1,2]，席 斌[1,2]

(1. 中国农业科学院兰州畜牧与兽药研究所，兰州 730050；2. 农业部动物毛皮及制品质量监督检验测试中心，兰州 730050)

摘 要：本文通过描述动物福利的概念及国内外现状，毛皮动物处死的方法，提出了毛皮动物处死过程中动物福利的必要性。

关键词：动物福利；毛皮动物；处死方法

（发表于《经济动物学报》）

微波辅助合成金丝桃素四磺酸衍生物

李兆周[1]，王学红[1]，李雪虎[2]，梁剑平[1]*

(1. 中国农业科学院兰州畜牧与兽药研究所农业部兽用药物创制重点实验室甘肃省/中国农业科学院新兽药工程重点实验室，兰州　730050；2. 中国科学院近代物理研究所，兰州　730000)

摘　要：利用微波辅助合成的方法，对金丝桃素四磺酸衍生物合成路线中的大黄素蒽酮缩合反应分别从微波加热温度、辐射时间和催化剂3个因素进行了优化，确定了最佳反应条件：温度150℃，反应时间1h，不使用催化剂；该步反应的产率可达82.2%，整个合成路线经放大后的总产率为65.3%。与常规加热方法相比，微波辅助合成法提高了反应效率和产率，减少了副反应的发生，使合成工艺更加低碳环保。
关键词：微波辅助合成；金丝桃素；四磺酸金丝桃素；医药与日化原料

(发表于《精细化工》)

不同预处理对羊毛纤维平均直径的影响及其相关性研究

郭天芬[1,2]，王宏博[1,2]，牛春娥[1,2]，席　斌[1,2]

(1. 中国农业科学院兰州畜牧与兽药研究所，兰州　730050；2. 农业部动物毛皮及制品质量监督检验测试中心，兰州　730050)

摘　要：为了研究不同预处理对羊毛纤维平均直径的影响及其相关性，用激光纤维直径分析仪分别检测了常温、烘箱、恒温恒湿3种不同预处理的标准毛条、同质羊毛、异质羊毛的纤维平均直径。结果表明：常温、烘箱、恒温恒湿3种不同预处理的8个标准毛条的平均纤维直径分别为24.88μm、24.84μm、25.11μm；10个同质毛的分别为17.66μm、17.79μm、18.16μm；10个异质毛的分别为27.91、27.76、29.24μm。经检验，3种预处理后羊毛纤维平均直径的差异均不显著，常温和烘箱预处理与恒温恒湿预处理的羊毛纤维直径之间存在较强的线性相关性。
关键词：预处理；羊毛；直径；影响；相关性

(发表于《毛纺科技》)

制定毛纤维直径成分分析仪检测方法标准重要性的探讨

郭天芬[1,2]，李维红[1,2]，高雅琴[1,2]，王宏博[1,2]

(1. 中国农业科学院兰州畜牧与兽药研究所，兰州 730050；2. 农业部动物毛皮及制品质量监督检验测试中心，兰州 730050)

摘 要：毛纤维直径和成分是评定毛绒品质和使用价值的重要指标，其检测方法和检测仪器的研究一直受到人们的关注。文章对我国研制开发的纤维直径成分分析仪与投影显微镜、激光细度仪、OFDA 仪、气流仪，从原理、价格、检测条件等方面进行了全面地分析比较。认为此仪器具有检测结果准确、速度较快、价格便宜、使用方便、用户多等特点，但其目前还没有相应的检测方法标准。因此，建议尽快制定毛纤维直径成分分析仪法检测方法标准，以促进我国毛纤维品质分析技术的普及和提高。

关键词：毛纤维直径成分分析仪；检测；标准；制定

(发表于《毛纺科技》)

奥司他韦快速鉴别及含量测定方法的建立

刘玉荣，杨亚军，李剑勇*，李 冰

(中国农业科学院兰州畜牧与兽药研究所，农业部兽用药物创制重点实验室·甘肃省新兽药工程重点实验室，甘肃兰州 730050)

摘 要：目的建立奥司他韦 (OS) 的快速鉴别和含量测定方法。方法硅胶 GF_{254} 薄层板，甲醇-醋酸乙酯-甲苯-三乙胺：2：4：2：1) 为展开剂，UV 254nm 下检视；C_{18} 柱，流动相甲醇-水 (75：25，V/V)，流速为 $0.4mL·min^{-1}$，以多重反应监测模式对 OS 进行进一步的鉴别和定量分析。结果展开后，254nm 下出现暗斑，Rf 约为 0.5；OS 准分子离子峰 [M+H] +m/z 313.30，子离子 m/z 166.20 和 208.20 分别为定量离子和辅助定性离子，在 $5\sim160ng·mL^{-1}$ 的范围内线性关系良好 ($R^2=0.9995$)，平均回收率为 100.94%，日内精密度和日间精密度分别为 99.89%和 98.90%。结论所建立的方法快速、简便，选择性强，灵敏度高，可用于 OS 的鉴别和含量测定。

关键词：奥司他韦；薄层色谱；液相色谱串联质谱

(发表于《时珍国医国药》)

肃北高寒草原不同放牧强度土壤养分变化特征

杨红善[1]，常根柱[1]，周学辉[1]，路　远[1]，那·巴特尔[2]

（1. 中国农业科学院兰州畜牧与兽药研究所，兰州　730050；2. 全球环境基金（GEF）项目肃北县项目办公室，肃北　736300）

摘　要：研究了肃北高寒草原不同放牧强度下不同土层土壤养分及 5 种微量元素有效态含量变化特征。结果表明：①高寒草原土壤物理性质的变化对土壤养分及微量元素具有重要的调控作用；②随着放牧强度的提高，0~10、10~20 cm 土层土壤体积质量均呈不同程度的增加，土壤孔隙度和土壤含水量则呈显著的递减趋势；③轻度放牧草地土壤有机质、土壤全 N 含量高于中度放牧和重度放牧草地；20~30 cm 土层有机质随放牧强度的增大呈明显下降趋势，即随放牧强度的增大深层土壤肥力呈退化趋势；肃北高寒草原的速效养分以多 N 少 P 富 K 为特点，土壤速效 N、P、K 含量在总体上随放牧强度的增加呈下降趋势；④肃北高寒草原 5 种微量元素的高低顺序依次是：Na>Fe>Mn>Cu>Zn，不同放牧强度下各微量元素的变化一致。顺序依次是：轻度放牧>对照>中度放牧>重度放牧；⑤放牧强度对 10~20 cm 的土层影响最大，随放牧强度的增大，地表植物营养吸收层土壤营养成分、微量元素呈降低，导致地表植被生长能力降低，最终导致地表土壤沙化，最后使草地大面积退化。

关键词：高寒草原；放牧强度；土壤养分；微量元素

（发表于《土壤》）

西藏拉萨地区不同施氮量对鸭茅产量的影响

张小甫[1]，李锦华[1]，田福平[1,2]，余成群[3]，黄秀霞[4]，江　措[5]

（1. 中国农业科学院兰州畜牧与兽药研究所，兰州　730050；2. 农业部兰州黄土高原生态环境重点野外科学观测试验站，兰州　730050；3. 中国科学院地理科学与资源研究所，北京　100101；4. 西藏自治区农业技术推广服务中心，拉萨　850000；5. 西藏自治区山南地区草原工作站，泽当　850000）

摘　要：在西藏拉萨地区进行不同施氮量（尿素）对鸭茅产量的影响研究。结果表明，鲜（干）草产量、种子产量、植株生长速度随施氮量的增加而明显提高，但当施氮达到一定量时，其产量不再提高，反而呈下降趋势。7 个处理的鲜（干）草产量均显著高于对照，尿素施量160~240kg/hm² 的鲜（干）草产量保持在较高水平，施量160kg/hm² 的鲜（干）草产量最高；施尿素的鸭茅种子产量与未施者差异显著，施量160~280kg/hm² 的鸭茅种子产量较高；施量120~200kg/hm² 的鸭茅生长速度明显高于其他施肥水平；对于牧草高效生产的经济效益而言，尿素施量为160 kg/hm² 的处理能够获得最大经济效益。

关键词：尿素；鸭茅；鲜（干）草产量；种子产量

（发表于《西北农学报》）

苜蓿引进品种半干旱、半湿润区适应性试验

杨红善，常根柱，周学辉，路　远

（中国农业科学院兰州畜牧与兽药研究所，兰州　730050）

摘　要：苜蓿王，Spreador-3，Pick 3006 和 Pick 8925 是从美国和加拿大引进的优良苜蓿品种，以国产中兰1号为对照品种，2002—2004年在兰州半干旱地区开展引种试验，2008年与2009年在半湿润区天水甘谷继续开展适应性试验，同时对其营养成分和抗旱指标进行测定。结果表明，引进的4个苜蓿品种，在半干旱、半湿润区均能正常生长生活，完成生育周期的全过程，无异常表现。未发现病虫害侵染，苜蓿王的抗旱性最强，鲜草产量在半干旱区兰州达68 715.8 kg/hm²，在半湿润区天水高达125 642.8 kg/hm²；Spreador-3 具叶大、杆中实，抗倒伏等优良特性，对照中兰1号具优良抗旱性能，建议3种苜蓿在中国黄土高原半干旱地区及半湿润区推广种植。

关键词：紫花苜蓿；半湿润区；半干旱区；栽培试验；

（发表于《西北农业学报》）

白术提取液急性毒性试验

王东升[1,2]，李锦宇[1,2]，罗超应[1,2]，郑继方[1,2]，张世栋[1,2]，胡振英[1,2]

（1. 中国农业科学院兰州畜牧与兽药研究所，兰州　730050；2. 中国农业科学院临床兽医学研究中心，兰州　730050）

摘　要：测定白术提取液对小鼠腹腔注射和肌内注射的急性毒性，评价其安全性，为临床安全用药提供依据。采用最大给药量法，将60只小鼠随机分为3组，每组20只，分为白术提取液100g/kg腹腔注射组、20g/kg肌内注射组和生理盐水100mL/kg腹腔注射

组,注射给药后连续观察7d,测定其1日最大给药量,确定白术提取液的急性毒性。结果表明,小鼠腹腔注射和肌内注射的1日最大给药量依次为100和20g/kg,分别相当于临床犬日用量的400和80倍。说明白术提取液较为安全,可在临床研究试用。

关键词:白术;急性毒性;小鼠;最大给药量

(发表于《西北农业学报》)

犬瘟热病毒甘肃株的分离和 N蛋白基因遗传进化分析

王旭荣[1,2],王小辉[1,2],韩富杰[3],李世宏[1,2],李建喜[1,2],孟嘉仁[1,2]

(1. 中国农业科学院兰州畜牧与兽药研究所,兰州 730050;2. 中国农业科学院临床兽医学研究中心,兰州 730050;3. 兰州恒泰动物医院,兰州 730050)

摘 要:采用Vero细胞从甘肃2008—2009年疑似犬瘟热感染病犬体内分离到8株病毒,通过RT-PCR方法扩增M蛋白基因初步鉴定所获得毒株均为犬瘟热病毒(CDV)。进一步通过RT-PCR方法扩增分离株的N蛋白基因,并进行序列分析。结果表明,8株分离株均为CDV,它们之间N蛋白基因的核苷酸同源性和推定的氨基酸同源性分别为98.4%~99.7%、98.3%~99.6%;8株CDV分离株的N蛋白基因和TN、A75/17相应序列的核苷酸同源性分别为98.5%~99.5%、97.0%~98.0%,推定的氨基酸同源性分别为98.3%~99.2%、98.7%~99.8%;8株CDV分离株N蛋白基因和疫苗株(Onderstepoort和Snyder Hill)的相应序列的核苷酸同源性和推定的氨基酸同源性分别为93.0%~94.0%、96.8%~98.3%。根据N蛋白基因所建立的遗传系统发育树可知这8株CDV分离株与TN的亲缘关系较近,而与疫苗株处于不同的分支上。说明从甘肃分离的CDV野毒株和TN由同一毒株演化而来,与疫苗株亲缘关系较远。

关键词:犬瘟热病毒;N蛋白基因;遗传进化

(发表于《西北农业学报》)

常山提取物急性毒性试验研究

雷宏东[1]，梁剑平[2]，郭志廷[2]，王学红[2]，华兰英[2]

(1. 甘肃农业大学动物医学院，兰州 730070；2. 中国农业科学院兰州畜牧与兽药研究所，兰州 730050)

摘 要：为了评价常山提取物的安全性，对其进行了小白鼠的急性毒性试验。预试验初步确定用药剂量，正式试验取60只昆明种小白鼠，灌胃给药后，连续观察7d，记录小白鼠的急性毒性反应，并通过改良寇氏法计算常山提取物对小白鼠的半数致死量（LD_{50}）及LD_{50}的95%可信限。结果表明，给药组部分小白鼠出现中毒反应并死亡，剖检主要脏器未见病理变化，测得LD_{50}为18.16g/kg体重，LD_{50}的95%可信限为15.35~21.49g/kg体重。试验结果初步提示，常山提取物的毒性很低，临床用药安全可靠。

关键词：常山提取物；急性毒性；小白鼠

（发表于《中国畜牧兽医》）

电针复合静松灵麻醉对山羊血液指标的影响

王贵波[1,2]，丁明星[2]，郑继方[1]，罗超应[1]，王东升[1]，李锦宇[1]

(1. 中国农业科学院兰州畜牧与兽药研究所，甘肃省中兽药工程技术研究中心，兰州 730050；2. 华中农业大学动物医学院，武汉 430070)

摘 要：本研究旨在探索新型的麻醉方法——电针复合静松灵麻醉对山羊血液指标的影响。将30只体重22~27kg的成年杂交山羊随机分为5组，分别为0.5mg/kg静松灵组、2mg/kg静松灵组、电针复合0.5mg/kg静松灵组、电针组和空白对照组，每组6只，公母各半。分别在静松灵给药后0、1.5、24、48、96、168h采集山羊血液，测定血液指标。电针复合0.5mg/kg静松灵对山羊的麻醉效果（痛阈值）与2mg/kg静松灵的麻醉效果相当时，对所测山羊血液指标的影响却比2mg/kg静松灵组小。电针复合静松灵麻醉是一种对山羊血液指标干扰较小的麻醉方法。

关键词：电针；静松灵；麻醉；血液指标；山羊

（发表于《中国畜牧兽医》）

发酵型黄芪提取物对肉仔鸡生产性能及血液生化指标的影响

张 凯，李建喜，杨志强，王学智，孟嘉仁，张景艳，秦俊杰，王 龙

(中国农业科学院兰州畜牧与兽药研究所，兰州 730050)

摘 要：将192只体重均一的21日龄肉仔鸡随机分成4组，分别饲喂正常日粮、含黄芪生药（RA）日粮、含发酵型黄芪提取物（FRAE）日粮和含黄芪多糖（AP）日粮，试验期28d。测定肉仔鸡全程饲料利用率、日增重；50日龄心脏采血制备血清，测定其中乳酸脱氢酶、丙氨酸氨基转移酶、碱性磷酸酶、天冬氨酸氨基转移酶、γ-谷氨酸转肽酶的酶活性，尿酸的含量。扑杀后测定器官指数，并做组织切片镜检。结果表明：①经过28d试验，RA组的平均日增重和日采食量明显下降（$P<0.05$），FRAE组饲料转化率明显下降（$P<0.05$）；②和RA组相比，FRAE组的乳酸脱氢酶活性明显降低（$P<0.05$），和对照组相比，RA组的γ-谷氨酸转肽酶活性明显降低（$P<0.05$）；③显微镜下观察各组织无病理变化。结果显示，在试验期内发酵型黄芪提取物对肉仔鸡的生产性能和健康状况有明显促进作用。

关键词：发酵型黄芪提取物；血清生化；肉仔鸡

(发表于《中国畜牧兽医》)

复合活菌制剂降解虾塘中亚硝酸盐的研究与应用

陶 蕾[1,2]，梁剑平[1]，郭 凯[1]，李兆周[1]

(1. 中国农业科学院兰州畜牧与兽药研究所/农业部兽用药物创制重点实验室/甘肃省新药工程重点实验室，兰州 730050；2. 甘肃农业大学，兰州 730070)

摘 要：本试验以真菌和反硝化细菌为基础菌种按一定比例制得复合活菌制剂，对虾塘养殖水中的亚硝酸盐进行降解或处理。结果表明，活菌制剂净化亚硝酸盐污染的虾塘养殖水效果好于单个反硝化细菌、某复合活菌制剂和硝氮综合剂，复合活菌制剂各菌种在虾塘养殖水中分解亚硝酸盐的作用上各有特点，可为形成共生长效的虾塘养殖水环境修复微生物种群提供基础。

关键词：复合活菌制剂；亚硝酸盐；虾塘养殖

(发表于《中国畜牧兽医》)

贵州地区牛源金黄色葡萄球菌分离株 16S rRNA 分子鉴定及 RAPD 分型研究

邓海平，蒲万霞，梁剑平，倪春霞，孟晓琴

(中国农业科学院兰州畜牧与兽药研究所/中国农业科学院新兽药重点开放实验室/甘肃省新兽药重点实验室，兰州 730050)

摘 要：利用 16S rRNA 保守序列对引起贵州地区奶牛乳房炎的金黄色葡萄球菌进行分离鉴定，建立随机多态性扩增体系对金黄色葡萄球菌分离株进行基因分型研究。结果表明，69 份奶样中共分离出金黄色葡萄球菌 37 株，DNA 随机多态性扩增（random amplified polymorphic DNA，RAPD）结果显示这 37 株金黄葡萄球菌扩增产物在 1~7 条带之间，产物大小为 400~4500bp。菌株可分为 7 个基因型，其中 I 型 7 株（18.9%）、II 型 2 株（5.4%）、III 型 2 株（5.4%）、IV 型 1 株（2.7%）、V 型 19 株（51.4%）、VI 型 1 株（2.7%）、VII 型 5 株（13.5%），V 型为该地区的流行优势菌群。地理和气候环境对病原菌流行传播的影响是病原菌基因型分布差异的重要原因。

关键词：奶牛乳房炎；金黄色葡萄球菌；16S rRNA；RAPD；基因型

(发表于《中国畜牧兽医》)

黄白口服液的质量控制研究

王胜义[1]，齐志明[1]，刘治岐[2]，荔 霞[1]，董书伟[1]，刘世祥[1]，刘永明[1]

(1. 中国农业科学院兰州畜牧与兽药研究所/农业部兽用药物创制重点实验室/甘肃省中兽药工程技术研究中心，兰州 730050；2. 兰州理工大学生命科学与工程学院，兰州 730050)

摘 要：为了制定黄白口服液质量标准，采用薄层色谱（TLC）法对组方中的黄连、金银花、黄芩、地榆炭进行定性鉴别；用高效液相色谱（HPLC）法对制剂中盐酸小檗碱和黄芩苷进行含量测定。结果 TLC 鉴别专属性强，重现性好；盐酸小檗碱线性范围在 0.139~0.834μg（$R^2=0.9999$），平均回收率为 97.04%，RSD 为 0.97%（n=6）；黄芩苷线性范围 0.235~1.41μg（$R^2=0.9999$），平均回收率为 97.64%，RSD 为 0.71%（n=6）。对黄连、金银花、黄芩、地榆炭的定性鉴别及盐酸小檗碱和黄芩苷的含量测定可用来进行制剂的质量控制。
关键词：黄白口服液；薄层鉴别；含量测定；质量标准

（发表于《中国畜牧兽医》）

六茜素的质量标准研究

王学红，刘　宇，郭文柱，郝宝成，郑红星，梁剑平

（中国农业科学院兰州畜牧与兽药研究所/农业部兽用药物创制重点实验室/甘肃省新兽药工程重点实验室，兰州　730050）

摘　要：试验考察了六茜素的质量要求，试验包括澄明度测定、无菌检查、渗透压测定、溶血性试验、酸度测定、局部刺激性试验、热原检查、过敏反应试验、含量测定。结果表明，六茜素符和注射剂的一般要求（中国兽药典委员会，2005）。

关键词：六茜素；注射剂；质量标准

（发表于《中国畜牧兽医》）

六茜素注射液的亚慢性毒性试验

王学红，梁剑平，郭文柱，冯艳丽，刘　宇，郝宝成

（中国农业科学院兰州畜牧与兽药研究所/农业部兽用药物创制重点实验室/甘肃省新兽药工程重点实验室，兰州　730050）

摘　要：为了研究六茜素注射液的安全性，进行了六茜素注射液的亚慢性毒性试验。试验每天给大白鼠灌服六茜素注射液80、160、320mg/kg，对各组大白鼠的体重变化、脏器系数、血液生理生化指标及心脏、肝脏、脾脏、肾脏、胸腺等脏器病理变化进行了测定和分析。结果表明，试验组大白鼠的各脏器无明显肉眼可见剖检病变；试验组大白鼠的各脏器系数与对照组差异不显著（$P>0.05$）；320mg/kg剂量组大白鼠的肝脏和肾脏器官出现轻微的病理变化。结果说明，六茜素注射液毒性低，可以安全用于动物。

关键词：六茜素注射液；亚慢性毒性试验；体重变化；脏器系数；病理切片

（发表于《中国畜牧兽医》）

芩连液与白虎汤对脂多糖致家兔肝损伤的疗效比较

张世栋[1]，王东升[1]，李世宏[1]，李宏胜[1]，荔 霞[1]，
李锦宇[1]，陈炅然[1]，龚成珍[2]，严作廷[1]

(1. 中国农业科学院兰州畜牧与兽药研究所/中国农业科学院临床兽医学研究中心/甘肃省中兽药工程技术研究中心，兰州 730050；2. 甘肃农业大学动物医学院，兰州 730070)

摘 要：为比较自拟方芩连液和白虎汤对高热病症的治疗效果，通过静脉注射脂多糖（LPS）复制了家兔高热病症，并在两种药物治疗前后分别进行了肝脏组织病理观察与血清生化指标分析。结果显示，15μg/kg LPS 能致家兔肝脏组织产生明显病理损伤，血清丙氨酸氨基转移酶（ALT）、天冬氨酸氨基转移酶（AST）和尿素（Urea）含量上升，而总胆固醇（TC）含量下降，经白虎汤和芩连液治疗后肝脏组织病理损伤显著减轻，4 项血清生化指标向正常水平恢复。结果表明，白虎汤和芩连液对 LPS 引起的肝损伤都具有显著的治愈效果，且两种组方的疗效各有优势。
关键词：LPS；肝损伤；血清生化指标；中药；组方

(发表于《中国畜牧兽医》)

犬瘟热病毒分离株 N、H、F 蛋白基因的变异分析

王旭荣，王小辉，张世栋，李世宏，潘 虎，严作廷

(中国农业科学院兰州畜牧与兽药研究所中国农业科学院临床兽医学研究中心，兰州 730050)

摘 要：通过 RT-PCR 方法扩增到犬瘟热病毒（canine distemper virus，CDV）分离株 GS0812-4 的 N、H、F 蛋白基因，然后克隆入 pGEM-T Easy 载体并进行序列分析和抗原表位预测分析。结果表明，GS0812-4 分离株的 N、H、F 蛋白基因与其他分离株相应序列的核苷酸同源性分别为 93.2%~97.8%、90.3%~97.3%、93.2%~97.8%，其推定的氨基酸同源性分别为 96.6%~98.7%、91.3%~97.9%和 96.6%~98.7%；GS0812-4 分离株 N、H、F 蛋白基因与疫苗株相应序列的核苷酸同源性分别为 93.1%~93.8%、89.4%~90.5%、93.1%~93.8%，其推定的氨基酸同源性分别为 96.4%~97.3%、

89.3%~91.3%、96.4%~97.3%。由基因系统发育树可知，GS0812-4 与疫苗株处于不同的分支，其中与 MKY-KM08 的亲缘关系最近，但 GS0812-4 在 N、H、F 3 个基因系统发育树上与其他同一毒株的亲缘关系远近不同。H、F 蛋白的抗原表位预测结果表明，GS0812-4 与 MKY-KM08 和疫苗株的抗原表位均有差异。说明 GS0812-4 是野毒株，与 MKY-KM08 属于同一基因型，来源于同一毒株，但两毒株间的 H 蛋白和 F 蛋白的抗原表位仍有差异。

关键词：犬瘟热病毒；H 蛋白；H 蛋白；F 蛋白；变异分析

（发表于《中国畜牧兽医》）

天祝白牦牛 *Leptin* 和 *SCD*1 基因单核苷酸多态性与肌肉脂肪酸含量的关联性分析

刘自增，阎 萍

（中国农业科学院兰州畜牧与兽药研究所畜牧研究室，兰州 730050）

摘 要：试验旨在研究 *Leptin* 和 *SCD*1 基因多态性对天祝白牦牛肌肉脂肪酸含量的影响。通过对 PCR 产物直接测序，在 *Leptin* 基因外显子 2 和 3 共发现 7 个 SNPs 位点（其中 4 个是错义突变）；在 *SCD*1 基因第 5 外显子区发现 1 个 SNP 位点（错义突变），对所有 SNPs 等位基因替代效应对 25 种单一脂肪酸、单一不饱和度指数、多不饱和度指数和去饱和效应指数的影响进行了检测。结果发现，*SCD*1 基因 SNP 和 *Leptin* 基因 3 个 SNPs 位点在不同程度上影响到由饱和脂肪酸 FA 到单一不饱和脂肪酸 MUFA 的去饱和作用。说明除了 *SCD*1 基因对 FA 去饱和作用的影响外，*Leptin* 基因错义突变也会对肌肉脂肪中 FA 的组成产生影响。

关键词：*Leptin* 基因；*SCD*1 基因；脱氢作用；脂肪酸；SNP

（发表于《中国畜牧兽医》）

猪戊型肝炎病毒 ORF2 部分片段原核表达载体的构建

郝宝成[1,2]，兰 喜[2]，胡永浩[3]，柳纪省[2,3]，梁剑平[1,3]

(1. 中国农业科学院兰州畜牧与兽药研究所农业部兽用药物创制重点实验室/甘肃省新兽药工程重点实验室，兰州 730050；2. 中国农业科学院兰州兽医研究所，兰州 730046；3. 甘肃农业大学动物医学院，兰州 730070)

摘 要：利用 Protean 软件对猪戊型肝炎病毒（hepatitis E virus，HEV）swCH189 株的 ORF2 进行潜在抗原位点分析，选取 381aa-623aa 段作为表达片段，运用 PCR 方法从 ORF2 全长质粒中扩增出目的片段 cp239，经纯化、连接等，成功构建了表达载体 pET32a（+）-cp239，为下一步进行基因原核表达及其生物学活性分析奠定了基础。

关键词：猪；戊型肝炎病毒；ORF2；表达载体构建

(发表于《中国畜牧兽医》)

塞拉菌素透皮制剂的眼刺激性试验

汪 芳，周绪正，张继瑜，李金善，李剑勇，
李 冰，牛建荣，魏小娟，杨亚军

(中国农业科学院兰州畜牧与兽药研究所/中国农业科学院新兽药工程重点开放实验室，兰州 730050)

摘 要：[目的] 观察塞拉菌素透皮制剂对日本大耳白兔眼的刺激性反应。[方法] 以《化学药物刺激性、过敏性和溶血性研究技术指导原则》为指导对日本大耳白兔进行眼刺激性试验，评价药物对眼的刺激性。[结果] 该制剂对兔子的结膜有轻度刺激性，对虹膜和角膜没有刺激性。[结论] 塞拉菌素制剂对眼的刺激为轻度。

关键词：塞拉菌素制剂；日本犬耳白兔；眼刺激性试验

(发表于《中国动物检疫》)

一起梅花鹿结核病的临床诊断与防治

郭志廷[1]，梁剑平[1]，罗晓琴[2]

(1. 中国农业科学院兰州畜牧与兽药研究所新兽药重点开放实验室，兰州 730050；
2. 兰州市动物卫生监督所，兰州 730050)

摘　要：鹿结核病是由牛型结核杆菌引起的一种慢性、消耗性人畜共患传染病。结核杆菌可通过唾液、粪便、尿液等传播，传播速度极快，在鹿群中常大批发生。本文对兰州市某新建梅花鹿养殖场于2007年12月发生的一起鹿以咳嗽、腹泻及渐行性消瘦为主要特征的鹿结核病从流行病学、发病情况、临床症状、剖检变化、实验室诊断和综合防治等方面进行了描述。
关键词：鹿结核病；诊断；治疗；剖检变化

(发表于《中国动物检疫》)

喹乙醇单克隆抗体制备中细胞融合条件的筛选

王　磊，李建喜，张景艳，杨志强，王学智，张　凯，孟嘉仁

(中国农业科学院兰州畜牧与兽药研究所中兽药工程中心，兰州 730050)

摘　要：为提高特异性杂交瘤细胞阳性得率，制备喹乙醇单克隆抗体，对PEG诱导细胞融合条件进行优化与筛选。

(发表于《中国毒理学会第三届中青年学者科技》)

喹乙醇免疫抗原的合成及评价

张景艳，李建喜，王　磊，杨志强，王学智，张　凯，孟　嘉

(中国农业科学院兰州畜牧与兽药研究所中兽药工程中心，兰州 730050)

摘　要：本研究旨在从喹乙醇分子上的-OH位点出发，合成喹乙醇全抗原，制备抗喹乙醇的血清，优化间接EILSA评价方法，为喹乙醇残留快速检测方法的建立提供依据。

(发表于《中国毒理学会第三届中青年学者科技》)

散养产蛋鸡群感染前殖吸虫的诊治

郝宝成[1]，王学红[1]，郭文柱[1]，郭志廷[1]，朱银萍[2]，胡永浩[3]，梁剑平[1]

(1. 中国农业科学院兰州畜牧与兽药研究所，兰州 730050；2. 甘肃省兰州市红古区王家口工农联合学校，兰州 730030；3. 甘肃农业大学，兰州 730070)

摘 要：2010年4月，兰州市红古区某蛋鸡养殖户745只甘肃本地柴鸡中，有135只蛋鸡突然出现产薄壳蛋，并间隔伴有软壳蛋，产蛋率下降症状。养殖户疑为缺钙，及时用维生素D、钙制剂等药物添加到饲料中进行治疗。20d后，蛋鸡产蛋停止，出现不食，腹部膨大等症状，并间歇有11只鸡陆续死亡。经过临床观察、病理解剖、粪便检查，确诊该鸡群患前殖吸虫病。

(发表于《中国家禽》)

人为干扰对玛曲高寒退化草地的影响

田福平[1,2]，陈子萱[3]

(1. 甘肃省农业科学院生物技术研究所，兰州 730070；2. 中国农业科学院兰州畜牧与兽药研究所/农业部兰州黄土高原生态环境重点野外科学观测试验站，兰州 730050；3. 甘肃农业大学草业学院，兰州 730070)

摘 要：近年来玛曲高寒草地退化日趋严重，位于黄河第一弯的甘肃省玛曲县是黄河径流重要的汇集区和黄河上游至源头的重要水源涵养区。补充黄河水量约达45%，但由于人口和牲畜的增加，掠夺式经营使草场退化严重，湿地萎缩。全县近90%的天然草地出现了不同程度的退化和沙化，其中沙化面积达$5 \times 10^4 hm^2$以上，且每年以10.8%的速度蔓延。文章综述了近年来人为干扰对玛曲高寒退化草地的影响，对玛曲县高寒草地生态环境恢复和沙漠化防治有重要意义。
关键词：玛曲；高寒草地；人为干扰；影响

(发表于《中国农学通报》)

微生物检测中糖量子点的制备及其与凝集素的相互作用

孔潇艺[1,2]，程水红[2]，刘雪峰[3]，李学兵[2]，吴培星[4]

(1. 天津科技大学食品工程与生物技术学院，天津 300457；2. 中国科学院微生物研究所/中科院病原微生物与免疫学重点实验室，北京 100101；3. 国家纳米科学中心，北京 100190；4. 中国农业科学院兰州畜牧与兽药研究所，兰州 730050)

摘 要：为了以量子点为探针建立简便快速的病原微生物检测方法，采用配体交换法将具有微生物特异性的糖链偶联到 CdSeS/ZnS 量子点表面，制备 2 种糖量子点复合材料（甘露糖–量子点和半乳糖–量子点）。通过对它们的结构、物性和光谱学特性进行表征，结果显示，2 种糖量子点均具有良好的水溶性、荧光发射和紫外可见光吸收性能。进一步以这 2 种糖量子点为探针与凝集素共孵育后，溶液颜色和荧光发射光谱的变化表明：甘露糖–量子点对刀豆蛋白具有特异性，最低可检测浓度为 95nmol/L；而半乳糖–量子点对蓖麻毒素具有特异性，最低可检测浓度为 3nmol/L。因为细菌和病毒表面的黏附素多为凝集素类蛋白，本研究为进一步建立病原微生物的检测方法奠定了基础。

关键词：微生物检测；量子点；糖；复合材料；凝集素

(发表于《中国农业大学学报》)

响应面法优化贯叶连翘中金丝桃素超声提取工艺

郭 凯[1,2]，梁剑平[1]，华兰英[1]，王学红[1]，石广亮[3]

(1. 中国农业科学院兰州畜牧与兽药研究所新兽药工程重点实验室，兰州 730050；2. 中国农业科学院研究生院，北京 100081；3. 甘肃农业大学动物医学院，兰州 730070)

摘 要：[目的] 以贯叶连翘为原料，应用响应面法探求超声提取金丝桃素的最优工艺。[方法] 在对影响金丝桃素提取率的单因素系统考察的基础上，采用部分析因设计分析筛选以上因素，并应用响应面法对其显著因素进行研究和优化。[结果] 建立了超声提取贯叶连翘中金丝桃素的工艺数学模型。超声提取金丝桃素的最佳工艺条件为超声提取时间 36.15min，丙酮/乙醇混合提取溶剂中丙酮体积分数为 68.14%。液料比为

11.87mL·g^{-1}，此时金丝桃素提取率理论值町达到1.454%。经验证，提取率为1.44%，与理论值相符。主效应关系为超声提取时间>丙酮体积分数>液料比。[结论]通过以上试验得到的提取工艺条件可靠，金丝桃素提取率远高于常规提取方法，可为金丝桃素的工业化放大生产提供了一定的理论依据。

关键词：贯叶连翘；金丝桃素；超声提取；部分析因设计；响应面法

（发表于《中国实验方剂学杂志》）

大蒜多糖对伴刀豆蛋白A致肝损伤小鼠肝组织抗氧化能力的影响

程富胜[1]，程世红[2]，张 霞[3]

(1. 中国农业科学院兰州畜牧与兽药研究所农业部兽用药物创制重点实验室甘肃省新兽药工程重点实验室，兰州 730050；2. 兰州大学实验室与设备管理处，兰州 730000；3. 甘肃农业大学生命科学技术学院，兰州 730070)

摘 要：为了探讨大蒜多糖（GAP）对肝脏氧化还原性的调节作用，试验就GAP对伴刀豆蛋白A（ConA）致肝损伤小鼠肝组织中超氧化物歧化酶（SOD）、丙二醛（MDA）、谷胱甘肽（GSH）影响进行了研究。50只小鼠随机平均分成5组：空白对照组，模型组，多糖不同剂量组（低、中、高）；多糖组分别以100mg/kg，200mg/kg，400mg/kg体重的大蒜多糖对小鼠进行灌胃，然后采用刀豆蛋白A（ConA）静脉注射致小鼠免疫性肝损伤，摘取肝脏，进行肝组织匀浆中SOD活力、MDA、GSH含量测定。结果显示，与ConA模型组比较，多糖低剂量组SOD活力提高不显著（$P>0.05$），MDA含量降低显著（$P<0.05$），GSH含量升高不显著（$P>0.05$）；多糖中剂量组SOD活力提高显著（$P<0.05$），MDA含量降低极显著（$P<0.01$），GSH含量升高显著（$P<0.05$）；多糖高剂量组SOD活力升高极显著（$P<0.01$），MDA含量降低极显著（$P<0.01$），GSH含量升高极显著（$P<0.01$）。试验表明，大蒜多糖能够有效提高机体抗氧化机能，对ConA所致的免疫性肝损伤有一定保护作用，并且在一定范围内存在着量效相关性。

关键词：大蒜多糖；ConA；SOD；MDA；GSH

（发表于《中国兽医杂志》）

牛源金黄色葡萄球菌 16S rRNA 分子鉴定与随机多态性扩增分型

邓海平，蒲万霞，梁剑平，倪春霞，孟晓琴

（中国农业科学院兰州畜牧与兽药研究所中国农业科学院新兽药工程重点开放实验室/甘肃省新兽药工程重点实验室，兰州 730050）

摘　要：利用 16S rRNA 保守序列对引起内蒙古地区奶牛乳房炎的金黄色葡萄球菌进行分离鉴定，并且建立引物随机多态性扩增（RAPD）体系对金黄色葡萄球菌分离株进行基因分型研究。结果表明，共分离出金黄色葡萄球菌 35 株，RAPD 结果显示这 35 株菌均可得到清晰的 RAPD 指纹图谱，扩增产物在 1~8 条带之间，产物大小在 350~4 500bp 之间。菌株分为 6 个基因型，Ⅳ型为该地区的流行优势菌群。不同牛场各基因型菌株分布有明显差异，这与牛场的环境和病原菌在牛场间流行传播情况有关。研究结果为地区性奶牛乳房炎的防治及新疫苗的研制提供了可靠的理论依据。

关键词：奶牛乳房炎；金黄色葡萄球菌；16S rRNA；RAPD；基因型

（发表于《中国兽医杂志》）

几种中草药对绵羊瘤胃发酵、抗氧化功能和营养物质消化率的影响

乔国华，杨　晓，周学辉，张怀山，王晓力，李锦华

（中国农业科学院兰州畜牧与兽药研究所）

摘　要：本研究包括 2 个试验。试验一，采用单因素随机区组试验设计研究了黄芪、党参、女贞子对绵羊屠宰体增重和抗氧化功能的影响。试验共 4 个处理，含 3 种中草药和 1 个对照组。每个处理 4 个重复。每个重复 1 只羊。中草药添加量为 6g/d。结果表明，女贞子组提高了绵羊屠宰体增重（$P<0.05$），改善了绵羊血液抗氧化功能，提高了谷胱甘肽还原酶和超氧化物歧化酶活性，降低了血浆丙二醛含量（$P<0.05$）。其他组对绵羊屠宰体增重和血液抗氧化功能影响不显著（$P>0.05$）。试验二，采用 4×4 拉丁方试验设计研究了女贞子提取物对绵羊瘤胃发酵和血液生化指标的影响。试验分 4 个处理，女贞子提取物的添加量分别为 0、100、300mg/kg 和 500mg/kg。结果表明，添加量为

300mg/kg 和 500 mg/kg 组在晨饲后 2、4、6、8h 显著提高了绵羊瘤胃液总挥发性脂肪酸浓度和丙酸的浓度，降低了瘤胃液氨态氮的浓度和血浆尿素氮的浓度。300mg/kg 和 500mg/kg 组显著提高了日粮有机物质和干物质表现消化率（$P<0.05$）。

关键词：中草药；瘤胃发酵；营养物质消化率；抗氧化功能；绵羊

（发表于《中国饲料》）

不同苜蓿品种叶面积与抗旱性的关联性研究

张怀山[1]，代立兰[2]，乔国华[1]，王春梅[1]，杨 晓[1]

(1. 中国农业科学院兰州畜牧与兽药研究所，兰州 730050；2. 甘肃省兰州市农业科技研究推广中心，兰州 730000)

摘 要：[目的] 研究不同叶面积苜蓿品种的抗旱性及其两者的关系。[方法] 通过研究紫花苜蓿在干旱条件下不同品种叶面积脯氨酸含量的差异性，探讨紫花苜蓿抗旱性与叶面积大小的关联性。[结果] 在旱作条件下，不同苜蓿品种的抗旱性与其叶面积大小有直接关系，叶面积越大的苜蓿品种脯氨酸含量越高，其抗旱性越强。[结论] 叶片面积可以作为衡量不同苜蓿品种抗旱性强弱的一个较为直观的辅助性指标。

关键词：紫花苜蓿；脯氨酸；叶面积；抗旱性

（发表于《安徽农业科学》）

流产母马血液流变学研究

严作廷[1,2]，荔 霞[1,2]，王东升[1,2]，张世栋[1,2]

(1. 中国农业科学院兰州畜牧与兽药研究所，兰州 730050；2. 中国农业科学院临床兽医学研究中心，兰州 730050)

摘 要：[目的] 研究流产对母马血液流变学的影响。[方法] 分别检测正常非孕母马、正常怀孕母马、流产母马、正常分娩母马的血液流变学指标，并作比较。[结果] 正常非孕母马与正常怀孕母马的各项血液流变学指标无明显差异（$P>0.05$）。正常分娩母马的全血比黏度、全血还原比黏度、血浆比黏度、血细胞压积、红细胞电泳时间均显著高于正常非孕母马（$P<0.001$）。而流产母马的各项血液流变学指标均显著高于正常分娩母马（$P<0.001$）。[结论] 母马流产后血液流变学存在异常。

关键词：流产母马；血液流变学；指标

(发表于《安徽农业科学》)

黄花矶松栽培驯化试验

常根柱，路 远，周学辉，杨红善，屈建民

(中国农业科学院兰州畜牧与兽药研究所，兰州 730050)

摘 要：对生长于北方沙漠、戈壁等干旱地区具有开发应用前景的园林花卉、药用及防风固沙植物，亦为荒漠地区的可食牧草黄花矶松进行了驯化栽培试验。结果表明，人工栽培的黄花矶松难度在于种子表层的腊质化，对其种子采用化学方法处理最为有效，育苗出苗率可达73%，田间出苗率可达54%；其次为机械处理方法，育苗和田间出苗率分别为58%、41%。经人工栽培驯化，黄花矶松可以育苗、移栽、盆栽和田间栽培。栽培方式的不同，可使其根系生长的形态结构有所变化，即田间栽培具有明显的主根，而盆栽则须根发达，说明该种植物人工栽培的适宜性较强。

关键词：栽培驯化；试验；黄花矶松

(发表于《北方园艺》)

12个引进苜蓿品种适应性栽培试验

杨红善，常根柱，周学辉，路 远

(中国农业科学院兰州畜牧与兽药研究所，兰州 730050)

摘 要：苜蓿王、菲尔兹、金字塔、里奥、辛普劳2 000、诺瓦、霍普兰德、多叶苜蓿、大叶苜蓿、Spreador-3、Pick3006和Pick8925是从美国和加拿大引进的优良紫花苜蓿(*Medicago sativa*) 品种，以国产中兰1号为对照品种，2003年在半干旱区兰州开展引种试验；2008年在半湿润区天水甘谷继续开展适应性试验，同时对营养成分和抗旱指标进行测定。结果表明，引进的12个苜蓿品种，在半干旱、半湿润区均能正常生长，完成生育周期的全过程，无异常表现，未发现病虫害侵染。各项试验得出，苜蓿王的抗旱性强 鲜草产量在兰州平均达54 207.00kg/hm²，在甘谷平均达80 883.65kg/hm³；PICK 3006农艺性状仅次于苜蓿王；Spreador-3具叶大、杆中实、抗倒伏等优良特性 对照中兰1号具优质、高产、抗旱特点；建议这4个苜蓿品种在我国黄土高原半干旱及半湿润地区推广种植。

关键词：紫花苜蓿；引种；栽培试验

(发表于《草业科学》)

牧区放牧管理传统乡土知识挖掘

杨红善，周学辉，苗小林，常根柱

(中国国农业科学院兰州畜牧与兽药研究所，兰州 730050)

摘　要：畜牧业生产力的提高往往被认为与先进的科学技术相关，而传统的放牧管理乡土知识却有逐渐被研究者遗弃的趋向。为搜集、挖掘一些长期以来牧区、半农半牧区及少数民族居住区的放牧管理乡土知识，总结提出一整套科学、合理、实用的放牧方法提供实践材料，以此提高畜牧业生产力。通过深入两个牧区县(肃北蒙古族自治县、肃南裕固族自治县)和两个半农半牧区县(甘肃永昌县、甘肃景泰县)进入多家农牧户家中采访、座谈，挖掘他们的先辈以及他们多年来的传统放牧经验。挖掘的乡土知识包括：放牧时间、放牧方式、不同季节的放牧方法、疾病预防与治疗等许多内容。认为应该认真总结传统的乡土知识对农牧区畜牧业发展的潜在积极作用。

关键词：乡土知识；放牧管理；牧区

(发表于《草业与畜牧》)

沙拐属植物种质资源描述规范和数据标准的研究

田福平，杨志强，时永杰，路　远，张　茜

(中国国农业科学院兰州畜牧与兽药研究所/农业部兰州黄土高原生态环境重点野外科学观测试验站，兰州 730050)

摘　要：沙拐枣属植物为灌木或半灌木，多分枝，是荒漠地区优良固沙和饲料植物。文章论述了沙拐枣属植物种质资源描述规范和数据标准制定的原则、方法及主要研究内容，对于规范全国沙拐枣属植物种质资源的收集、整理、保存、鉴定、评价、信息与实物共享具有重要意义。

关键词：沙拐枣属；牧草种质资源；描述规范；数据标准

(发表于《草业与畜牧》)

我国毛皮产业风险预警系统建设

郭天芬[1,2], 王宏博[1,2], 李维红[1,2]

(1. 中国农业科学院兰州畜牧与兽药研究所,兰州 730050; 2. 农业部动物毛皮及制品质量监督检验测试中心,兰州 730050)

摘 要:我国是毛纤维及毛皮生产及加工大国。这是由于我国的毛、皮产业的生产、加工、流通等各环节之间相互脱节,存在一定的不规范性,且动物毛、皮产业是高风险的产业,为了使我国的毛纤维及毛皮产业健康稳定地发展,本文建议建立毛皮产业风险预警系统,并阐述了毛、皮产业风险预警系统的构成及建设步骤。

关键词:毛;皮;产业;风险预警系统

(发表于《畜牧兽医科技信息》)

旱獭、麝鼠、兔狲、青鼬、石貂毛绒纤维超微结构比较

李维红[1,2,3], 吴建平[1]

(1. 甘肃农业大学动物科技学院,兰州 730060; 2. 中国农业科学院兰州畜牧与兽药研究所,兰州 730050; 3. 农业部动物毛皮及制品质量监督检验测试中心(兰州),兰州 730050)

摘 要:以旱獭、麝鼠、兔狲、青鼬、石貂为试验材料,观察其毛绒纤维的超微结构,利用扫描电镜法比较其鳞片层结构特征。结果显示,旱獭针毛翘角平均值为37.0°,鳞片高度平均值为7.76μm,鳞片厚度平均值为0.43μm,旱獭绒毛翘角平均值为25.4°,鳞片高度平均值为14.57μm,鳞片厚度平均值为0.68μm;麝鼠针毛鳞片高度平均值为4.58μm,鳞片厚度平均值为0.22μm,麝鼠绒毛翘角平均值为24.7°,鳞片高度平均值为11.14μm,鳞片厚度平均值为0.40μm;兔狲针毛翘角平均值为31.9°,鳞片高度平均值为10.65μm,鳞片厚度平均值为0.52μm,兔狲绒毛翘角平均值为24.8°,鳞片高度平均值为9.30μm,鳞片厚度平均值为o.46μm;青鼬针毛翘角平均值为37.8°,鳞片高度平均值为4.33μm,鳞片厚度平均值为0.24μm,青鼬绒毛鳞片高度平均值为13.88μm,鳞片厚度平均值为0.65μm;石貂针毛翘角平均值为33.6°,鳞片高度平均值

为 23.93μm，鳞片厚度平均值为 0.74μm，石貂绒毛翘角平均值为 25.2°，鳞片高度平均值为 29.87μm，鳞片厚度平均值为 0.64μm。不同种的动物纤维具有独特的形态特征，在超微结构上存在明显的差别。

关键词：动物纤维；扫描电镜；超微结构

（发表于《畜牧兽医学》）

阿拉善沙拐枣引种驯化栽培试验

常根柱，张 茜，路 远，杨红善，周学辉

（中国国农业科学院兰州畜牧与兽药研究所，兰州 730050）

摘 要：：通过 10 年（2001~2010）的驯化栽培，将具有高抗旱、耐风沙的野生灌木阿拉善沙拐枣从民勤沙区引种至兰州获得成功。试验结果表明，经过驯化栽培，该种植物在兰州地区人工栽培条件下，能够正常生长、结籽和成熟，从播种（移栽）出苗（萌发）至结籽成熟生长 160d 以上。植株高度 80~113cm，5 月出苗（萌发），8 月开花，10 月种子成熟。干鲜比 1:4.4，茎叶比 1:2.0。种子产量 538.4kg/hm² （千粒重 64.46g），青草产量 12 809.8kg/hm²。根系发达，生长速度快，种植第 5 年的植株地上和地下生物量各接近 50%，其根系具有很强的固土护坡能力。营养含量中等，可消化吸收率高。阿拉善沙拐枣在我国黄土高原干旱、半干旱地区可作为水土保持和生态建设灌木树种使用，亦可饲用。种植方式以育苗移栽效果最佳，耐粗放管理，种苗成活后不再需浇水、施肥。

关键词：阿拉善沙拐枣；引种；驯化栽培

（发表于《中国草地学报》）

兰州大尾羊微卫星 DNA 多态性研究

郎 侠[1]，吕潇潇[2]

（1. 中国农业科学院兰州畜牧与兽药研究所，兰州 730050；2. 甘肃农业大学动物科技学院，兰州 730070）

摘 要：研究旨在分析兰州大尾羊的微卫星 DNA 多态性，以期了解该品种的遗传多样性，为该品种保种选育和品质的进一步提高提供资料。选择 15 对微卫星引物，通过计算基因频率、多态性信息含量、有效等位基因数和杂合度，评估其品种内的遗传变异。结果表明，在 15 个座位中，共检测到 153 个等位基因，每个座位平均为 10.2 个等位基因；座位平均杂合度为 0.814 9；平均有效等位基因数为 6.089 1；平均多态性信息含量为 0.776 2。结果提示，兰州大尾羊群体存在丰富的遗传多样性，所选微卫星标记可用于绵羊遗传多样性评估。

关键词：微卫星；遗传多样性；兰州大尾羊

(发表于《中国畜牧杂志》)

鸡传染性喉气管炎及其防制

严作廷[1]，刘家彪[2]，严建鹏[3]

(1. 中国农业科学院兰州畜牧与兽药研究所中国农业科学院临床兽医学研究中心，兰州 730050；2. 甘肃省家禽业协会，兰州 730040；3. 甘肃省兰州市动物疾控中心，兰州 730050)

摘　要：鸡传染性喉气管炎是由鸡传染性喉气管炎病毒引起的鸡的一种传播快速的急性接触性上部呼吸道传染病。以呼吸困难、气喘、咳出血样渗出物为特征。本病主要侵害鸡，育成鸡和成年蛋鸡多发。褐羽蛋鸡发病后较为严重。本病主要通过呼吸道传播，本病康复鸡可带毒一年以上，因此病鸡和康复后带毒鸡是本病的主要传染来源。本病自然感染的潜伏期为 6~12 天，病程一般 10~14 天，部分鸡群可延长到 3~4 周，疫情易于反复。经统计，产蛋鸡群发病后因发病程度不同产蛋率可下降 20%~40%，产出畸形蛋、薄皮蛋、沙皮蛋、软壳蛋等不正常蛋的比例可达 2%~10%，严重鸡群可达 5%~14.8%，多数鸡群病愈后经过 2~4 周产蛋率仍可恢复接近至发病前的水平。经统计，鸡群的发病率可达 5.7%~68.4%，死亡率为 5.6%~56.3%，病死率可达 70.4%~89.3%，其中多数产蛋鸡群的死亡率一般在 7.7%~68.4%，商品肉鸡群、肉杂鸡、肉公鸡及其它鸡群的死亡率较蛋鸡群低，一般在 4.6%~34.2%。若传染性喉气管炎继发感染其它疫病时，死亡率可达 40.2%~80.4%。

(发表于《中国畜禽种业》)

阿司匹林丁香酚酯的高效解热作用及作用机制

叶得河[1]，于远光[1]，李剑勇[2]，杨亚军[2]，张继瑜[2]，
周旭正[2]，牛建荣[2]，魏小娟[2]，李 冰[2]

(1. 甘肃农业大学动物医学院，兰州 730070；2. 中国农业科学院兰州畜牧与兽药研究所/农业部兽用药物创制重点实验室/甘肃省新兽药工程重点实验室，兰州 730050)

摘 要：[目的] 探讨阿司匹林丁香酚酯 (AEE) 解热作用及其作用机制。[方法] 采用 sc 给予 Wistar 大鼠15%酵母混悬液10mL·kg^{-1}制备发热模型。体温升高> 0.8℃的大鼠按分组分别 ig 给予阿司匹林 0.27g·kg^{-1}、丁香酚 0.24 g·kg^{-1}、AEE 0.32，0.48，0.65 g·kg^{-1}观察给药后 2h，4h 和 6h 后大鼠体温，6h 后采血取脑，应用酶联免疫法 (ELISA) 测定致热大鼠腹中隔区及血浆中精氨加压素 (AVP) 的含量和下丘脑中及血浆中环磷酸腺苷 (cAMP) 的含量。结果与发热模型组 2h 和 6h 自然降温相比较，ig 给予 AEE 0.32，0.48 和 0.65 g·kg^{-1}、阿司匹林 0.27 g·kg^{-1}、丁香酚 0.24 g·kg^{-1}后 2 h，大鼠体温分别降温 (-1.2±28)，(-1.14±35)，(-2.09±45)，(-2.19±32)，(-0.94±0.42)℃；6h 后分别降温 (-1.32±34)，(-1.45±0.41)，(-2.49±49)，(-1.78±51)，(-1.21±29)℃，差异显著 ($P<0.05$)。比较同组药后 6h 与 2h 的降温作用发现，阿司匹林解热药效作用下降较快，差异显著 ($P<0.05$)；丁香酚药效作用时间较为持久，但无增加或下降差异；AEE 0.65g·kg^{-1}药效快速持久，降温作用增加，差异显著 ($P<0.05$)。药后 6h，AEE 0.65 g·kg^{-1}解热降温作用明显强于阿司匹林和丁香酚，差异极显著 ($P<0.01$)。发热模型组中下丘脑及血浆中 cAMP 的含量与腹中隔区、血浆中 AVP 含量较正常对照组高。与发热模型组相比，AEE 0.32，0.48 和 0.65g·kg^{-1}组、阿司匹林 0.27g·kg^{-1}组、丁香酚 0.24g·kg^{-1}组中腹中隔区中 AVP 含量明显下降，而血浆中 AVP 含量明显升高，下丘脑中 cAMP 的含量明显降低，血浆中 cAMP 的含量变化不明。结论 AEE 的解热作用药效快速持久，明显优于阿司匹林与丁香酚，其解热机制可能通过改变下丘脑中 cAMP 的含量和腹中隔区、血浆中 AVP 含量而发挥作用。

关键词：阿司匹林丁香酚酯；镇痛药；非麻醉

(发表于《中国药理学与毒理学杂志》)

阿司匹林丁香酚酯的抗炎作用及其可能的作用机制

李剑勇[1]，王棋文[2]，于远光[3]，杨亚军[1]，

牛建荣[1]，周旭正[1]，张继瑜[1]，魏小娟[1]，李 冰[1]

(1. 中国农业科学院兰州畜牧与兽药研究所农业部兽用药物创制重点实验室甘肃省新兽药工程重点实验室，兰州 730050；2. 毕节学院试验区研究院草地生态研究所，毕节 551700；3. 甘肃农业大学动物医学院，兰州 730070)

摘 要：[目的] 研究阿司匹林丁香酚酯（AEE）的抗炎作用及其可能的作用机制。[方法] 阿司匹林 0.2 g·kg^{-1} 组、丁香酚 0.18 g·kg^{-1} 组、AEE0.18 和 0.36 g·kg^{-1} 组小鼠按体重 ig 给予相应药物，每天 1 次，连续 3d。用二甲苯致小鼠耳肿胀模型、小鼠腹腔毛细血管通透性增高模型、角叉菜胶致小鼠足跖肿胀模型及小鼠棉球肉芽肿模型观察 AEE 的抗炎作用，并测定血清中一氧化氮（NO）和足爪组织中前列腺素 E$_2$（PGE$_2$）、丙二醛（MDA）和 5-羟色胺（5-HT）的含量。[结果] 二甲苯致小鼠耳肿胀实验，模型组肿胀度为 (7.0±1.2) mg，阿司匹林、丁香酚、AEE0.18 和 0.36 g·kg^{-1} 明显抑制耳肿胀度，肿胀度分别为 3.9±1.2，4.7±1.9，3.9±1.9 和 (3.8±1.1) mg（$P<0.05$，$P<0.01$）。小鼠腹腔毛细血管通透性实验结果表明，模型组伊文思蓝的渗出（A$_{590}$ nm）为 0.56±0.11，阿司匹林、丁香酚、AEE0.18 和 0.36 g·kg^{-1} 组腹腔毛细血管通透性均降低，伊文思蓝的渗出（A$_{590}$ nm）分别为 (0.29±0.15) mg，(0.34±0.12) mg 和 (0.35±0.18) mg（$P<0.05$，$P<0.01$）。小鼠棉球肉芽肿实验结果表明，模型组肉芽肿增生 (12.5±2.4) mg，阿司匹林、丁香酚、AEE0.18 和 0.36 g·kg^{-1} 组明显降低，分别为 (7.0±2.1) mg，(9.0±1.7) mg，(9.9±1.4) mg 和 (8.8±1.8) mg（$P<0.01$）；角叉菜胶致小鼠足跖肿胀实验结果表明，AEE0.18 g·kg^{-1} 显著抑制角叉菜胶致炎小鼠足跖肿胀（$P<0.05$），降低致炎足组织中 PGE2，5-HT 和血清 NO 含量，作用与阿司匹林和丁香酚较强或相当。[结论] AEE 具有明显的抗炎作用，与阿司匹林和丁香酚作用相当；其抗炎机制可能与抑制炎性介质释放有关。

关键词：阿司匹林；丁香酚；阿司匹林丁香酚酯；炎症

（发表于《中国药理学与毒理学杂志》）

非抗生素疗法在防治奶牛子宫内膜炎上的研究概况

严作廷，荔 霞，王东升，陈炅然，张世栋

(中国国农业科学院兰州畜牧与兽药研究所/中国农业科学院临床兽医学研究中心，兰州 730050)

摘 要：奶牛子宫内膜炎是造成奶牛不孕症的主要原因之一，对奶牛业的危害较大。国内外学者对奶牛子宫内膜炎的防治作了大量研究，提出了许多行之有效的治疗方法。目前治疗本病的非抗生素治疗方法包括生物疗法、激素疗法、物理疗法、中草药疗法等。但最有应用前景的是生物疗法和中草药疗法。今后应加强生物疗法和中草药方面的研究，同时应加强中草药制剂的质量控制和安全性评价，以及中草药的作用机理方面的研究。

关键词：子宫内膜炎；非抗生素疗法；奶牛；防治

(发表于《畜牧与兽医》)

西门塔尔牛在甘肃省的利用与发展

郭 宪，梁春年，丁学智，阎 萍

(中国国农业科学院兰州畜牧与兽药研究所甘肃省牦牛繁育工程重点实验室，兰州 730050)

摘 要：本文在分析甘肃省肉牛产业现状的基础上，从西门塔尔牛的外貌特征、生产性能、培育以及在甘肃省的适应性等方面讨论了西门塔尔牛的中国化，并从产肉性能、产乳性能、繁殖性能、役用性能等方面分析了西门塔尔牛在甘肃省的利用，最后讨论了西门塔尔牛在甘肃省的进一步发展与应用方向，旨在促进甘肃省肉牛产业持续、健康发展。

关键词：西门塔尔牛；甘肃省；利用

(发表于《畜牧与兽医》)

针刺麻醉机制研究进展

王贵波[1]，罗超应[1]，郑继方[1]，李锦宇[1]，王东升[1]，丁明星[2]

(1. 中国农业科学院兰州畜牧与兽药研究所，兰州 730050；2. 华中农业大学动物医学院，武汉 430070)

摘 要：针刺麻醉是针灸与麻醉学相结合的产物，具有简便安全、可操作性强、对动物生理干扰小、术后恢复快等优点。近年来，为更好的将针刺麻醉应用于临床，人们对针刺麻醉的机理进行了进一步研究。本文对针刺镇痛的神经机制，神经化学递质和分子水平及经络的作用机制进行了简要回顾。

关键词：钟刺；麻醉；机翻

(发表于《畜牧与兽医》)

现代生物技术在动物营养中的应用

郭天芬[1,2]，王宏博[1,2]，李维红[1,2]

(1. 中国农业科学院兰州畜牧与兽药研究所，兰州 730050；2. 农业部动物毛皮及制品质量监督检验测试中心，兰州 730050)

摘 要：现代生物技术是在传统生物技术基础上发展起来的，以DNA重组技术的建立为标志，以现代生物学研究成果为基础，以基因或基因组为核心，现已辐射到各个生物科技领域。从现代生物技术在动物营养中的作用、应用及营养调控等几个方面进行了概述。

关键词：生物技术；动物营养；营养调控

(发表于《畜牧与饲料科学》)

羊毛纤维的结构及影响羊毛品质的因素

郭天芬[1,2]，李维红[1,2]，牛春娥[1,2]，王宏博[1,2]

（1. 中国农业科学院兰州畜牧与兽药研究所，兰州 730050；2. 农业部动物毛皮及制品质量监督检验测试中心，兰州 730050）

摘 要：目前，我国是世界上最大的羊毛加工国。通过介绍羊毛的结构及羊毛的形成过程。分析了影响羊毛品质的因素，以便使羊毛生产人员了解羊毛品质的有关制约因素。并利用科学技术提高羊毛产量和品质。

关键词：羊毛；结构；品质；因素

（发表于《畜牧与饲料科学》）

非甾体抗炎药物研究进展

王海为[1,2]，李剑勇[1]，杨亚军[1]，于远光[1,2]

（1. 中国农业科学院兰州畜牧与兽药研究所/甘肃省新兽药工程重点实验室/中国农业科学院新兽药工程重点实验室，兰州 730050；2. 甘肃农业大学动物医学院，兰州 730046）

摘 要：非甾体抗炎药（NSAIDs）包括传统的非甾体类抗炎药（COX-1 抑制剂）和非传统的非甾体类抗炎药（COX-2 抑制剂）。论文分析了国内外市场状况，并指出临床应用所存在的问题。介绍了常用非甾体抗炎药的作用机制，总结了常用抗炎药不良反应，提出了应用时应注意的事项。

关键词：非甾体抗炎药物；作用机制；临床应用；不良反应

（发表于《动物医学进展》）

预防围产期奶牛酮病的营养策略

乔国华,李锦华,周学辉,杨　晓,张　茜,王春梅

(中国国农业科学院兰州畜牧与兽药研究所草业饲料研究室,兰州　730050)

摘　要:本文综述了围产期奶牛的脂肪代谢变化特点,对处在围产期奶牛酮病预防的营养调控策略进行了总结,以期为奶牛生产提供理论基础和实践参考。

关键词:围产期;奶牛;酮病;营养策略

(发表于《动物营养学报》)

STATs家族基因多态性在奶牛育种中的研究进展

褚　敏,阎　萍,裴　杰,梁春年,郭　宪,曾玉峰,包鹏甲

(中国国农业科学院兰州畜牧与兽药研究所,兰州　730050)

摘　要:信号转导及转录激活因子家族(STATs)是信号转导途径JAK-STAT的重要组成部分,该家族蛋白被激活以后和相应的受体结合,随后穿过核膜与相应的靶基因位点结合,从而调控基因的转录。研究表明,该家族部分成员与奶牛的泌乳性能有密切的关系。本文将近几年该家族基因多态性的研究进展作一综述。

关键词:信号转导及转录激活因子家族;基因多态性;奶牛育种

(发表于《黑龙江畜牧兽医》)

大环内酯类抗寄生虫药物的研究进展

汪 芳，周绪正，李 冰，张继瑜

(中国国农业科学院兰州畜牧与兽药研究所/农业部兽用药物创制重点实验室/甘肃省新兽药工程重点实验室，兰州 730050)

摘 要：大环内酯类抗寄生虫药物具抗动物体内外寄生虫的双重作用，对该类药物的了解有助于进行深入研究和指导临床应用。文章综述了大环内酯类抗寄生虫药物理化性质、作用机制、药物代谢动力学特征、临床药效学的研究概况，并对研发趋势进行了展望，旨在为开发适用于肉食动物和宠物的体内外抗寄生虫药物及以药物作用靶标为基础的新药研制提供参考。

关键词：大环内酯类；抗寄生虫药物；理化性质；作用机制；药物代谢动力学；药效学

(发表于《黑龙江畜牧兽医》)

环形泰勒虫主要膜蛋白的研究进展

蔺红玲，张继瑜，魏小娟，李 冰

(中国国农业科学院兰州畜牧与兽药研究所新兽药工程重点开放实验室，兰州 730050)

摘 要：环形泰勒虫病是由蜱传播引起的一种血液原虫病，为世界各国公认的危害养牛业最重要的寄生虫病之一。环形泰勒虫各发育阶段表面抗原是防制本病的研究基础，目前已经鉴定和正在研究的环形泰勒虫表面抗原有子孢子表面抗原、裂殖体表面抗原、裂殖子表面抗原及热休克蛋白等，它们不但具有很好的免疫原性，而且具有很重要的生理功能，可作为药物靶点和疫苗的候选位点。本文就上述几种膜蛋白抗原的结构、功能及应用作一概述。

关键词：环形泰勒虫；膜蛋白；结构；药物靶点

(发表于《黑龙江畜牧兽医》)

家畜 H-FABP 基因的研究进展

曾玉峰，阎　萍，梁春年，郭　宪，包鹏甲，裴　杰，褚　敏

（中国国农业科学院兰州畜牧与兽药研究所，兰州　730050）

摘　要：肌内脂肪含量与肉的风味和嫩度等有关，心型脂肪酸结合蛋白基因作为影响肌内脂肪含量的候选基因之一，已经引起国内外研究人员的关注。本文就心型脂肪酸结合蛋白的生物学功能、心型脂肪酸结合蛋白基因的结构、性质及其在家畜中的相关研究进展等作一详细阐述，并提出今后的研究前景，为深入研究该基因提供参考。

关键词：家畜；心型脂肪酸结合蛋白；候选基因

（发表于《黑龙江畜牧兽医》）

牦牛数量性状基因的研究进展

包鹏甲，阎　萍，梁春年，郭　宪，裴　杰，褚　敏，朱新书

（中国国农业科学院兰州畜牧与兽药研究所，兰州　730050）

摘　要：牦牛在青藏高原具有广泛分布，对高寒、高海拔地区有较好的适应性，集肉、乳、毛、皮、役等生产性能于一身，是青藏高原上特有家畜和"全能家畜"；但是牦牛是比较原始的品种，饲养管理粗放，生产性能没有完全发挥出来。为提高牦牛生产性能已开展了一系列的研究，并取得了一些成果。随着现代生物技术研究的深入，如何从基因角度诠释和高效利用牦牛，已经成为近年牦牛研究的热点，文章仅就牦牛生产性能相关数量性状基因的研究进展作一综述。

关键词：牦牛；生产性能；基因；数量性状

（发表于《黑龙江畜牧兽医》）

饲料添加剂喹胺醇的研究进展

郭文柱，梁剑平，王学红，郭志廷，郝宝成，尚若锋，华兰英

（中国国农业科学院兰州畜牧与兽药研究所/农业部新兽药创制重点实验室/甘肃省新兽药工程重点实验室，兰州 730050）

摘　要：为了更加系统地了解喹胺醇作为饲料添加剂方面的研究进展，通过查阅大量文献资料，总结了喹胺醇的药理学、毒理学、药代动力学、残留试验以及临床试验的研究成果，最后推断出喹胺醇是一种高效、低毒的饲料添加剂。

关键词：喹胺醇；喹乙醇；毒性；残留

（发表于《黑龙江畜牧兽医》）

抑制素免疫及其在肉牛繁殖中的应用

郭宪，阎萍，梁春年，包鹏甲，朱新书，褚敏，裴杰

（中国国农业科学院兰州畜牧与兽药研究所，兰州 730050）

摘　要：为了研究抑制素（INH）免疫及其在肉牛繁殖中的应用，试验从 INH 的特性、来源、免疫原性、免疫方式等方面进行了综述，结果 INH 免疫在肉牛繁殖中有广阔的应用前景。

关键词：抑制素（INH）；免疫；肉牛；繁殖性能

（发表于《黑龙江畜牧兽医》）

青蒿琥酯治疗泰勒焦虫病的研究进展

张 杰[1,2]，张继瑜[1]，李 冰[1]，周绪正[1]

(1. 中国农业科学院兰州畜牧与兽药研究所/中国农业科学院新兽药工程重点开放实验室，兰州 730050；2. 甘肃农业大学动物医学院，兰州 730070)

摘 要：青蒿琥酯（Artesunate）是抗疟药青蒿素（Artemisinin）的衍生物，在兽医临床上主要用于治疗牛、羊的泰勒焦虫病（Theileriosis）以及双芽焦虫病（Double buds piroplasmosis）。作者对近20年来青蒿琥酯在不同动物的药物代谢动力学特征、泰勒焦虫病的临床治疗以及泰勒焦虫病对中国畜牧业的危害进行了总结。

关键词：青蒿琥酯；泰勒焦虫病；临床症状；临床治疗

(发表于《湖北农业科学》)

我国乳品质量标准在安全管理中存在的问题及对策

王宏博，高雅琴，牛春娥，郭天芬，李维红

(中国国农业科学院兰州畜牧与兽药研究所/农业部动物毛皮及制品质量监督检验测试中心，兰州 730050)

摘 要：论述了我国乳品质量标准的发展现状及存在的问题，并对我国乳品质量标准的安全管理提出了相应的控制对策。以便为我国乳品质量标准的修订以及乳品业的健康发展提供参考。

关键词：乳品质量标准；安全管理；污染；检验检疫

(发表于《湖北农业科学》)

药动学-药效学结合模型及其在兽用抗菌药物中的应用

杨亚军，李剑勇，李 冰

(中国农业科学院兰州畜牧与兽药研究所/中国农业科学院新兽药工程重点实验室，兰州 730050)

摘 要：药动学-药效学（PK-PD）结合模型，综合研究体内药物浓度的动态过程与其药效消长之间关系，被广泛用于优化抗菌药物的给药方案。简要介绍了抗菌药物PK-PD模型的基本概念、研究方法等，综述了氟喹诺酮类、β-内酰胺类及其他抗菌药的PK-PD结合模型研究应用进展，以期为兽用抗菌药物的研究开发及临床合理应用提供帮助。

关键词：药动学-药效学结合模型；突变选择窗；氟喹诺酮类药物；β-内酰胺类抗生素

(发表于《湖北农业科学》)

中药鸭跖草的研究进展

王兴业[1,2]，李剑勇[1]，李 冰[1]，杨亚军[1]

(1. 中国农业科学院兰州畜牧与兽药研究所/农业部兽用药物创制重点实验室/甘肃省新兽药工程重点实验室，兰州 730050；2. 甘肃农业大学动物医学院，兰州 730046)

摘 要：鸭跖草（*Commelina communis* L）具有祛风行水、清热、凉血、解毒的功效，主治寒热瘴疟、水肿尿少、高热不退、咽喉肿痛等症。阐述了鸭跖草的化学成分、药理作用及临床应用，以期使新型高效、副作用小的中药鸭跖草越来越多地应用于临床，达到疗效最大化，副作用最小化的目的。

关键词：鸭跖草；化学成分；药理作用；临床应用

(发表于《湖北农业科学》)

非生物学药物筛选方法及其应用

刘玉荣，李剑勇，杨亚军

(中国农业科学院兰州畜牧与兽药研究所新兽药工程重点实验室，兰州 730050)

摘 要：分子印迹技术、高通量筛选技术、生物芯片技术、生物信息学技术是4种常用的非生物学药物筛选技术。本文要介绍了每项技术的概念、原理和应用，以期使其更广泛地应用于抗病毒等安全性低的药物筛选领域。

关键词：药物筛选；分子印迹技术；高通量筛选技术；生物芯片技术；生物信息学技术

(发表于《时珍国医国药》)

黄花补血草醇提物的抗炎作用研究

刘 宇，蒲秀英，梁剑平，华兰英，王学红，尚若峰

(中国国农业科学院兰州畜牧与兽药研究所新兽药工程重点开放实验室，兰州 7a0050)

摘 要：采用不同类型的炎症动物犊型评价黄花补血草提取物的抗炎作用。通过二甲苯致小鼠耳肿胀实验、抗棉球内芽肿实验现察黄花补血草的抗炎作用。黄花补血草醇提物具有明显的抑制小鼠耳肿胀作用，且对小鼠内芽组织增生也有明显的抑制作用。

关键词：黄花补血草；抗炎作用

(发表于《时珍国医国药》)

中兽药防治动物疫病研究现状

郝宝成[1]，王学红[1]，郭志廷[1]，郭文柱[1]，胡永浩[2]，梁剑平[1,2]

(1. 中国农业科学院兰州畜牧与兽药研究所新兽药工程重点实验室，兰州　730050；
2. 甘肃农业大学动物医学院，兰州　730070)

摘　要：近年来，随着科学技术的进步和畜牧业水平的提高，我国传统的中兽医理论（正气存内，邪不可干）和中草药的物性（阴阳寒凉温热）、物味（酸苦甘辛咸）、物间关系及辅以现代技术和生产工艺生产的纯天然药物和纯天然中草药饲料添加剂，已在世界各地悄然兴起。我国无公害安全食品的发展，使得人们对动物源性食品的安全标准要求越来越高，畜禽产品生产的安全问题和中兽药产品的开发也越来越受到人们的重视。中药的优势和特色除了毒副作用小、残留低外，主要体现在药效的整体性和药源的天然性两方面Ⅲ。由于中药绝大多数来自天然，因而具有许多天然属性。每种中药都含有复杂的化学成分，由多味中药组成的复方，其化学成分就更加复杂，一旦作用于机体，往往产生多元组合效应。因此，我们可在动物疫病免疫预防的薄弱领域充分发挥中兽药的优势。动物疫病所涉及的患病动物种类较多，但主要是禽、猪和牛。本文主要从中兽药对禽、猪和牛疫病的防治研究现状及中兽药其他方面的应用进行简要阐述。

（发表于《现代畜牧兽医》）

我国主要毛皮市场及其规范对策

郭天芬[1,2]，席　斌[1,2]，李维红[1,2]，常玉兰[1,2]

(1. 中国农业科学院兰州畜牧与兽药研究所，兰州　730050；2. 农业部动物毛皮及制品质量监督检验测试中心，兰州　730050)

摘　要：介绍了我国几个主要的毛皮市场，阐述了规范我国毛皮市场的对策，以期促进我国毛皮市场的健康发展。
关键词：毛皮市场；规范对策；中国

（发表于《现代农业科技》）

牛繁殖性状候选基因的研究进展

梁春年，阎 萍，郭 宪，包鹏甲，裴 杰，褚 敏

(中国农业科学院兰州畜牧与兽药研究所，兰州 730070)

摘 要：繁殖性状是牛的重要性状之一，繁殖性能高低直接关系到生产成本和生产效率，因此，提高牛的繁殖效率是目前国内外养牛业研究的热点课题。该文综述了牛繁殖性状相关功能基因的研究进展，并对其应用前景进行了展望。

关键词：繁殖性状；功能基因；研究进展

(发表于《中国畜牧兽医》)

山羊痘的流行及防治措施

郝宝成[1]，梁剑平[1,2]，王学红[1]，郭志廷[1]，郭文柱[1]，胡永浩[2]

(1. 中国农业科学院兰州畜牧与兽药研究所中国农业科学院新兽药工程重点开放实验室/甘肃省新兽药工程重点实验室，兰州 730050；2. 甘肃农业大学动物医学院，兰州 730070)

摘 要：山羊痘是由山羊痘病毒属的痘病毒引起的羊的一种急性、热性、接触性传染病。文章从山羊痘流行病学、症状、诊断和防治措施等方面对山羊痘进行综述。

关键词：山羊痘病毒；流行病学；诊断；防治措施

(发表于《中国畜牧兽医》)

基因组学和蛋白组学在纳米药物毒理研究中的应用

高昭辉[1,2,4]，董书伟[1,4]，薛慧文[2]，荔 霞[1,4]，申小云[3]，
刘永明[1,4]，王胜义[1,4]，刘世祥[1,4]，齐志明[1,4]

(1. 中国农业科学院兰州畜牧与兽药研究所/农业部兽用药物创制重点实验室，兰州 730050；2. 甘肃农业大学动物医学院，兰州 730030；3. 贵州省毕节学院，毕节 551700；4. 甘肃省中兽药工程技术中心，兰州 730050)

摘 要：近年来，纳米材料药物因其独特的生物学特性而被广泛应用于医疗领域，但其安全性问题已成为广大学者所关注的焦点。基因组学和蛋白组学的快速发展，为研究纳米药物对机体的毒性作用机理提供了相关的技术支持。本文介绍了基因组学和蛋白组学的概念及其核心技术、纳米药物毒性，并对基因组学和蛋白组学在纳米药物毒性研究中的应有进行了综述。

关键词：基因组学；蛋白组学；纳米药物；毒理

(发表于《中国奶牛》)

如何进行奶牛乳房炎金黄色葡萄球菌疫苗的合理评估

王旭荣，李建喜，王小辉，李宏胜，杨志强，孟嘉仁

(中国农业科学院兰州畜牧与兽药研究所/中国农业科学院临床兽医学研究中心/甘肃省中兽药工程技术研究中心，兰州 730050)

摘 要：奶牛乳房炎是造成全球奶业经济损失的主要原因，而金黄色葡萄球菌是最主要的病原菌。本文系统、直观的对研究奶牛乳房炎金黄色葡萄疫苗的科研结论进行量化总结和评估，目的是为奶牛乳房炎疫苗的研发提供新思路，为设计合理的奶牛乳房炎疫苗本动物评估试验方案提供科学依据。本文在 pubmed、science 数据库和 cnki 数据库对"奶牛乳房炎疫苗"类的文章进行了合理的电子检索，然后对细菌苗、细菌类毒素疫苗、DNA-重组蛋白疫苗、单一的重组蛋白疫苗等方面的科研论文从试验设计、方法、疫苗类型和研究结果 4 个方面进行了全面的分析。结果表明，采用 DNA、重组蛋白新技术的疫苗和传统的细菌苗已取得良好效果，疫苗可以成为预防和控制金黄色葡萄球菌性奶

牛乳房炎的最好或最有前景的一种途径。但是研究方法差异和双盲试验的缺乏阻碍对疫苗效果的合理评估。

我国奶牛普通病防治现状和发展对策

严作廷，杨志强，荔　霞

(中国国农业科学院兰州畜牧与兽药研究所/中国农业科学院临床兽医学研究中心，兰州　730050)

摘　要：本文主要介绍了我国奶牛普通病防治现状和存在的问题，分析了奶牛普通病及其防治的特殊性，提出了今后的研究对策。
关键词：奶牛；普通病；防治；对策

(发表于《中国奶牛》)

治疗奶牛卵巢疾病性不孕症中药制剂研究进展

王东升，严作廷，李世宏，张世栋，谢家声

(中国国农业科学院兰州畜牧与兽药研究所/中国农业科学院临床兽医学研究中心/甘肃省中兽药工程技术研究中心，兰州　730050)

摘　要：奶牛卵巢疾病包括持久黄体、卵巢囊肿、卵巢静止及卵巢萎缩等，是引起奶牛不孕症的主要病因，常用的疗法以药物疗法为主，主要有激素疗法和中药疗法。本文综述了治疗奶牛卵巢疾病性不孕症的中药制剂散剂、汤剂、子宫灌注剂、酊剂和颗粒剂5种剂型的组方、疗效等研究进展。
关键词：奶牛；中药；卵巢疾病；不孕症；治疗

(发表于《中国奶牛》)

基因芯片技术及其在兽医学中的应用

龚成珍[1,2]，胡永浩[2]，王东升[1]，张世栋[1]，严作廷[1]

(1. 中国农业科学院兰州畜牧与兽药研究所，兰州 730050；2. 甘肃农业大学，兰州 730070)

摘 要：基因芯片技术是由生命科学与物理学、微电子技术、生化技术等众多科学领域交叉综合的一项高新技术，具有强大的类比性、重复性、微型化和自动化等特点。对该技术的原理、分类、技术流程、优点及其在兽医学中的应用及存在的问题进行概述，并对其应用前景进行展望。

关键词：基因芯片技术；兽医学；基因组；基因表达

(发表于《安徽农业科学》)

植物精油对奶牛和肉牛瘤胃发酵的影响研究进展

乔国华，张怀山，周学辉，常根柱，张 茜，路 远

(中国农业科学院兰州畜牧与兽药研究所草业饲料研究室，兰州 730050)

摘 要：本文综述了近几年关于植物精油对反刍动物瘤胃影响及其机理的报道。旨在探讨植物精油作为新型绿色饲料添加剂的可行性，并探讨了植物精油应用在生产实践中所需要解决的问题。

关键词：植物精油；瘤胃发酵；研究进展；奶牛；肉牛

(发表于《中国畜牧杂志》)

羔羊腹泻综合防治技术

严作廷，王 东

(中国农业科学院兰州畜牧与兽药研究所中国农业科学院临床兽医学研究中心，兰州 730050)

摘　要：羔羊腹泻是出生 1~20 日龄的羔羊的常发病，发病后羔羊生长缓慢，体质下降，抵抗力减弱，不及时预防和治疗就会引起死亡，给养羊业造成严重的经济损失。从 2008 年至 2009 年对甘肃省武威市、白银市、临夏州等地区羔羊腹泻发病情况进行了调查，并采取中西医综合防治措施，共治疗腹泻羔羊 6 355 只，治愈 5 451 只，治愈率达 85.77%，好转 484 只，无效 420 只，总有效率达到 93.39%。现将防治情况报道如下。

(发表于《中国畜禽种业》)

抗血小板药物研究及临床应用状况

张泉州[1]，于远光[2]，李剑勇[2]，杨亚军[2]

(1. 甘肃省医疗器械检测中心，兰州　730030；2. 中国农业科学院兰州畜牧与兽药研究所新兽药工程重点实验室，兰州　730050)

摘　要：[目的] 介绍近年来国内外研发的抗血小板凝集药及临床应用情况。[方法] 根据各类药作用机制的不同，分别从抑制花生四烯酸代谢药、影响环核苷酸代谢药、作用于血小板膜特异激动剂和受体药、一氧化氮供体药等几个方面进行分析。[结果] 抗血小板药物可显著抑制血小板的黏附、聚集和释放，有效控制血小板活化，阻止或延缓血栓的形成，随着对该药品的认识提高，心脑血管不良事件的发生不断减少。[结论] 为寻找疗效更强、抑制作用更广谱的血小板聚集抵制剂提供更加宽阔的视野，为减少药品的不良反应提供更多帮助。

关键词：抗血小板药；花生四烯酸；环核苷酸；膜特异激动剂和受体药；一氧化氮供体

(发表于《中国医院药学杂志》)

高速公路绿化工程系统构建概述

常根柱，周学辉，杨红善，路　远，屈建民，石天欢

(中国国农业科学院兰州畜牧与兽药研究所，兰州　730050)

摘　要：作者于1996—1998年实施"陕西省高等级公路绿化研究"课题，2009年完成《高速公路绿化》一书。本研究将我国高速余路绿化工程划分为7大组成部分：中央分隔带防眩绿化；路堤边坡防护绿化；景观路树栽植绿化；刺篱植物封闭绿化；路堑土、石质坡面立体垂直绿化；立交区景观再造绿化；服务区、收费站环境绿化。对绿化理念、设计与施工、绿化技术和工程监理等方面进行简要阐述，研究提出我国高速公路绿化理论体系和工程技术标准。

关键词：高速公路；绿化工程；研究

(发表于《中国园艺文摘》)

紫草的药理研究及其在兽医临床上的应用概况

严作延，杨国林，巩忠福

(中国农业科学院兰州畜牧与兽药研究所，兰州　730050)

摘　要：紫草为多年生草本植物，《中华人民共和国药典》2000年版（一部）收载的紫草（Radix Arnebiae, RadixLithospermi）为紫草科植物新疆紫草 Arnebia euchroma (Royle) Johnst.、紫草 Lithospermum erythrorhizon Sieb.et Zucc.或内蒙紫草 Arnebia guttata Bunge 的干燥根。其性味苦寒，归心包络、肝经，有凉血、活血、清热、解毒的作用，主要用于温热斑疹、湿热黄疸、紫癜、吐、衄、尿血、淋浊和血痢、热结便秘、烧伤、湿疹、丹毒、痈疡等病的治疗。《本草纲目》云：紫草"治斑疹、痘毒，活血凉血、利大肠"以往研究表明紫草具有抗病原微生物、抗炎及抗过敏、解热、抗肿瘤、保肝、止血、降血糖、镇静和影响免疫功能等作用[1]。近几年来紫草的现代药理作用与临床应用研究取得了新的进展。

(发表于《中兽医学杂志》)

Effects of Arecoline Hydrobromide on Taeniasis and Cysticercosis in Domestic Animals

ZHOU Xu-zheng, ZHANG Ji-yu, LI Jian-yong, LI Jin-shan,
WEI Xiao-juan, NIU Jian-rong, LI Bing

(Key Laboratory of New Animal Medicine Project, Lanzhou Institute of Animal Husbandry and Veterinary Medicine, Chinese Academy of Agricultural Sciences, Lanzhou 730050, China)

Abstract: Taeniasis and cysticercosis in domestic animals belong to zoonosis and seriously threaten the public health security. Especially the cysticercosis and echinococcosis caused by the tapeworm eggs have great harms to bodies because they can attack many organs of body. According to the combination of experimental results and literature materials, the morphology and transmission mode of taenia and cysticercus, the prevalence status and monitoring of taeniasis and cysticercosis as well as the antitapeworm mechanism, comparative analysis to other drugs, expelling tapeworm tests *in vitro*, dose determining tests and usage notes of arecoline hydrobromide were expounded in detail. It provides a theoretical basis for prevention of taeniasis and cysticercosis and more scientific usage of arecoline hydrobromide and thus relieves the harms of taeniasis and cysticercosis and ensuring the public health security.

Key words: Taeniasis; Cysticercosis; Arecoline hydrobromide

(发表于《Animal Husbandry and Feed Science》)

Study on the Correlation Between Leaf Area and Drought Resistance in Different Alfalfa Varieties

ZHAN GHuai-shan[1], DAI Li-lan[2], QIAO Guo-hua[1],
WANG Chun-mei[1], YANG Xiao[1]

(1. Lanzhou Institute of Animal Sciences and Veterinary Pharmaceutics, Chinese Academy of Agriculture Science, Lanzhou 730050; 2. Lanzhou Agricultural Scienceand Technology Research Promotion Center, Lanzhou 730030)

Abstract: [Objective] This study was to reveal the corre lation bet ween leafarea and drough tresistance in different varieties of alfalfa.

[Method] Using variousal falfavarie tiesasex perimen talmaterials, the droughtresistanceo fleave sofdrought‐stressedal falfaplan tswasas sessed bymeasuring thecontento ffreeproline foranal yzingitscorrela tion withlea farea.

[Result] Under drought condition, the drough tresistance of alfalfaisdirectlyrelatedtoleafareainapositivecorrelation.

[Conclusion] Leafarea could beuseda sanins titutiona lassistan tindextore flect theresis tanceo fdifferen talfalfa varieties.

Key words: Alfalfa; Proline; Leafarea; Droughtresistance

(发表于《Agricultural Science & Technology》)

Progress in Determination Methods of Degradation Rate in Ruminants

WANG Hong-bo[1,2]*, GAO Ya-qin[1,2], NIU Chun-e[1,2], LI Wei-hong[1,2], GUO Tian-fen[1,2]

(1. Lanzhou Institute of Animal Science and Veterinary Pharmaceutics, Chinese Academy of Agricultural Sciences (CAAS), Lanzhou 730050, China; 2. Quality Supervising, Inspecting and Testing Center for Animal Fiber, Fur, Leather and Products (Lanzhou) of the Ministry of Agriculture, Lanzhou 730050, China)

Abstract: Degradation rate of feed proteins in rumen is a basic indicator of new intestinal protein system of ruminants. In this paper, determination methods of degradation rate in rumen including *in-vivo* method, nylon bag method and artificial rumen method are compared in order to provide a reference for animal nutrition.

Key words: Rumen; Degradation rate; Determination method; Progress

(发表于《Animal Husbandry and Feed Science》)

Comparison on Two Different Instrumental Methods for Determination of Fat Content in Washed Cashmere

WANG Hong-bo, CHANG Yu-lan, LI Wei-hong,
DU Tian-qing, LIANG Li-na

(Lanzhou Institute of Animal Science and Veterinary Pharmaceutics of CAAS, Quality Supervising, Inspecting and Testing Center for Animal Fiber, Fur, Leather and Products (Lanzhou) of MOA of China, Lanzhou 730050, China)

Abstract: The aim of this study was to compare two different instrumental methods for determination of fat content in washed cashmere. The fat content in washed cashmere was determined using Soxhlet fat extraction unit and fiber grease extractor quickly, respectively. The results show that no significant difference was found between these two instrumental methods, and reliability of fiber grease extractor quickly was confirmed for determination of fat content in washed cashmere. Therefore, the fiber grease extractor quickly can meet current standard for determination of fat content in washed cashmere.

Key words: Cashmere; Fat content; Fiber grease extractor quickly; Soxhlet fat extraction unit

(发表于《Animal Husbandry and Feed Science》)

4个美国标准秋眠型苜蓿对自然光周期的生长反应

冯长松[1]，李锦华[2]，李绍钰[1]，卢欣石[3]

(1. 河南省农业科学院畜牧所，郑州 450002；2. 中国农业科学院兰州畜牧与兽药研究所，兰州 730050；3. 北京林业大学林学院，北京 100082)

摘 要：以美国分属4个秋眠类型的4个标准级紫花苜蓿 Medicago sativa L. 品种为材料，研究了北京地区春季第一茬草和秋季再生草的植株生长性状的动态变化。结果表明，在春季光照延长和日辐射增强的条件下，盛花期前4个秋眠型品种的植株高度持续提高，植株生长速度有逐步加快趋势，但不同秋眠型间生长速度变化的差异不显著($P>0.05$)；盛花期后植株生长趋缓至停滞，秋眠型的最终株高顺序为：Archer（半秋眠型) >UC-1465 (极非秋眠型) >Pierce (非秋眠型) > Maverick (秋眠型)。在秋季光照时间缩短过程中,不同秋眠型品种的再生株高顺序为:UC-1465 (极非秋眠型) > Pierce (非秋眠型) > Archer (半秋眠型) > Maverick (秋眠型);其再生速度与春季的变化曲线不同,Maverick (秋眠型) 和 Archer (半秋眠型) 持续下降,而 UC-1465 (极非秋眠型) >Pierce (非秋眠型) 呈先高后低的单峰曲线。

关键词:紫花苜蓿;秋眠性;光周期;生长反应

(发表于《中国草食动物》)

不同断奶日龄对羔羊育肥效果的影响

张鹏俊[1],郎 侠[2]

(1. 甘肃省靖远县农牧局,靖远 730600;2. 中国农业科学院兰州畜牧与兽药研究所,兰州 730050)

摘 要:选择出生重及出生日期相近、生长发育正常的滩羊羔羊60只,随机分为4组,分别为30、45和60日龄断奶组,120日龄哺乳组为对照,各选取15只羔羊进行育肥试验,直至120日龄。结果表明:在120日龄时,试验组羔羊的体重和体尺均极显著高于对照组($P<0.01$)。其中体重最高的是60日龄断奶组羔羊,为36.25 kg,日增重(75~120d)达278.67g,极显著高于对照组的23.27kg和136.0g,经济效益比对照组高出80.12元/只。

关键词:断奶;育肥;日增重;经济效益;滩羊羔羊

(发表于《中国草食动物》)

复方杨黄灌注液对奶牛子宫内膜炎的药效试验

倪鸿韬[1,2],王 瑜[3]

(1. 甘肃农业大学动物医学院,兰州 730070;2. 甘肃省兽医局,兰州 730030;3. 中国农业科学院兰州畜牧与兽药研究所,兰州 730050)

摘 要:采用常规方法对奶牛子宫内膜炎进行细菌学检查;以液体培养基对倍稀释法测定复方杨黄灌注液对奶牛子宫内膜炎主要病原菌的最低抑菌浓度(MIC),并以双黄连注射液作对照,观察复方杨黄灌注液对奶牛子宫内膜炎的临床疗效。[结果]37例患子宫内膜炎奶牛子宫中共分离到49株细菌,其中链球菌、大肠埃希氏菌和葡萄球菌分别占30.61%、28.57%和24.49%，其他菌的比例较低。复方杨黄灌注液对奶牛子宫内膜炎的主要病原菌均有不同程度的抑制作用，与对照药物双黄连注射液作用相仿。复方杨黄灌注液对奶牛子宫内膜炎的治愈率为88.06%，疗效优于对照药物（$P<0.05$）。

关键词：复方杨黄灌注液；奶牛；子宫内膜炎；临床试验

(发表于《中国草食动物》)

甘肃省绒毛生产销售情况调查研究

牛春娥[1,2]，杨博辉[1]，曹藏虎[3]，岳耀敬[1]，张　力[3]，郭　健[1]，
孙晓萍[1]，郎　侠[1]，刘建斌[1]，程胜利[1]，焦　硕[1]，冯瑞林[1]

(1. 中国农业科学院兰州畜牧与兽药研究所，兰州　730050；2. 农业部动物毛皮及制品质量监督检验测试中心（兰州），兰州　730050；3. 甘肃省农牧厅畜牧处，兰州　730000)

摘　要：文章对2009年甘肃省毛绒用羊养殖，毛绒生产、质量控制、销售渠道及2008—2009年度销售情况等进行了全面调查研究，分析了甘肃省绒毛产业存在的问题，提出了发展甘肃省绒毛产业的建议和措施。

关键词：甘肃省；绒毛；生产；销售

(发表于《中国草食动物》)

狼尾草属牧草种质资源的开发利用

张怀山

(中国农业科学院兰州畜牧与兽药研究所，甘肃　730050)

摘　要：狼尾草属（*Pennisetum* Rich）是个大属，中国是狼尾草属牧草资源最为丰富的国家之一。至今已发现并发表的包括常见种、引进种以及新发现种、新发现变种共计12种3变种（包括4个引进种）。狼尾草属牧草的广泛适应性、高产草量、高蛋白含量及无融合生殖特性等受到世界多国农业部门与育种专家的高度关注，在狼尾草饲用型、生态型、观赏型、能源型牧草开发应用方面取得了显著成就。狼尾草属优良草种的引种驯化、杂交育种和品种引进具有广阔的发展前景。

关键词：狼尾草属；牧草资源；开发利用

(发表于《中国草食动物》)

青海省湟源县奶牛养殖存在问题与对策

严作廷[1]，荔 霞[1]，齐志明[1]，王胜义[1]，刘永明[1]，王国仓[2]，付明善[2]

(1. 中国农业科学院兰州畜牧与兽药研究所，兰州 730050；2. 青海省湟源县畜牧兽医工作站，湟源)

摘 要：湟源县位于我国黄土高原最西端的日月山下，属祁连山系，分川水、浅山、半浅山、脑山4类地区，海拔2 470~4 898m，年均气温3.7℃。2009年底全县共存栏各类草食畜29.24万头（只、匹），同比增长1.3%，其中存栏牛4.38万头（奶牛1.35万头）。奶牛养殖主要分布在城关镇、大华乡、申中乡、波航乡、和平乡、塔湾乡、日月乡等乡镇。其中大部分为改良牛，目前全县奶牛良种化比率达88%，日产奶量由改良前的黄牛2.5kg增加到现在改良牛的9.6kg，且良种化程度不断提高，高代牛已占改良牛数的45%。奶牛养殖小区和奶牛基地村不断增多，目前全县20头以上奶牛养殖户达到26户，百头以上奶牛养殖村达到9个，奶牛收入已成为一些养殖户家庭收入的主要来源。

(发表于《中国草食动物》)

我国奶牛主要疾病研究进展

严作廷，王东升，王旭荣，张世栋，李宏胜

(中国农业科学院兰州畜牧与兽药研究所/甘肃省中兽药工程技术研究中心/中国农业科学院临床兽医学研究中心，兰州 730050)

摘 要：文章主要分析了影响我国奶牛业发展的主要疾病，如乳房炎、子宫内膜炎、卵巢疾病、营养代谢疾病、胎衣不下、犊牛腹泻、肢蹄病等的防治现状，并提出了今后的研究方向与对策。

关键词：奶牛；疾病；中国

(发表于《中国草食动物》)

西藏达孜箭筈豌豆西牧 324 播种期试验

李锦华[1]，张小甫[1]，田福平[1]，乔国华[1]，黄秀霞[2]，苗小林[1]

（1. 中国农业科学院兰州畜牧与兽药研究所，兰州 730050；2. 西藏自治区农业技术推广服务中心，拉萨 850000）

摘　要：西牧 324 为春箭筈豌豆晚熟品种，在西藏达孜县 4 月 13 日播种时，生育期约 150d；种子千粒重 66.07g；种子产量达到 2442.1kg/hm²，极显著高于其他播期（$P<0.01$）。4 月 20 日以后播种，则不能开花或开花结荚后大部分籽粒不能成熟，种子产量急剧下降到 795.2kg/hm² 以下。如果 4 月中旬作为西牧 324 的播期，稍早于当地油菜等作物的传统播期，所以出苗期极易受低温和晚霜的影响，是需要进一步研究解决的问题。种子收获期的秸秆产量性状测定表明，试验设计范围的不同播期间株高随播期的延迟下降，但产量没有显著差异。

关键词：西藏；箭筈豌豆；种子；播种期

（发表于《中国草食动物》）

西藏山南地区微肥对苜蓿种子产量影响的研究

马海兰[1]，刘学录[1]，李锦华[2]，江措[3]，田福平[2]，乔国华[2]

（1. 甘肃农业大学，兰州 730070；2. 中国农业科学院兰州畜牧与兽药研究所，兰州 730050；3. 西藏自治区山南地区草原站，山南）

摘　要：在分析西藏山南地区乃东县种子基地微量元素有效成分含量的基础上，初步研究了 7 种微量元素对苜蓿种子产量的效应。结果表明，苜蓿开花期喷施不同种类的微肥，硫酸锰处理的种子产量为 353.3kg/hm²，与对照组差异显著（$P<0.05$）；千粒重以硫酸锰、硼肥、硫酸亚铁处理组最高，与对照组差异达极显著水平（$P<0.01$）；同时硫酸锰处理组的种子长度、两侧宽、背腹宽在各处理之间相对较高。以不同浓度的钼酸铵进行喷施时，随喷施浓度的提高，苜蓿种子产量有提高的趋势。分析不同微肥处理组种子产量与鲜草产量的相关性，结果均不显著（$P>0.05$），但不同钼酸铵浓度处理条件下的种子产量与鲜草产量呈极显著负相关（$P<0.05$），相关系数为 -0.958。

关键词：西藏山南地区；微肥；苜蓿；种子产量

（发表于《中国草食动物》）

应用不对称竞争性 PCR 技术确定藏羊 Agouti 基因拷贝重复数

杨树猛[1,2]，岳耀敬[1]，杨博辉[1]，郭 健[1]，孙晓萍[1]，
牛春娥[1]，冯瑞林[1]，郭婷婷[1]

(1. 中国农业科学院兰州畜牧与兽药研究所，甘肃 730050；2. 甘肃省甘南州畜牧科学研究所，合作 747000)

摘 要：实验采用不对称竞争性 PCR 技术，测定 16 只不同毛色藏羊 Agouti 基因的拷贝重复数，对记录所得 Agouti 基因连接点的比值进行分析。对样品（$n=16$）的 Agouti 基因拷贝数进行分析，区分未知样品 Agouti 基因的拷贝数。10 只全白藏羊的拷贝数在 3~6 个之间，其余不同毛色 6 只藏羊均为 2 个拷贝数。实验证明：不对称竞争性 PCR 技术检测 Agouti 基因拷贝重复数变异的方法，可以直接对产物进行定量，具有良好的稳定性和特异性，而且操作简单、安全，自动化程度高，不易产生污染，不需昂贵设备，适合在普通实验室检测。

关键词：不对称竞争性 PCR；藏羊；Agouti 基因；拷贝重复数

（发表于《中国草食动物》）

中草药防治奶牛子宫内膜炎的现状及其药理作用机制

严作廷，王东升，荔 霞，张世栋，李世宏

(中国国农业科学院兰州畜牧与兽药研究所/中国农业科学院临床兽医学研究中心，甘肃 730050)

摘 要：奶牛子宫内膜炎是奶牛的常见疾病之一，严重影响奶牛业的健康持续发展。文章综述了中草药防治奶牛子宫内膜炎的制剂类型及其作用机制，并提出了中草药防治奶牛子宫内膜炎需要进一步研究的问题。

关键词：奶牛；中草药；子宫内膜炎

（发表于《中国草食动物》）

鸡大肠杆菌病及其中草药防治的概况

严作延，王东升，陈炅然，张世栋

（中国农业科学院兰州畜牧与兽药研究所/农业部兽用药物创制重点实验室/甘肃省中兽药工程技术中心，兰州 730050）

摘 要：禽大肠杆菌病是由埃希氏大肠杆菌引起的多种病的总称，包括大肠杆菌性肉芽肿、腹膜炎、输卵管炎、脐炎、滑膜炎、气囊炎、眼炎、卵黄性腹膜炎等疾病，对养禽业危害严重。

（发表于《中国畜禽种业》）

甘南牧区耗牛产业发展与生态环境建设浅议

丁学智，郭 宪，梁春年，阎 萍

（1. 中国国农业科学院兰州畜牧与兽药研究所，兰州 730050；2. 甘肃省耗牛繁育工程重点实验室，兰州 730050）

摘 要：通过对甘南牧区耗牛产业发展现状及生态环境进行分析，指出了甘南牧区畜牧业发展中存在的问题，并指出指出人为和自然因素导致草地严重退化，对黄河、长江源区的生态安全与甘南牧区民族经济和社会的可持续发展产生了重大影响。提出落实草地产权制度，摸清草地状况，加强草地载畜量管理，加强草畜科学技术的推广，加大资本投入，对饲草季节矛盾进行深入研究和解决等，是提高甘南地区草畜平衡及可持续发展的措施。
关键词：甘南草原；生态环境；耗牛生产；可持续发展

（发表于《中国牛业科学》）

甘肃省牛种业现状与"十二五"科技发展方向

梁春年；阎 萍；丁学智；郭 宪；包鹏甲；装 杰；朱新书

(中国农业科学院兰州畜牧与兽药研究所，兰州 730070)

摘 要：牛产业是甘肃省农业中的重要产业，近年来，甘肃省牛产业发展迅速，牛产业链也日臻完善。本文从甘肃省牛产业生产现状和牛种业现状分析入手，分析指明了甘肃省牛种业"十二五"科技发展方向。

关键词：甘肃省；牛；种业；现状；科技发展方向

(发表于《中国牛业科学》)

高寒牧区耗牛繁育综合技术措施

郭 宪，丁学智，梁春年，包鹏甲，阎 萍*

(中国国农业科学院兰州畜牧与兽药研究所/甘肃省耗牛繁育工程重点实验室，兰州 73050)

摘 要：托牛是青藏高原特有的遗传资源，是高寒牧区的基本生产生活资料。耗牛繁育是耗牛生产中的关键环节，是改良耗牛品质、增加良种数量、提高经济效益的重要步骤。本文在分析讨论的基础上，提出了高寒牧区耗牛繁育综合技术措施，包括加强托牛种质特性研究、开展托牛系统选育，建立高效繁育技术体系、提高托牛生产性能，综合应用生态高效草原收养技术、提高耗牛生产效率等方面，为进一步开展托牛选育与改良提供技术参考。

关键词：托牛；高寒牧区；繁育

(发表于《中国牛业科学》)

高寒牧区牦牛冷季暖棚饲养技术要点

包鹏甲，阎　萍，梁春年，郭　宪，丁学智

(中国国农业科学院兰州畜牧与兽药研究所，兰州　730050；甘肃省牦牛繁育工程重点实验室，兰州　730050)

摘　要：牦牛是青藏高原特有的放牧家畜，是唯一能充分利用青藏高原牧草资源进行动物性生产的牛种。但长期以来由于掠夺式的经营和粗放的管理模式，使牦牛始终处于"夏饱、秋肥、冬瘦、春乏"的恶性循环中，加之近年来牦牛数量迅速增加，草畜矛盾日益突出，加速了天然草地的退化，严重影响着高寒草地生态系统的平衡与稳定。本文就高寒牧区牦牛冬季暖棚饲养技术要点进行归纳总结，以求寻求解决草畜矛盾及季节不平衡、提高牦牛的商品率、增加牧民收入、减轻天然草地压力、恢复天然草地植被的新途径，同时建议相关部门给予资金和政策方面的支持，使牦牛冷季暖棚饲养技术发挥更大效益。

关键词：高寒牧区；牦牛；暖棚；饲养技术

(发表于《中国牛业科学》)

牦牛 ATP6 和 ATP8 基因生物信息学分析

吴晓云[1,2]，阎　萍[1,2]，梁春年[1,2]，郭　宪[1,2]

(1. 中国农业科学院兰州畜牧与兽药研究所，兰州　730050；2. 甘肃省牦牛繁育工程重点实验室，兰州　730050)

摘　要：对牦牛 ATP6 和 ATP8 基因序列进行初步的分析，结果表明，牦牛 ATP6 基因长 68lbp。编码 226 个氨基酸残基的蛋白质，其蛋白分子质量为 24839Da，等电点为 9.99；ATP8 基因长 201bp，编码 66 个氨基酸残基的蛋白质，其蛋白分子质量为 7966.5Da，等电点为 9.46。牦牛 ATP6 基因和 ATP8 基因与其他 5 种牛（家牛、欧洲野牛、美洲野牛、瘤牛和亚洲水牛）的该基因具有很高的同源性。通过分析牦牛 ATP6 基因和 ATP8 基因的遗传进化关系，发现 ATP6 和 ATP8 基因有较快的进化速率。

关键词：牦牛；ATP6 基因；ATP8 基因；生物信息学分析

(发表于《中国牛业科学》)

GPA 对小鼠肝损伤血清及肝组织中 GOT、GPT 的影响

张 霞[1]，程富胜[2]

(1. 甘肃农业大学生命科学技术学院，兰州 730070；2. 中国农业科学院兰州畜牧与兽药研究所/农业部兽用药物创制重点实验室/甘肃省新兽药工程重点实验室，兰州 730050)

摘 要：采用刀豆蛋白 A（ConA）尾静脉注射方法建立小鼠免疫性肝损伤模型，分别用低（100mg/kg）、中（200mg/kg）、高（400mg/kg）3 个不同剂量大蒜多糖（GPA）对小鼠进行灌胃，灌服 7d 后，小鼠尾静脉注射刀豆蛋白 A，采血及取肝组织，用赖氏法测定天冬氨酸转氨酶（GOT）及丙氨酸转氨酶（GPT）变化。结果模型组和多糖组与空白组比较，除高浓度多糖组差异不显著（$P>0.05$）外，其他各组均差异显著（$P<0.05$），且多糖组与模型组比较差异显著（$P<0.05$），并呈现一定的量效关系。表明不同浓度的大蒜多糖均对 ConA 致小鼠免疫性肝损伤具有保护作用。

关键词：大蒜多糖；刀豆蛋白 A（ConA）；GOT；GPT

(发表于《中兽医医药杂志》)

动物疫病发展与防控趋势分析

倪鸿韬[1,2]，王 瑜[3]

(1. 甘肃农业大学动物医学院，兰州 730070；2. 甘肃省兽医局，兰州 730070；3. 中国农业科学院兰州畜牧与兽药研究所，兰州 730050)

摘 要：随着我国与世界各国贸易往来不断增加，国内畜禽及其产品流通渠道增多，动物疫病传播途径和机会也在增加，疫病传播的速度加快，重大动物疫病不断发生，新的动物疫病不断出现，给动物防疫带来前所未有的难题。另外，随着人类生活方式不断改善以及各种新技术的广泛应用和动物饲养环境的变化，动物疫病变得复杂和多样化，防制难度加大，尤其是一些人畜共患病对人类健康和生命安全带来严重威胁。

(发表于《中兽医医药杂志》)

附红清治疗小白鼠人工感染猪附红细胞体试验观察

苗小楼[1]，李 芸[2]，李宏胜[1]，潘 虎[1]，尚小飞[1]

(1. 中国农业科学院兰州畜牧与兽药研究所/中国农业科学院临床兽医学中心，兰州 730050；2. 甘肃中医学院)

摘 要：将人工感染附红细胞体小白鼠50只分为附红清治疗高、中、低剂量组和三氮脒治疗对照组及生理盐水对照组，采用腹腔注射给药，1次/d，连用3d，连续6d采血观察红细胞感染率。结果附红清高、中剂量组与三氮脒对照组疗效接近。

关键词：附红细胞体；人工感染；附红清

(发表于《中兽医医药杂志》)

复方茜草灌注液质量标准研究

王学红[1]，王作信[2]，郭文柱[1]，刘 宇[1]，郝宝成[1]，梁剑平[1]

(1. 中国农业科学院兰州畜牧与兽药研究所/农业部兽用药物创制重点实验室/甘肃省新兽药工程重点实验室/中国农业科学院新兽药工程重点开放实验室，兰州 730050；2. 临夏州畜牧局，临夏)

摘 要：[目的] 制定复方茜草灌注液质量标准。[方法] 采用薄层色谱法对制剂中的丹参、茜草、苦参进行鉴别；采用高效液相色谱法测定制剂中丹参酮II_A和隐丹参酮的含量。[结果] 隐丹参酮的回归方程 $Y = 1×10^6 X + 744\ 462$，$r = 0.999\ 5$，平均回收率为98.5%，RSD为1.61%。丹参酮II_A的回归方程 $Y = 2×10^6 X + 2×10^6$，$r = 0.999\ 5$，平均回收率为97.8%，RSD为2.05%。[结论] 建立的鉴别方法直观、简单、专属性强；含量测定方法准确、重现性好。可作为制定复方茜草灌注液质量标准的试验依据。

关键词：复方茜草灌注液；丹参酮II_A；隐丹参酮；薄层色谱；高效液相色谱法

(发表于《中兽医医药杂志》)

黑曲霉发酵对黄柏中游离态小檗碱含量的影响

陶 蕾[1,2]，王宏伟[3]，许冠英[3]，殷劲松[3]，梁剑平[1]，郭 凯[1]

(1. 中国农业科学院兰州畜牧与兽药研究所/中国农业科学院新兽药工程重点开放实验室/甘肃省新兽药工程重点实验室，兰州 730050；2. 甘肃农业大学动物医学院；3. 保定冀中药业有限公司)

摘 要：中药材黄柏经黑曲霉菌液发酵15d后，将发酵后药材烘干、氨醇水浸泡、索式提取等方法处理，与未发酵原药材比较小檗碱含量变化。检测结果表明，经黑曲霉菌液发酵过的黄柏中游离态小檗碱含量较高。该结论可为进一步开发中药发酵饲料添加剂提供依据。

关键词：黄柏；小檗碱；黑曲霉；发酵

(发表于《中兽医医药杂志》)

家畜脉诊研究进展

严作廷[1]，王东升[1]，万玉林[2]，荔 霞[1]

(1. 中国农业科学院兰州畜牧与兽药研究所/农业部兽用药物创制重点实验室/甘肃省新兽药工程重点实验室，兰州 730050；2. 中农威特生物科技股份有限公司)

摘 要：中兽医学是我国文化宝库中富有民族特色的珍贵遗产，脉学是中兽医学有代表性的学科之一。近50年来，家畜脉诊研究尤其是马属动物的脉诊研究取得较大进展，内容涉及到家畜脉理、脉诊部位、脉的实质、脉象分类等，提出了一些新的学术见解，现将主要成就简述如下。

(发表于《中兽医医药杂志》)

六茜素稳定性影响因素研究

王学红，郭文柱，刘　宇，郝宝成，郑红星，梁剑平

(中国农业科学院兰州畜牧与兽药研究所/甘肃省新兽药工程重点实验室/中国农业科学院新兽药工程重点开放实验室，兰州　730050)

摘　要：[目的] 探讨六茜素的固有稳定性、了解影响其稳定性的因素及可能的降解途径与分解产物，为制剂生产工艺、包装、贮存条件提供科学依据。[方法] 分别通过高温试验、高湿度试验、强光照射试验等，考察六茜素制剂的稳定性。[结果] 六茜素在强光照射和高温条件下稳定，湿度和水分对六茜素的稳定性影响较大。[结论] 六茜素应密封贮存，使用时宜现配现用。

关键词：六茜素；影响因素试验；稳定性

(发表于《中兽医医药杂志》)

鸦胆子苦木素对小鼠血清胆碱酯酶和总胆红素的影响

程富胜[1]，程世红[2]，张　霞[3]，王华东[1]

(1. 中国农业科学院兰州畜牧与兽药研究所/农业部兽用药物创制重点/实验室甘肃省新兽药工程重点实验室，兰州　730050；2. 兰州大学实验室与设备管理处，兰州　730000；3. 甘肃农业大学生命科学技术学院，兰州　730000)

摘　要：为了探讨鸦胆子提取物苦木素的毒性，实验以小白鼠为研究对象，对血清中胆碱酯酶活性和总胆红素含量进行了测定。70只健康昆明系雌性小白鼠随机分为1个空白对照组和6个药物组，药物组分别以1.25、1.80、2.80、4.00、6.00、8.50g/kg灌胃，在第7天和第14天分别进行采血，分离血清，测定血清胆碱酯酶活性和总胆红素含量。结果鸦胆子苦木素类提取物具有抑制小鼠血清胆碱酯酶活性的作用，与空白组相比差异显著（$P<0.05$，$P<0.01$），并能使血清中总胆红素升高（$P<0.05$，$P<0.01$），对两种酶的抑制与促生作用，第14天较第7天明显增强（$P<0.01$）。实验表明鸦胆子苦木素对动物机体有一定的毒性作用，并随着浓度的增加而加强，同时呈现出一定的时效性。

关键词：鸦胆子；苦木素；胆碱酯酶；总胆红素

（发表于《中兽医医药杂志》）

射干麻黄地龙散对小白鼠的止咳祛痰作用

罗永江，谢家声，辛蕊华，郑继方，胡振英，邓素平

（中国农业科学院兰州畜牧与兽药研究所/农业部兽用药物创制重点实验室/甘肃省新兽药工程重点实验室，兰州 730050）

摘 要：目的：为了解射干地龙颗粒对动物的止咳祛痰作用的效果。方法：选用浓氨水致咳小鼠实验和小白鼠酚红祛痰法。结果：生理盐水对照组和药物组引起半数小白鼠咳嗽所需的浓氨水喷雾时间（ET50）分别为19.2秒和30.43秒，R值为158.50%；以酚红标准溶液（X）对其吸收度（A值）（Y）作标准曲线为：$Y=0.0747X-0.0036$，浓度在 $0.1\sim10.0\mu g/ml$ 范围内呈线性关系，其相关系数为0.9998，小鼠气管酚红排泌量3个药物剂量组与生理盐水对照组比较均呈显著性差异；药物组之间比较，3个剂量组之间没有显著性差异。结论：射干地龙颗粒对小白鼠具有明显的止咳作用；对小白鼠气管分泌功能有显著性增强作用。

关键词：射干地龙颗粒；小白鼠；止咳作用；祛痰作用

（发表于《中兽医学分会论文集》）

紫花苜蓿航天诱变田间变异观察研究

杨红善[1]，常根柱[1]，李红民[2]，柴小琴[3]，周学辉[1]，路 远[1]

（1. 中国农业科学院兰州畜牧与兽药研究所 兰州 730050；2. 甘肃省航天育种工程技术中心，天水；3. 天水市农业科学研究所，天水）

摘 要：紫花苜蓿经过空间搭载后，表现出较强的诱变效应，本文以搭载于我国神舟三号飞船上的4个航天诱变紫花苜蓿品种：德宝、德福、阿尔金冈、三得利，种植而成的苜蓿航天诱变原原种为研究对象。天水原原种田及兰州原原种田，连续几年观测记载，初步确定了突变类型和单株，并进行了集团分类，大致分为以下7种类型：多叶突变单株（6株）；大叶突变单株（11株）；速生单株（14株）、白花突变单株（1株）；抗病单株（7株）；早熟单株（7株）；矮生、分蘖性强单株（2株）。可见航天诱变对苜蓿形态指标具有变异频率高，变异幅度大、有益变异多，优势明显等特点。

关键词：紫花苜蓿；航天诱变；田间观察

(发表于《〈2011年航天工程育种论坛〉论文集》)

甘南藏系绵羊饲养管理现状及发展对策

王宏博[1,2]，阎 萍[1]，梁春年[1]，郎 侠[1]，丁学智[1]

(1. 中国农业科学院兰州畜牧与兽药研究所，兰州 730050；2. 农业部动物毛皮及制品质量监督检验测试中心（兰州），兰州 730050)

摘 要：<正>草地型藏系绵羊属于我国西藏绵羊品种之一，主要分布在海拔2 880~3 500m的青藏高原，分布广，具有抗寒、耐粗饲、遗传性稳定、产肉性能好的特点，是青藏高原的主体畜种。依其生态环境，结合生产、经济特点，可分为高原型（草地型）、山谷型和欧拉型3类。在甘南地区饲养的优良品种有草地类型的欧拉羊、甘加羊、乔科羊。草地型藏羊占甘南州绵羊总数的87.44%，构成了甘南州养羊业的主体。但由于甘南州地处青藏高 更多还原

关键词：藏系；绵羊饲养

(发表于《〈2011中国羊业进展〉论文集》)

饲草青贮系统中的乳酸菌及其添加剂研究进展

王晓力[1]，张慧杰[2]，孙启忠[2]，玉 柱[3]，郭艳萍[3]

(1. 中国农业科学院兰州畜牧与兽药研究所，兰州 73002；2. 中国农业科学院草原研究所，呼和浩特 0100103；3. 冲国农业大学草地研究所，北京 100193)

摘 要：在饲草青贮系统中有许多微生物存在，其中乳酸菌是最重要的微生物。本文着重介绍了饲草青贮系统中乳酸菌的种类及作用、乳酸菌青贮添加剂，并对近些年国内外的研究现状进行了分析。

关键词：青贮；乳酸菌；添加剂

(发表于《〈第三届全国微生物资源学术暨国家微生物资源平台运行服务研讨会〉论文集》)

母牦牛的繁殖特性与人工授精

郭 宪，裴 杰，包鹏甲，梁春年，丁学智，褚 敏，阎 萍

(中国国农业科学院兰州畜牧与兽药研究所/甘肃省牦牛繁育工程重点实验室，兰州 730050)

摘 要：牦牛是青藏高原特有的遗传资源，是高寒牧区的基本生产生活资料。牦牛繁殖是牦牛生产中的关键环节，深入了解母牦牛的繁殖特性是实施牦牛繁殖调控的基础。本文从初情期、性成熟、发情、妊娠期、分娩、繁殖力等方面综述了牦牛的繁殖特性，并探析了提高牦牛人工授精效率的技术措施，旨在为牦牛品种改良提供技术参考。

关键词：牦牛；繁殖特性；人工授精

(发表于《2011年国家肉牛牦牛产业技术交流大会论文汇编》)

动物用抗球虫药-常山酮

王学红，梁剑平，郭文柱，刘 宇，郝宝成

(中国农业科学院兰州畜牧与兽药研究所/农业部兽用药物创制重点实验室/甘肃省新兽药工程重点实验室，兰州 730050)

摘 要：本文综述了动物用抗球虫药物-常山酮的理化性质、作用机理、化学合成等，从而为常山酮的进一步研究提供理论指导和参考。

关键词：抗球虫药物；常山酮；理化性质、作用机理；化学合成

(发表于《2011年中国畜牧兽医学会小动物医学分会（乌鲁木齐）》)

一例雏鸭曲霉菌病的临床诊治

郭志廷[1]，梁剑平[1]，罗晓琴[2]

(1. 中国农业科学院兰州畜牧与兽药研究所新兽药重点开放实验室，兰州 730050；
2. 兰州市动物卫生监督所，兰州 730050)

摘　要：2008 年 8 月，兰州某小型养鸭场发生一起以腹泻、伴发神经症状为主的死亡病例。据畜主口述，养鸭场共有 12 日龄雏鸭 400 余只，第 1 天死亡 15 只，第 3 天死亡 34 只。第 3 天应畜主请求前往调查，最终经临床诊断和实验室检查确诊为曲霉菌病。

(发表于《中国家禽》)

奶牛乳房炎元乳链球菌抗生素耐药性研究

李宏胜，罗金印，李新圃，王　玲，苗小楼，王旭荣

(中国农业科学院兰州畜牧与兽药研究所/中国农业科学院临床兽医学研究中心/甘肃省中兽药工程技术研究中心，兰州 730050)

摘　要：为了查明引起我国奶牛乳房炎的元乳链球菌抗生素耐药情况，指导临床合理用药。对从我国部分地区奶牛场采集的临床型乳房炎病乳中分离鉴定出的 78 株无乳链球菌，采用 K-B 纸片法测定了这些菌株对抗生素的耐药情况。结果表明，无乳链球菌对目前临床上使用的大部分抗生素，头孢唑啉、头孢噻肟、丁胺卡那霉素、卡那霉素、庆大霉素、四环素、强力霉素、麦迪霉素、林可霉素、氟苯尼考、多粘菌素 B、环丙沙星、氟哌酸、氨苄青霉素/舒巴坦、头孢噻肟/棒酸和头孢他啶/棒酸均比较敏感；但对青霉素 G、氨苄青霉素、链霉素、恩诺沙星、阿莫西林/棒酸和复方新诺明，有一定的耐药性，其耐药率达 50%~100%。本项研究对指导临床合理用药具有重要的意义。
关键词：奶牛乳房炎；无乳链球菌；抗生素；敏感性

(发表于《第三届全国微生物资源学术研讨会》)

我国奶牛乳房炎病原菌区系调查及抗生素耐药性检测

李宏胜，罗金印，李新圃，王　玲，王旭荣，李建喜，杨　峰

(中国国农业科学院兰州畜牧与兽药研究所/中国农业科学院临床兽医学研究中心/农业部新兽药创制研究中心/甘肃省中兽药工程研究中心，兰州　730050)

摘　要：为了查明我国奶牛乳房炎病原菌区系分布状况及主要病原菌对常见抗生素的耐药情况，为进一步更好的预防和治疗奶牛乳房炎提供科学数据。对从我国40个城市150多个奶牛场采集的临床型乳房炎、隐性乳房炎和健康奶样共计3 941份，进行了细菌分离和鉴定，同时对分离鉴定出的乳房炎4种主要病原菌（无乳、停乳、金葡和大肠）采用药敏纸片法进行了抗生素耐药性检测。细菌分离和鉴定结果表明，从3 941份奶样中分得细菌和霉菌24种4 337株，无菌奶样496份，细菌总检出率为87.41%。在4 337株细菌中，分离鉴定出与乳房炎有密切关系的病原菌12种2 386株，病原菌总检出率为62.12%。各种病原菌主要为：无乳链球菌（35.54%）、停乳链球菌（23.55%）、金黄色葡萄球菌（16.89%）、大肠杆菌（12.99%）、乳房链球菌（5.28%）、克雷伯氏菌（1.63%）、变形杆菌（1.59%）、绿脓杆菌（1.22%）、星状诺卡氏菌（0.59%）、化脓棒状杆菌（0.38%）、白色念株菌（0.17%）和化脓链球菌（0.17%）。细菌耐药性结果表明，奶牛乳房炎4种常见病原菌对所有的抗生素均产生了不同程度的耐药性，其耐药率达10%~100%。对4种菌有较强抑制作用的药物主要有氟苯尼考、左氟沙星、洛美沙星、氟哌酸和头孢类药物，其敏感度达60%~100%。对4种菌产生较强耐药性的药物主要有青霉素G、链霉素和复方新诺明，其耐药率达30%~100%。本试验查明了我国奶牛乳房主要病原菌区系分布及常见病原菌对抗生素的耐药情况，对进一步预防和治疗奶牛乳房炎，指导临床合理用药奠定了基础。

关键词：奶牛乳房炎；病原菌；区系分布；抗生素；敏感性

（发表于《中国畜牧兽医协会家畜内科学分会第七届代表大会暨学术研讨会》）

细胞因子在奶牛乳房炎生物学防治中的应用

王 玲[1], 李宏胜[1], 陈炅然[1], 苗小楼[1], 王正兵[2], 尚若锋[1]

(1. 中国农业科学院兰州畜牧与兽药研究所/农业部新兽药工程重点实验室,甘肃省新兽药工程重点实验室, 兰州 730050; 2. 甘肃农业大学动物医学院, 兰州 730070)

摘 要：鉴于细胞因子对乳房免疫机能独特有效的操纵和调节,在奶牛乳房炎控制中具有一定独特优势,作者就细胞因子的生物学活性、作用机制和特点进行了综述,并重点对国内外应用奶牛重组细胞因子防治奶牛乳房炎进行了总结、分析,有望为控制奶牛乳房炎提供安全、有效、实用的生物学防治途径。

关键词：奶牛乳房炎；重组细胞因子；生物学防治

(发表于《中国畜牧兽医学会家畜内科学分会》)

马属动物感冒的辨证施治

谢家声，严作廷

(中国农业科学院兰州畜牧与兽药研究所甘肃省中兽药工程技术中心, 兰州 730050)

摘 要：感冒,中兽医称为伤风,临床常分为风寒感冒和风热感冒,患病动物主要临床表现为头低耳聋,上眼角翘起呈三角形皱（系头痛眩晕的表现）,发热恶寒,发抖,被毛粗乱逆立,脉象现浮,鼻流清涕或粘浆液,时有吹鼻、咳嗽、食少、精神不振等。因动物体制的强弱、季节的变化,临床表现常有寒热虚实的不同。动物感冒虽不像人感冒临床症状明显、脉搏清晰,但临诊只要掌握感冒动物的特有症状,辨清寒热虚实,进行合理治疗,常能收俘鼓之效。如果诊断差错,误用药物,会导致引邪入内,使病情加重甚至死亡。笔者临诊常将其分为风寒与风热辨证施治,取效甚捷,现介绍如下。

(发表于《中国畜牧兽医学会家畜内科学分会第七届代表大会暨学术研讨会论文集》)

牦牛 ATP6 和 Cytb 基因的分子进化研究

吴晓云，阎 萍，梁春年，郭 宪，包鹏甲，
裴 杰，丁学智，褚 敏，焦 斐，刘 建

(中国国农业科学院兰州畜牧与兽药研究所/甘肃省牦牛繁育工程重点实验室，兰州 730050)

摘 要：线粒体基因组（Mitoehondrial Gonome）具有区别于核基因组的诸多特性，文章通过对牦牛线粒体基因组的 ATP6 和 Cytb 基因进行分子进化分析发现，ATP6 和 Cytb 基因均受到纯化选择，且 ATP6 基因受到纯化选择的作用大于 Cytb 基因，选择的主要因素应该是该群体所处的特殊的地理环境，即地理环境具有直接的选择作用。

关键词：ATP6；Cytb；分子进化

(发表于《中国畜牧兽医学会养牛学分会第八届全国会员代表大会暨 2011 年学术研讨会》)

牛 FAS 基因的生物信息学分析

焦 斐，阎 萍，梁春年，郭 宪，包鹏甲，
裴 杰，丁学智，褚 敏，吴晓云，刘 建

(中国国农业科学院兰州畜牧与兽药研究所/甘肃省牦牛繁育工程重点实验室，兰州 730050)

摘 要：对牛 FAS 基因进行生物信息学分析，以预测 FAS 基因编码产物的理化性质、结构与功能。结果表明，FAS 基因编码产物为较稳定疏水性蛋白，不具有信号肽，为非分泌蛋白。二级结构主要以 α 螺旋和无规则卷曲为主，主要在质膜中发挥生物学作用。FAS 基因编码产物可能具有结构蛋白功能的可能性是 1.735，还具有信号转导、生长因子等功能。

关键词：牛；FAS 基因；蛋白质结构；生物信息学

(发表于《中国畜牧兽医学会养牛学分会第八届全国会员代表大会暨 2011 年学术研讨会》)

牛 FTO 基因功能结构生物信息学分析

刘 建，阎 萍，梁春年，郭 宪，包鹏甲，
裴 杰，丁学智，褚 敏，吴晓云，焦 斐

(中国农业科学院兰州畜牧与兽药研究所/甘肃省牦牛繁育工程重点实验室，兰州 730050)

摘 要：FTO (fat mass and obesity associated) 基因也称肥胖基因，是一种与肥胖相关的等位基因，Peters[1]等在首次在小鼠中克隆出这个基因。近年来，随着对FTO基因的研究，多个单核苷酸多态性位点经证实与肥胖有关，并确定FTO基因通过调节食物摄入和机体内脂肪代谢活动发挥其功能[2]。FTO基因定位于牛的18号染色体上，Wei等[3]人检测了五个中国地方黄牛品种FTO基因的多态性，发现第五外显子C1071T突发变位点与牛的背膘厚和背最长肌面积有显著的关联性。柴继杰等人[4]解析了肥胖基因的蛋白质结晶结构，发现了该蛋白质是一种具有双加氧酶特征的大分子。本研究用生物信息学方法对牛FTO基因其编码产物的蛋白质结构进行预测和分析，旨在为进一步开展牛核基因编码蛋白结构及其生理功能等方面的实验研究提供基础研究资料。

(发表于《中国畜牧兽医学会养牛学分会第八届全国会员代表大会暨2011年学术研讨会》)

牛 PCR 性别鉴定研究进展

裴 杰[1,2]，阎 萍[1,2]，程胜利[1]，梁春年[1,2]，
郭 宪[1,2]，包鹏甲[1,2]，褚 敏[1,2]

(1. 中国农业科学院兰州畜牧与兽药研究所，兰州 730050；2. 甘肃省牦牛繁育工程重点实验室，兰州 730050)

摘 要：动物的性别控制一直是性别研究领域的热点，尤其是对具有较高经济价值的牛的性别控制的研究。目前通过牛早期胚胎的性别鉴定来控制出生牛性别的技术应用最为广泛。本文综述了近年来应用于牛早期胚胎性别鉴定的各种基因及DNA序列，主要包括SRY基因、ZFY/ZFX基因、AMEL基因、Y染色体重复序列，并对各种性别鉴定方法的原理和特点作了进一步的说明，力图为将要从事牛性别鉴定的研究人员和基层工作者提供技术参考。

关键词：牛；性别鉴定；早期胚胎

(发表于《中国畜牧兽医学会养牛学分会第八届全国会员代表大会暨2011年学术研讨会》)

奶牛乳腺的生理防御及乳房炎病理学研究

李新圃，罗金印，李宏胜

(中国农业科学院兰州畜牧与兽药研究所，兰州 730050)

摘　要：从物理屏障、细胞防御、体液抗体、乳汁抑菌剂及细胞因子方面，对奶牛乳腺的生理防御功能进行了综述，并且在此基袖上分析了乳房炎的发病机理，认为乳房炎的复杂性已经限制了人们对奶牛乳腺生理防御及病理学过程的全面了解和掌握，进而影响了防治乳房炎新方法——生物疗法的研究进程因此，进一步深入开展奶牛乳腺生理及病理学研究非常重要。

关键词：奶牛；乳腺；乳房炎

(发表于《中国奶业协会第26次繁殖学术年会暨国家肉牛牦牛/奶牛产业技术体系第3届全国牛冰防治学术研讨会》)

奶牛乳房炎以疫苗预防为主的综合防控措施应用效果研究

李宏胜，罗金印，李新圃，苗小楼，王　玲，王旭荣，王小辉

(中国农业科学院兰州畜牧与兽药研究所/中国农北科学院临床兽医学研究中心/甘肃省中兽药工程研究中心，兰州 730050)

摘　要：为了进一步提高奶牛乳房炎综合防控效果，我们先后在兰州地区的3个规模化奶牛场和7个个体奶牛场开展了以奶牛乳房炎疫苗预防为主，药物防治和改善环境卫生为辅的综合防控措施试验，结果表明，10个奶牛场经过1~2年的试验，隐性乳房炎乳区发病率比试验前平均降低63.97%（$P<0.01$），临床型乳房炎头发病率比试验前平均降低66.36%（$P<0.01$），奶产量与试验前相比平均提高10.10%。达到了良好的防治效果。表明以奶牛乳房炎疫苗预防为主的综合防控措施是一项有效的预防奶牛乳房炎的有效方法。

关键词：奶牛乳房炎；多联苗；免疫；综合防控

(发表于《中国奶业协会第26次繁殖学术年会暨国家肉牛牦牛/奶牛产业技术体系第3届全国牛冰防治学术研讨会》)

奶牛乳房炎综合防控技术研究进展

李宏胜，李新圃，罗金印，苗小楼，王 玲，李建喜

(中国农业科学院兰州畜牧与兽药研究所/中国农业科学院临床兽医学研究中心/甘肃省中兽药工程研究中心，兰州 730050)

摘 要：奶牛乳房炎是一种复杂的给奶牛养殖业造成极大经济损失的常见多发病。本文综述了奶牛乳房炎造成的危害，近年来国内外引起奶牛乳房炎的主要病原菌及其区系分布状况、乳房炎发病原因、发生规律、诊断治疗及疫苗研制等研究进展，以期为进一步更好的综合防控奶牛乳房炎提供借鉴和参考。

关键词：奶牛乳房炎；病原菌；诊断治疗；疫苗

(发表于《中国奶业协会第26次繁殖学术年会暨国家肉牛牦牛/奶牛产业技术体系第3届全国牛病防治学术研讨会》)

奶牛隐性子宫内膜炎的生物学诊疗想法

张世栋，严作廷，王东升，李世宏

(中国农业科学院兰州畜牧与兽药研究所/中国农业科学院临床兽医学研究中心/甘肃省中兽药工程技术研究中心，兰州 730050)

摘 要：本文简要综述了奶牛隐性子宫内膜炎的病理学特征和当前具有代表性的几种诊断与治疗方法。基于对奶牛隐性子宫内膜炎的病理研究现状，认为子宫内膜组织持续受到氧化损伤应该是该病的主要原因之一，研究中性粒细胞在子宫组织中持续浸润的机制，寻找其细胞信号转导中有效的生物学标记可以为隐性子宫内膜炎新的诊断和治疗方法奠定理论基础。

关键词：隐性子宫内膜炎；生物学诊断和治疗；氧化损伤

(发表于《中国奶业协会第26次繁殖学术年会暨国家肉牛牦牛/奶牛产业技术体系第3届全国牛病防治学术研讨会》)

清宫助孕液治疗奶牛子宫内膜炎临床试验

严作廷[1]，王东升[1]，李世宏[1]，张世栋[1]，谢家升[1]，王雪郦[2]，杨明成[2]，朱新荣[3]，陈道顺[3]

(1. 中国农业科学院兰州畜牧与兽药研究所/甘肃省中兽药工程技术研究中心，兰州 730084；2. 兰州市城关奶牛场，兰州 730020；3. 甘肃荷斯坦奶牛繁育示范中心，兰州 730086)

摘 要：为研究"清宫助孕液"治疗奶牛子宫内膜炎的临床疗效，本试验将临床确诊为子宫内膜炎的奶牛80头，随机分为"清宫助孕液"治疗组和药物对照组，每组分别为60头和20头，治疗组采用直肠把握法将清宫助孕液子宫灌注，每次100mL，隔日一次，4次为一疗程；对照药物将青霉素100万单位、链霉素100万单位，用生理盐水50mL稀释后，子宫灌注，隔日一次，4次为一疗程。结果表明，试验组治疗60头，治愈52头，治愈率86.67%，显效3头，显效率5.0%；有效2头，有效率3.33%；无效3头，无效率5.0%，总有效率95.0%。三个情期受胎率86.67%。对照组治疗20头，治愈15头，治愈率75.0%，显效1头，显效率5.0%；有效2头，有效率10%；无效2头，无效率10%，总有效率90.0%。三个情期受胎率75.0%。可以得出，"清宫助孕液"对奶牛子宫内膜炎具有较好的治疗效果，可以提高情期受胎率。

关键词：奶牛；清宫助孕液；子宫内膜炎

(发表于《中国奶业协会第26次繁殖学术年会暨国家肉牛牦牛/奶牛产业技术体系第3届全国牛病防治学术研讨会》)

如何进行奶牛乳房炎金黄色葡萄球菌疫苗的合理评估

王旭荣，李建喜，王小辉，李宏胜，杨志强，孟嘉仁

(中国国农业科学院兰州畜牧与兽药研究所/中国农业科学院临床兽医学研究中心/甘肃省中兽药工程技术研究中心，兰州 730050)

摘 要：奶牛乳房炎是造成全球奶业经济损失的主要原因。而金黄色葡萄球菌是最主要的病原菌。本文系统、直观地对研究奶牛乳房炎金黄色葡萄球菌疫苗的科研结论进行量化总结和评估，目的是为奶牛乳房炎疫苗的研发提供新思路。为设计合理的奶牛乳房炎疫苗给动物评估试验方案提供科学依据。本文在Pubmed、Science数据库和CNKI数据

库对"奶牛乳房炎疫苗"类的文章进行了合理的电子检索。然后对细菌疫苗、细菌类毒素疫苗、DNA-重组蛋白疫苗、单一的重组蛋白疫苗等方面的科研论文从试验设计、方法、疫苗类型和研究结果4个方面进行了全面的分析。结果表明,采用DNA、重组蛋白新技术的疫苗和传统的细菌疫苗已取得良好效果,疫苗可以成为预防和控制金黄色葡萄球菌性奶牛乳房炎的最好或最有前景的一种途径。但是研究方法差异和双盲试验的缺乏阻碍对疫苗效果的合理评估。

关键词:奶牛乳房炎;金黄色葡萄球菌;疫苗;合理评估

(发表于《中国奶业协会第26次繁殖学术年会暨国家肉牛牦牛/奶牛产业技术体系第3届全国牛病防治学术研讨会》)

中兽药在奶牛健康养殖中的作用

李建喜

(中国农业科学院兰州畜牧与兽药研究所/甘肃省中兽药工程技术研究中心,兰州 730050)

摘 要:健康养殖模式是提高我国奶牛养殖水平、改善奶品质和增强市场竞争力的新型支撑技术,疾病防控是健康养殖模式建设的关键环节,现代养殖中的疾病防控是一个立体概念,但药物和诊断技术仍然是疾病防控的两大要素。临床上可使用的兽药种类很多,但抗生素和化学合成药物的用量比较大,因此导致的药物残留和细菌耐药性问题倍受人们重视为解决该类问题,开展药物创新研究势在必行。中兽药防治奶牛疾病的有效性已被历史实践证实,尽管因其剂型单一等问题的存在而限制了市场发展,但如果能加大研发力度,中兽药定能成为奶牛健康养殖疾病防控的保障产品。

关键词:中兽药;奶牛;健康养殖

(发表于《中国奶业协会第26次繁殖学术年会暨国家肉牛牦牛/奶牛产业技术体系第3届全国牛病防治学术研讨会》)

中药子宫灌注剂治疗奶牛不孕症综述

王东升,严作廷,张世栋,谢家声,李世宏

(中国农业科学院兰州畜牧与兽药研究所/中国农业科学院临床兽医学研究中心/甘肃省中兽药工程技术研究中心,兰州 730050)

摘 要:中药子宫灌注剂在奶牛子宫内膜炎和卵巢疾病等繁殖疾病的防治中发挥着巨大

的作用，临床用于防治奶牛主要繁殖障碍疾病的子宫灌注剂主要有溶液型、混悬型和乳浊型3种。本文对这3种剂型的组方、疗效及作用机理等的研究概况进行综述，以期为新型中药子宫灌注剂的研制及其在临床上的应用提供参考。

关键词：奶牛；中药；子宫灌注剂；子宫内膜炎；卵巢疾病

(发表于《中国奶业协会第26次繁殖学术年会暨国家肉牛牦牛/奶牛产业技术体系第3届全国牛冰防治学术研讨会》)

蛋白质组学研究概况及其在中兽医学研究中的应用探讨

董书伟，荔 霞，刘永明，王胜义，王旭荣，刘世祥，齐志明

(中国农业科学院兰州畜牧与兽药研究所/中国农业科学院临床兽医学研究中心/甘肃省中兽药工程技术中心，兰州 730050)

摘 要：随着生命科学研究进入后基因组时代，蛋白质组学作为重要的实验技术，已经成为筛选重大疾病的特异生物标志物和研究发病机制的新途径本文总结了蛋白质组学研究的主要内容和方法，简述了蛋白质组学在生命科学的应用概况，并探讨了其在中兽医学研究中的应用前景。

关键词：蛋白质组；蛋白质组学；生物标志物；中兽医学

(发表于《中国奶业协会第26次繁殖学术年会暨国家肉牛牦牛/奶牛产业技术体系第3届全国牛冰防治学术研讨会》)

微量元素硒在奶牛上的研究进展

王胜义，刘永明，齐志明，荔 霞，董书伟，刘世祥

(中国农业科学院兰州畜牧与兽药研究所/农业部兽用药物创制重点实验室/甘肃省中兽药工程技术研究中心，兰州 730050)

摘 要：硒是奶牛营养中的必需微量元素之一，具有抗氧化、提高免疫力、调节激素代谢等多种生物学功能。文章主要阐述了硒的生物学功能及其在奶牛营养上的研究进展。

关键词：硒；生物学功能；奶牛健康

(发表于《中国奶业协会第26次繁殖学术年会暨国家肉牛牦牛/奶牛产业技术体系第3届全国牛冰防治学术研讨会》)

动物性食品中兽药残留分析检测研究进展

李 冰，李剑勇，周绪正，杨亚军，牛建荣，魏小娟，李金善，张继瑜

(中国农业科学院兰州畜牧与兽药研究所/农业部兽用药物创制重点实验室/甘肃省新兽药工程重点实验室，兰州 730050)

摘　要：兽药作为动物疾病防治的重要保障，对改善动物生产性能和产品品质、保持生态平衡等多方面具有重要功能，是畜牧业健康发展、食品安全和公共卫生的重要保障。但不合理使用和滥用兽药及饲料药物添加剂的情况普遍存在，既造成了动物食品中有害物质残留，又对人类健康造成损害，同时还威胁到环境和畜牧业持续健康发展。为保证人类健康，迫切需要开发简洁、快速、高灵敏度、高通量且低成本的兽药残留检测技术。本文对动物性食品中兽药残留分析的样品前处理方法和检测技术进行了研究探讨。
关键词：动物性食品；兽药残留；样品前处理；检测技术

(发表于《畜牧与兽医》)

规模化肉牛养殖场苍蝇的防控策略

周绪正，张继瑜，李 冰，魏小娟，牛建荣，李金善，李剑勇，杨亚军

(中国农业科学院兰州畜牧与兽药研究所/农业部兽用药物创制重点实验室/甘肃省新兽药工程重点实验室，兰州 730050)

摘　要：养殖场蚊蝇滋生是影响奶牛生产性能和环境卫生的一个令人头痛的问题。本文对苍蝇的生活习性进行了详细叙述，以便为有效防控、消灭苍蝇提供依据。同时，本文从环境防治、生物防治和化学防治三方面对苍蝇进行防控，根据苍蝇的生活周期采取一种、二种或三种并用的方式，达到最有效防控苍蝇的目的，并对环境污染最低化、最小化，最好达到无害化。
关键词：苍蝇；防控策略；生活习性

(发表于《中国奶牛》)

银翘蓝芩注射液中绿原酸、黄芩苷和连翘苷的含量测定

杨亚军，王兴业，李剑勇，李 冰，周绪正，刘希望

(中国农业科学院兰州畜牧与兽药研究所/农业部兽用药物创制重点实验室/甘肃省新兽药工程重点实验室，兰州 730050)

摘 要：[目的] 对银翘蓝芩注射液中的绿原酸、黄芩苷和连翘苷进行含量测定。[方法] 高效液相色谱法，Kromasil ODS-1 C18柱；绿原酸，流动相为甲醇-pH3.0磷酸水溶液 (20∶80，v/v)，检测波长为324nm；黄芩苷，流动相为甲醇-pH3.0磷酸水溶液 (50∶50，v/v)，检测波长为274nm；连翘苷，流动相为乙腈-纯水 (25∶75，v/v)，检测波长为230nm。[结果] 3种成分在一定的浓度范围内，线性关系良好；日内和日间精密度及加样回收率试验结果均符合实验要求。[结论] 该方法简单迅速，结果准确，精密度好，能够用于银翘蓝芩注射液质量标准（草案）的制定。

关键词：银翘蓝芩注射液；高效液相色谱法；绿原酸；黄芩苷；连翘苷

(发表于《中国奶业协会第26次繁殖学术年会暨国家肉牛牦牛奶牛产业技术体系第3届全国牛病防治学术研讨会论文集2011年》)

作者简介

张继瑜（1967—），中共党员，博士研究生，三级研究员，博（硕）士生导师，中国农业科学院三级岗位杰出人才，中国农业科学院兽用药物研究创新团队首席专家，国家现代农业产业技术体系岗位科学家。兼任中国兽医协会中兽医分会副会长，中国畜牧兽医学会兽医药理毒理学分会副秘书长，农业部兽药评审委员会委员，农业部兽用药物创制重点实验室常务副主任，甘肃省新兽药工程重点实验室常务副主任，中国农业科学院学术委员会委员，黑龙江八一农垦大学和甘肃农业大学兼职博导。主要从事兽用药物及相关基础研究工作，重点方向包括兽用化学药物的研制、药物作用机理与新药设计、细菌耐药性研究。带领的研究团队在动物寄生虫病、动物呼吸道综合症防治药物研究上取得了显著进展。在肠杆菌耐药机理、血液原虫药物作用靶标筛选的研究处于领先地位。先后主持完成国家、省部重点科研项目20多项，获得科技奖励6项，研制成功4个兽药新产品，其中国家一类新药一个，以第一完成人申报10项国家发明专利，取得专利授权5项。培养研究生21名，发表论文170余篇，主编出版《动物专用新化学药物》和《畜牧业科研优先序》等著作2部。

阎萍（1963—）中共党员，博士，三级研究员，博士生导师。2012年享受国务院特殊津贴，是中国农业科学院三级岗位杰出人才，甘肃省优秀专家，甘肃省"555"创新人才，甘肃省领军人才。曾任畜牧研究室副主任、主任等职务，2013年3月任研究所副所长职务。兼任国家畜禽资源管理委员会牛马驼品种审定委员会委员，中国畜牧兽医学会牛业分会副理事长，全国牦牛育种协作组常务副理事长兼秘书长，中国畜牧兽医学会动物繁殖学分会常务理事和养牛学分会常务理事等。阎萍研究员主要从事动物遗传育种与繁殖研究，特别是在牦牛领域的研究成绩卓越，先后主持和参加完成了科技部支撑计划、科技部基础性研究项目、科技部"863"计划、"948"计划、农业部行业科技项目、国家肉牛产业技术体系岗位专家、人事部回国留学基金项目、科技部成果转化项目、甘肃省科技重大专项计划、甘肃省农业生物技术项目等20余项课题。现为国家肉牛牦牛产业技术体系牦牛选育岗位专家，甘肃省牦牛繁育工程重点实验室主任。作为高级访问学者多次到国外科研机构进行学术交流。培育国家牦牛新品种1个，填补了世界上牦牛没有培育品种的空白。获国家科技进步奖1项，省部级科技进步奖5项及其他科技奖励3项。培养研究生15名，发表论文180余篇，出版《反刍动物营养与饲料利用》、《现代动物繁殖技术》、《牦牛养殖实用技术问答》、《Recend Advances in Yak Reproduction》、《中国畜禽遗传资源志-牛志》等著作。

包鹏甲（1980—），男，甘肃武威人，助理研究员，动物遗传育种与繁殖专业硕士。2007年毕业于甘肃农业大学，主要从事生物技术与草食动物遗传繁育研究工作，主持完成中央级公益性科研院所基本科研业务费2项，参加完成国家、省部级及其他科研课题多项，参编著作3部，第一作者发表国内外文章10余篇，获专利十余项。

常根柱（1956—），甘肃甘谷县人。研究员，中共党员，硕士研究生导师，中国草学会理事，甘肃省一层次领军人才，甘肃省草品种审定委员会委员。先后参加、主持完成国家、省部级课题12项；获省、部级奖3项，地、厅级奖4项；主编、副主编学术专著4部，发表论文68篇；获国家授权发明专利1项（第二）。选育成功牧草新品种4个（2个第一，2个第二），通过甘肃省草品种审定委员会审定登记，其中2个通过国家草品种审定委员会评审，参加全国区域试验。独立培养硕士研究生5名，合作培养博士研究生1名。研究方向：牧草育种。现承担项目：甘肃省科技支撑计划-牧草航天诱变品种选育。

丁学智（1979—），博士、副研究员，甘肃省杰青年金获得者、2012年度中国农业科学院"青年文明号"。主要从牦牛高寒低氧适应方面的研究工作。目前主持国家自然基金青年科学基金、甘肃省杰出青年科学基金、国家自然科学基金国际（地区）合作重大项目等，发表SCI收录论文10余篇、国家发明专利了2项、实用新型专利1项。

董书伟（1980—），男，汉族，在读博士，助理研究员，毕业于西北农林科技大学。主要从事奶牛营养代谢病与中毒病的蛋白质组学研究。先后主持参加国家科技支撑计划、中央级公益性科研院所基本科研业务费专项资金项目、中国科学院西部之光项目、国家自然科学青年基金，发表学术论文20余篇，申请专利18项。

郭天芬（1974—），学士、副研究员。主要从事动物营养、毛皮质量检测及标准研究制定工作。期间参加十余项国家级省部级科研项目及撰写制定十余项标准制定项目，主持完成一项国家标准制定项目和二项中央公益类基本科研项目。以第一完成人获发明专利1项、实用新型专利6项，获甘肃省科学技术进步奖1项（第3完成人），酒泉市科技进步奖1项（第4完成人）。正式发表学术期刊论文100余篇，其中第一作者的30余篇。参与撰写专著5部，其中副主编的2部。

郭宪（1978—），中国农业科学院硕士生导师，博士，副研究员。中国畜牧兽医学会养牛学分会理事，中国畜牧业协会牛业分会理事，全国牦牛育种协作组理事。主要从事牛羊繁育研究工作。先后主持国家自然科学基金、国家支撑计划子课题、中央级公益性科研院所基本科研业务费专项资金项目等5项。科研成果获奖3项，参与制定农业行业标准3项，授权专利3项。主编著作2部，副主编著作5部，参编著作2部。主笔发表论文30余篇，其中SCI收录5篇。

郭志廷（1979—），男，内蒙古人，助理研究员，执业兽医师，九三学社兰州市青年委员会委员，中国畜牧兽医学会中兽医分会理事。2007年毕业于吉林大学，获中兽医硕士学位。近年主要从事中药抗球虫、免疫学和药理学研究。先后主持或参加国家、省部级科研项目5项，包括中央级公益性科研院所专项基金。作为参加人获得兰州市科技进步一等奖1项，兰州市技术发明一等奖1项，完成甘肃省科技成果鉴定4项，授权国家发明专利5项（1项为第一完成人），参编国家级著作2部。在国内核心期刊上发表学术论文80余篇（第一作者30篇）。

郝宝成（1983—），甘肃古浪人，硕士研究生，助理研究员，研究方向为新型天然兽用药物研究与创制。先后主持和参与了中央级公益性科研院所基本科研业务费专项资金、国家支撑计划、863项目子课题等项目6项，以第一作者发表论文23篇（其中SCI论文2篇，一级学报2篇），参与编写著作3部《中兽药学》、《兽医中药学及试验技术》、《天然药用植物有效成分提取分离与纯化技术》，以第一发明人获得国家发明专利3项，荣获2012年度兰州市九三学社参政议政先进工作者。

李宏胜（1964—），九三学社社员，博士，研究员，硕士生导师，甘肃省"555"创新人才。中国畜牧兽医学会家畜内科学分会常务理事。多年来主要从事兽医微生物及免疫学工作，尤其在奶牛乳房炎免疫及预防方面有比较深入的研究。先后主持和参加完成了国家自然基金、国家科技支撑计划、国际合作、甘肃省、兰州市及企业横向合作等20多个项目。先后获得农业部科技进步三等奖1项；甘肃省科技进步二等奖2项、三等奖2项；中国农业科学院技术成果二等奖3项；兰州市科技进步一等奖2项、二等奖2项。获得发明专利4项，实用新型专利14项，培养硕士研究生5名，在国内外核心期刊上发表论文160余篇，其中主笔论文60余篇。

李剑勇（1971—），研究员，博士学位，硕士和博士研究生导师，国家百千万人才工程国家级人选，国家有突出贡献中青年专家。现任中国农业科学院科技创新工程兽用化学药物创新团队首席专家，农业部兽用药物创制重点实验室副主任，甘肃省新兽药工程重点实验室副主任，甘肃省新兽药工程研究中心副主任，农业部兽药评审专家，甘肃省化学会色谱专业委员会副主任委员，中国畜牧兽医学会兽医药理毒理学分会理事，国家自然基金项目同行评议专家，《黑龙江畜牧兽医杂志》常务编委，《PLOS ONE》、《Medicinal Chemistry Research》等SCI杂志审稿专家。多年来一直从事兽用药物创制及与之相关的基础和应用基础研究工作。曾先后完成药物研究项目40多项，主持16项。获省部级以上奖励10项，2011年度获第十二届中国青年科技奖；2011年度获第八届甘肃青年科技奖；获2009年度兰州市职工技术创新带头人称号。获国家一类新兽药证书，均为第2完成人。申请国家发明专利22项，获授权9项。发表科技论文200余篇，其中SCI收录22篇，第一作者和通讯作者15篇。出版著作4部，培养研究生15名。

李维红（1978—），博士，副研究员。主要从事畜产品质量评价技术体系、畜产品检测新方法及其产品开发利用研究等。先后主持和参加了中央级公益性科研院所基本科研业务费专项资金项目、甘肃省自然基金项目等。主笔发表论文20余篇。主编《动物纤维超微结构图谱》，副主编《绒山羊》，参加编写《动物纤维组织学彩色谱》和《甘肃高山细毛羊的育成和发展》著作2部。获得专利4项，作为参加人获甘肃省科技进步一等奖一项、三等奖一项。

李新圃（1962—），博士，副研究员。主要从事兽医药理学研究工作。已主持完成省、部、市级科研项目7项。参加完成国际合作、省、部、市级科研项目三十余。参加项目"奶牛重大疾病防控新技术的研究与应用"获2010年甘肃省科技进步二等奖；"奶牛乳房炎主要病原菌免疫生物学特性的研究"在2008年获兰州市科技进步一等奖、中国农科院科学技术成果二等奖和甘肃科技进步三等奖；"绿色高效饲料添加剂多糖和寡糖的应用研究"获2005年兰州市科技进步一等奖；"奶牛乳房炎综合防治配套技术的研究及应用"获2004年甘肃省科技进步二等奖。已发表研究论文40余篇，其中5篇被SCI收录。获实用新型专利授权3个。

梁春年（1973—），硕士生导师，博士，副研究员，副主任，兼任全国牦牛育种协作组常务理事，副秘书长，中国畜牧兽医学会牛学会理事，中国畜牧业协会牛业分会理事，中国畜牧业协会养羊学分会理事等职。主要从事动物遗传育种与繁殖方面的研究工作。现主持国家科技支撑计划子课题"甘肃甘南草原牧区牦牛选育改良及健康养殖集成与示范"和国家星火计划项目子课题"牦牛高效育肥技术集成示范"等课题4项；参加完成国家及省部级科研项目30余项，获得省部级科技奖励7项。参与制定农业行业标准5项。参加国内各类学术会议30余次，国际学术会议4次。主编著作2本，副主编著作5本，参编著作6本，发表论文90余篇，其中SCI文章5篇。

刘希望（1986—），硕士，助理研究员。2007年毕业西北农林科技大学环境科学专业，2010年获西北农林科技大学化学生物学专业硕士学位，同年参加工作至今。曾主持农业部兽用药物创制重点实验室开放基金项目1项，现主持中央公益性科研院所基本科研业务费1项。以第一作者发表SCI论文4篇，主要从事新兽药研发，药物合成方面的研究工作。

罗永江（1966—），硕士生导师，学士，副研究员。主要从事中兽药的研制与开发工作。主持或参加了十一五国家科技支撑项目子课题、甘肃省自然科学基金、兰州市农业科技攻关课题、中央级公益性科研院所基本科研业务费专项以及横向课题等。发表论文40余篇，参加编写专著8部，获得发明专利11项，实用新型专利5项。获得过农业部科技进步三等奖1次，中国农科院科技进步一等奖、二等奖各1次，甘肃省科技进步二等奖1次、三等奖2次，兰州市技术发明一等奖1次，北京市科学技术奖三等奖1次。

孙晓萍（1962—），硕士生导师，学士，副研究员。主要从事绵羊遗传育种工作，先后主持的项目有：甘肃省自然科学基金：绵羊毛生长机理研究；甘肃省农委推广项目：绵羊双高素推广应用研究；甘肃省支撑计划项目：奶牛产奶量的季节性变化规律研究；甘肃省星火项目：肉羊高效繁殖技术研究；甘肃省支撑计划项目：肉用绵羊高效饲养技术研究。先后发表学术论文40余篇，主编参编著作11部，实用新型专利3个。获甘肃省畜牧厅科技进步二等奖1项，甘肃省科技进步二等奖1项，中国农业科学院科技进步一等奖1项，中华农业科技二等奖1项。

田福平（1976—），甘肃武山人，副研究员。中国草学会会员，2003年获草业科学王栋奖学金，中国草学会青年工作委员会理事，一直从事草地生态、牧草种质资源与育种、牧草栽培等方面的研究。先后参加国家、部、省级相关研究项目40项，主持项目6项。获甘肃省科技进步奖二等奖2项，中国农业科学院科学技术成果奖二等奖1项。发表学术论文60余篇，其中SCI论文4篇。参编著作7部，其中主编著作4部。

王东升（1979—），农学硕士，助理研究员。主要从事奶牛繁殖疾病的研究。主持"奶牛子宫内膜中天然抗菌肽的分离、鉴定及其生物学活性研究"和"狗经穴靶标通道及其生物效应的研究"2个课题，参加"十二五"国家科技支撑计划项目"奶牛健康养殖重要疾病防控关键技术研究"和"十一五"国家科技支撑计划项目"奶牛主要繁殖障碍疾病防治药物研制"等10多个项目。参与申请并获得发明专利4项，实用性新专利5项，取得三类新兽药证书1个，参加的成果获甘肃省科技进步二等奖2项和兰

州市科技进步二等奖 2 项。参编著作《兽医中药配伍技巧》、《兽医中药学》、《奶牛围产期饲养与管理》和《奶牛常见病综合防治技术》等 5 部，主笔发表论文 20 余篇。

王贵波（1982—），临床兽医学硕士，助理研究员，中国生理学会会员，主要从事兽医针灸及中兽医药学研究。主持完成了甘肃省青年基金项目"针刺镇痛对犬脑内 Jun 蛋白表达的影响研究"，中央级公益性事业单位专项资金项目"针刺镇痛对中枢 Fos 蛋白表达的影响研究"、"电针对犬痛阈及中枢强啡肽基因表达水平的研究"及横向委托项目"银翘双解颗粒/饮与灵丹草饮临床疗效验证委托试验"；还参与完成了国家自然科学基金、财政部中央级公益性科研院所基本科研业务费项目、"十一五"国家科技支撑项目和农业科技成果转化资金项目等。第一作者发表论文 20 余篇，副主编著作一部，参编著作一部，第一发明人获得的专利 7 项。

王宏博（1977—），博士，助理研究员，博士。主要从事动物营养与饲料科学。先后主持甘肃省科技厅项目 2 项，中央级公益性科研院所科研业务费专项资金项目 3 项，先后参加农业部公益性行业（农业）专项 3 项，"948"项目 1 项，获得国家发明专利 2 项，实用新型专利 1 项，参与完成国家发明专利和实用新型专利总计 10 余项。参与制定国家和农业部标准 10 余项。主编 1 部，副主编著作 1 部，参编著作 3 部。发表学术论文 30 余篇。

王磊（1985—），硕士，研究实习员。主要从事中兽医药理学、奶牛疾病防治和药物残留研究。先后参加各类研究课题 8 项，主持课题 1 项。2013 年参加的"非解乳糖链球菌发酵黄芪转化多糖的研究与应用"获甘肃省科技进步三等奖、2014 年参加的"重金属镉/铅与喹乙醇抗原合成、单克隆抗体制备及 ELISA 检测技术研究"获中国农业科学院科学技术成果二等奖；主笔发表科技论文 4 篇、其中 SCI 收录 1 篇，获得授权专利 2 项。

王晓力（1965—），副研究员。主要从事牧草资源开发利用方面的研究工作。近年来参加省部级课题多项，主持农业行业科研专项子课题 2 项，获全国农牧渔业丰收二等奖 1 项、内蒙古自治区农牧业丰收一等奖 1 项、2011 年贵州省科学技术成果转化二等奖 1 项。发表论文 30 余篇，出版著作 4 部，参编 4 部，授权专利 5 项。

王旭荣（1980—），博士，助理研究员。主要从事分子病毒学和分子细菌学方面的研究。主持完成"犬瘟热病毒"的基本科研业务费项目1项；现主持"奶牛乳房炎病原菌"方面的农业行业标准项目1项和甘肃省农业生物技术项目1项；参与国家奶牛产业技术体系项目、948项目、国家科技支撑项目、国家自然基金、科技基础性工作专项等10余项。参与的研究项目在2011—2014年期间获得奖项3个，其中甘肃省科技进步三等奖1项，甘肃省农牧渔丰收奖一等奖1项、中国农业科学院科学技术成果二等奖1项。发表科技论文40余篇，主笔15篇；参编著作1部；实用新型授权20余项，第一发明人授权5项；申请发明专利（第一发明人）3项。

王学红（1975—），硕士，高级兽医师。主要从事天然药物提取研究。先后主持及参加国家级省部级药物研究项目20余项，包括"十二五"农村领域国家科技计划项目、国家支撑计划项目、甘肃省科技支撑计划项目及中央级公益性科研院所基本科研业务费专项资金项目等。获省部级科技进步奖4项，院厅级奖2项，国家发明专利10余个。在国内外学术刊物上发表论文20余篇，出版著作4部。

吴晓云（1986—），博士，研究实习员，主要从事牦牛低氧适应机制和牦牛肉品质形成遗传机制的研究。目前以第一作者发表论文11篇，其中7篇被SCI收录。

肖玉萍（1979—），动物遗传育种与繁殖专业硕士，助理研究员，主要从事《中国草食动物科学》的编辑出版工作。发表论文15余篇，副主编出版专著2部。

严作廷（1962—），硕导，博士，研究员。现为第五届农业部兽药评审专家、中国畜牧兽医学会家畜内科学分会常务理事、中国畜牧兽医学会中兽医学分会理事。主要从事工作奶牛疾病和中兽医学研究（先后主持国家自然科学基金、国家科技支撑计划等项目10多项，现主持"十二五"国家科技支撑计划课题"奶牛健康养殖重要疾病防控关键技术研究"课题）。主要成就（主编或副主编《家畜脉诊》、《奶牛围产期饲养与管理》等著作3部，参编《中兽医学》、《奶牛高效养殖技术及疾病防治》和《犬

猫病诊疗技术及典型医案》等著作6部。发表论文70余篇，获省部级科技进步奖6项，院厅级奖7项。获得国家新兽药证书2个，国家发明专利10个，实用新型专利5个）。

张景艳（1980—），硕士，助理研究员。从事兽药残留快速检测技术和细胞药理学等研究工作。先后参加各类研究课题10余项，主持课题1项。2011年参加的"新型中兽药饲料添加剂'参芪散'的研制与应用"获中国农业科学院科学技术成果二等奖；2013年参加的"非解乳糖链球菌发酵黄芪转化多糖的研究与应用"获得甘肃省科技进步三等奖；2014年参加的"重金属镉/铅与喹乙醇抗原合成、单克隆抗体制备及ELISA检测技术研究"获中国农业科学院科学技术成果二等奖；2014年参加的"奶牛乳房炎联合诊断和防控新技术研究及示范"获甘肃省农牧渔丰收奖一等奖；发表科技论文10余篇，授权发明专利1项、实用新型2项。

张凯（1983—）江西新余人，2008年毕业于西北民族大学，获得临床兽医学硕士学位，现就读于意大利墨西拿大学临床兽医学博士学位。同时现为兰州畜牧与兽医研究所助理研究员，研究方向为中兽药生物转化与畜禽传染病防控。发表论文20余篇，其中主笔7篇。曾获得中国农业科学院发明成果"三等奖"。

张世栋（1983—），硕士，助理研究员。主要从事奶牛疾病与中兽医药研究工作。近年来，主持中央级基本科研业务费及其增量项目共3项，国家自然基金1项；参与国家"十一五"、"十二五"课题3项，甘肃省、兰州市科技项目多项。参与获得兰州市科技进步二等奖1项。发表论文10篇，获得专利6项，参与申请专利10多项。

周绪正（1971—），副研究员，硕士生导师。先后主持和参加国家、省、部级科研课题近40项，其中主持3项；参与项目获得国家科技进步二等奖1项，甘肃省科技进步一等奖2项，兰州市技术发明一等奖1项、临夏回族自治州科技进步一等奖1项、兰州市科技进步二等奖1项、中国农业科学院科学技术成果二等奖2项；参与研制成功国家一类新药（喹烯酮原料及预混剂）2个，三类兽药新制剂4个（多拉菌素、替米考星、黄霉素），取得8项国家发明专利；在国内和国际学术刊物上共发表学术论文100余篇，其中主笔发表论文40余篇；主编、参编著作10部。

李建喜（1971—），研究员，博士学位，硕士研究生导师。现任中国农业科学院兰州畜牧与兽药研究所中兽医（兽医）研究室主任，中国农业科学院科技创新工程中兽医与临床创新团队首席专家，甘肃省中兽药工程技术研究中心副主任，农业部新兽药中药组评审专家，国家自然基金项目同行评议专家，国家现代农业（奶牛）产业技术体系后备人选等。从事兽医病理学、动物营养代谢病与中毒病、兽医药理与毒理、奶牛疾病防治、中兽医药现代化等研究工作。完成国家和省部级科研项目40余项，其中主持24项，包括国家自然基金面上项目、国家科技支撑计划课题、国家公益性行业（农业）科研专项课题、国家科技基础性工作专项课题、农业部948计划项目、甘肃省农业科技重大专项、甘肃省生物技术专项、中国西班牙科技合作项目、中泰科技合作项目等。先后获科技奖励6项，其中获省部级奖励2项，院厅级奖励4项。研发新产品6个，获授权发明专利9项，以第一导师培养硕士研究生6名，共同培养硕士研究生17名，参与培养博士研究生8名。在国内和国际学术刊物上共发表学术论文99篇，其中以第一作者和通讯作者发表论文54篇，SCI收录6篇，编写著作7部。

梁剑平（1962—），博士，三级研究员，博士生导师。2005年获"农业部有突出贡献的中青年专家"称号。2005年分别获中国科协"西部开发突出贡献奖"、中央统战部"为全面建设小康社会做出贡献的先进个人"、甘肃省"陇上骄子"、九三学社甘肃省委、"十佳青年"称号。现任兰州畜牧与兽药研究所兽药研究室副主任、农业部兽用药物创制重点实验室副主任、中国农业科学院新兽药工程重点开放实验室副主任。是中国农业科学院二级岗位杰出人才，甘肃"555"创新人才。兼任中国毒理学会兽医毒理学分会及中国兽医药理学分会理事，农业部新兽药评审委员会委员，农业部兽药残留委员会委员，中国兽药典委员会委员，中国农业科学院学术委员会委员，中国农科院研究生院教学委员会委员、政协兰州市委常委，九三学社兰州市七里河区主任委员，九三学社兰州市副主委。梁剑平研究员主要从事兽药化学合成和中草药的提取及药理研究，先后主持和参加国家和省部级重大科研项目20余项，获奖8项，其中获国家科技进步二、三等奖各1项，省（部）级二等奖2项、三等奖2项，发明专利10项。

牛春娥（1968—），高级实验师，农业部动物毛皮及制品质量监督检验测试中心（兰州）检测室主任。主要从事动物生产及畜产品加工，主持各类科研项目7项：科技部公益类研究项目"动物纤维毛皮产品质量评价技术的研究"；公益类科研院所专项基金项目"甘肃优质细羊毛质量控制关键技术的研究"。主持制定国家标准5项：其中GB/T 25243-2010甘肃高山细毛羊已颁布实施，《西藏羊》等4项完成报批稿。参与完成项目19项："国家毛绒用羊产业技术体系分子育种岗位"、"十一五"科技支撑计划重大项目"优质肉羊新品种选育"、甘肃省科技攻关"中国美利奴高山型细毛羊超细品系培育"、"中国美利奴高山型细毛羊超细品系配套体系研究"等。参与制定国家标准4项、农业行业标准3项。

裴杰（1979—），硕士研究生、助理研究员，研究方向为动物遗传资源与动物蛋白质结构功能研究。主要从事牦牛遗传资源和牦牛乳铁蛋白的结构功能研究工作，主持在研项目2项，包括国家自然科学基金青年基金项目1项、中央级公益性科研院所基本科研业务费专项资金项目1项，参加国家支撑计划子课题等各类项目共计5项。工作期间，获甘肃省科学技术进步二等奖1项（第五完成人）。参加编写《适度规模肉牛场高效生产技术》（副主编）。在各类期刊发表论文50篇，其中主笔16篇。

蒲万霞（1964—）博士后，四级研究员，硕士生导师，甘肃省微生物学会理事，政协兰州市七里河区第六届委员。从事兽医微生物与微生物制药研究，重点方向为兽用微生态制剂的研制及细菌耐药性研究，在金黄色葡萄球菌耐药性研究方面取得一定进展。先后主持农业部公益性行业科研专项"新型动物专用药物的研制与应用"子专题，国家科技支撑计划"奶牛主要疾病综合防控技术研究及开发"子专题，中央级公益性科研院所基本科研业务费专项资金项目、甘肃省科技成果重点推广计划项目、甘肃省农业科技创新项目、兰州市院地校企合作项目、兰州市科技局项目、农业部畜禽病毒学开放实验室基金项目等各级各类项目15项。获得各级政府奖7项。取得国家发明专利2项。主编《食品安全与质量控制技术》等著作5部，副主编著作1部，发表论文70多篇，培养硕士生11名。

王春梅（1981—），硕士研究生、助理研究员，研究方向为植物抗逆机理与饲用牧草研究。主要从事野生牧草耐盐碱生理机制与分子生物学研究，主持在研国家自然基金1项，已主持完成中央级公益性科研院所基本科研业务费专项资金项目2项，参加公益性行业专项、省重大等各类项目10余项。工作期间发表论文56篇，主笔14篇（其中一作/并列一作IF>5.0 SCI论文2篇）；参编著作2部。获甘肃省自然科学一等奖1项（第5完成人），甘肃省科技情报二等奖1项（第5完成人）。

朱新书（1957—），学士，副研究员，主要从事牦牛藏羊品种资源开发和利用工作，主要主持和参加的项目有：公益性行业（农业）科研专项《放牧牛羊营养均衡需要研究与示范》；农业部重点攻关项目《大通牦牛选育与杂交利用》；甘肃省重大科技项目《甘南牦牛改良与选育技术研究示范》；国家肉牛牦牛产业技术体系《牦牛繁育技术研究与示范岗位科学家团队》等。发表学术论文20余篇，主编参编著作3部，荣获农业部科技进步三等奖1项。

岳耀敬（1980—），在读博士。主要从事绵羊繁殖、羊毛和高原适应性等重要经济、抗逆性状的分子调控机制研究。国家绒毛用羊产业技术体系分子育种岗位团队成员，兼任中国畜牧兽医学会养羊学分会副秘书长、世界美利奴育种者联盟成员，曾先后到法国、澳大利亚、新西兰等国学习考察细毛羊育种工作。现主持国家自然青年基金、甘肃省青年基金等项目3项；申报发明专利5项，授权发明专利2项、实用新型2项；参编著作5部，发表学术论文38篇，其中SCI5篇。

谢家声（1956—），大学专科学历，高级实验师。主要从事猪、禽、奶牛等动物疾病的中兽医临床防治研究，主持完成了甘肃省自然基金——"复方中草药防治畜禽病毒性传染病新制剂的研究"、甘肃省科技攻关项目——"治疗奶牛胎衣不下天然药物的研制"等项目。完成国家"七五"、"八五"攻关项目、国家"十五"及"十一五"奶业专项、国家"十一五"、"十二五"支撑计划以及省市各类研究项目30余项。发表论文60余篇，合编专著8部，获得国家发明专利3项。1990年以来先后获甘肃省科技进步三等奖1项；甘肃省科技进步二等奖3项；中国农业科学院科技进步二等奖4项；兰州市科技进步二等奖3项；甘肃省农牧厅二等奖1项；甘肃省天水市科技进步二等奖2项。

王胜义（1981—），硕士学位，助理研究员。主要从事中兽药新药研发和微量元素代谢病研究，先后主持和参与了国家科技支撑计划项目、科技部农业成果转化项目、公益性行业（农业）科研专项、农业部"948"项目等。发表中文核心21篇，参与著作1部，以第一完成人或参与人获得专利13个，以第二完成人获得全国农牧渔业丰收奖二等奖1项。

尚小飞（1986—），硕士，助理研究员。主要从事藏兽医药的现代化研究。目前主持中央级公益性院所基本科研业务费项目一项，参与多项国家及省部级课题。在 Journal of Ethnopharmacology，Veterinary Parasitology 等SCI杂志发表文章8篇，参与编写著作一部。

苗小楼（1972—），学士，副研究员。主要从事兽药研发、传统兽医药物研究工作，主持参加多个省部课题，发表论文20余篇，获得授权专利3项，主持研制的中兽药"益蒲灌注液"获得国家三类新兽药证书，参与研制的一类兽药"喹烯酮"曾先后获国家科技进步二等奖和甘肃省科技进步一等奖。

　　杨红善（1981—），硕士，助理研究员。主要从事牧草种质资源搜集与新品种选育研究工作，主持在研项目3项，其中甘肃省青年基金项目1项、甘肃省农业生物技术研究与应用开发项目1项、中央级公益性科研院所基本科研业务费专项资金项目1项，参加国家支撑计划子课题等各类项目共计3项。工作期间以第一完成人或参加人审定登记甘肃省牧草新品种4个。获中国农业科学院科技进步二等奖1项（第三完成人）。参加编写《高速公路绿化》著作1本（副主编）。在各类期刊发表论文20篇，其中主笔13篇。

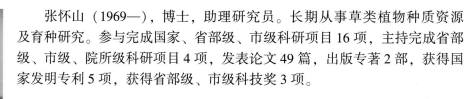
　　张怀山（1969—），博士，助理研究员。长期从事草类植物种质资源及育种研究。参与完成国家、省部级、市级科研项目16项，主持完成省部级、市级、院所级科研项目4项，发表论文49篇，出版专著2部，获得国家发明专利5项，获得省部级、市级科技奖3项。

　　刘宇（1981—），硕士，助理研究员。2007年7月来中国农业科学院兰州畜牧与兽药研究所工作至今，主要从事有机及天然药物化学研究。先后主持及参与国家级省部级药物研究项目10余项，包括"十二五"农村领域国家科技计划项目、国家支撑计划项目、甘肃省科技支撑计划项目、甘肃省自然科学基金项目及中央级公益性科研院所基本科研业务费专项资金项目等。获省部级科技进步奖1项，院厅级奖2项，国家发明专利4个。在国内外学术刊物上发表论文20余篇，出版著作2部。

　　郭文柱（1980—），男，汉族，硕士，助理研究员。2007年11月来中国农业科学院兰州畜牧与兽药研究所工作至今，主要从事新兽药的研究开发工作。先后主持及参与国家级省部级药物研究项目10余项，包括"十二五"农村领域国家科技计划项目、国家支撑计划项目以及中央级公益性科研院所基本科研业务费专项资金项目等。获兰州市科技进步奖1项，国家发明专利2个。在国内外学术刊物上发表论文10余篇，出版著作1部。

　　杨亚军（1982年），男，硕士研究生、助理研究员，研究方向为兽医药理与毒理学。主要从事药物筛选、新制剂的研究与开发工作，主持在研项目3项，其中国家自然科学基金青年基金项目1项，甘肃省青年基金项目1项，中央级公益性科研院所基本科研业务费专项资金项目1项，参加国家自然科学基金面上项目1项，甘肃省科技重大专项1项，公益性行业专项子课题2项等。获中国农业科学院科技进步二等奖2项、甘肃省科技进步一等奖1项，在各类期刊发表论文30篇，其中SCI论文10篇（第一作者2篇，通讯作者1篇），参编著作1部，参译著作1部，申请发明专利6项，授权3项。

程富胜（1971—）博士，副研究员，硕士生导师。主要从事天然药物活性成分免疫药理学研究及其制剂开发研制。先后主持和参加国家、省、部级科研课题 20 项。参加完成的国家支撑项目"中草药饲料添加剂'敌球灵'的研制"获农业部科技进步二等奖；国家自然基金项目"免疫增强剂-8301 多糖的研究与应用"成果获中国农业科学院科技进步二等奖；"蕨麻多糖免疫调节作用及其机理研究与临床应用"成果已分别获中国农科院、兰州市科技进步二等奖。主持完成的"富含活性态微量元素酵母制剂的研究"获兰州市科技进步二等奖。在国内和国际学术刊物上共发表学术论文 50 余篇，其中主笔发表论文 30 余篇。

李冰（1981 年—），女，硕士研究生、助理研究员，研究方向为兽药新制剂研制与安全评价。主要从事新兽药制剂质量标准制定、体内药代动力学研究与残留研究。中国畜牧兽医学会兽医药理毒理学分会理事。主持在研项目 2 项，其中公益性行业专项子课题 1 项、中央级公益性科研院所基本科研业务费专项资金项目 1 项；参加国家支撑计划项目、国家高技术研究发展计划（863 计划）、甘肃省科技重大专项、国家自然基金项目、国家基础专项等各类项目十余项。以主要完成人申报国家新兽药 1 个。前三以前作者发表论文 30 余篇，主笔核心期刊 3 篇，主笔 SCI 收录 2 篇，参编著作 2 部。获 2011 年兰州市科学技术进步奖一等奖，第 6 完成人；2012 年中国农业科学院科技进步二等奖，第 3 完成人；2013 年甘肃省科学技术发明奖一等奖，第 3 完成人；2013 年兰州市科学技术发明奖一等奖，第 3 完成人。

褚敏（1982—），中共党员，在读博士，助理研究员。主要从事牦牛分子遗传与育种研究，先后参加国家科技支撑计划、农业部行业科技、甘肃省重大科研项目、"863""948"项目、甘肃省生物技术和甘肃省科技支撑等重大课题十余项，现主持中央级公益性科研院所基本科研业务费专项资金项目 1 项。主笔发表 SCI 学术论文 2 篇，中文 10 多篇，参与编写著作《牦牛养殖实用技术问答》和《适度规模肉牛场养殖技术示范》2 部，翻译并编辑第五届国际牦牛会议英文学术论文集 1 部，发明实用新型专利 6 项。

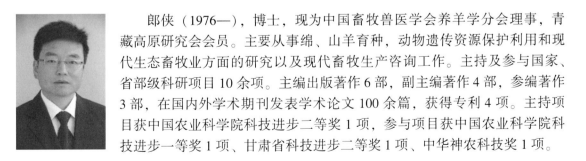
郎侠（1976—），博士，现为中国畜牧兽医学会养羊学分会理事，青藏高原研究会会员。主要从事绵、山羊育种，动物遗传资源保护利用和现代生态畜牧业方面的研究以及现代畜牧生产咨询工作。主持及参与国家、省部级科研项目 10 余项。主编出版著作 6 部，副主编著作 4 部，参编著作 3 部，在国内外学术期刊发表学术论文 100 余篇，获得专利 4 项。主持项目获中国农业科学院科技进步二等奖 1 项，参与项目获中国农业科学院科技进步一等奖 1 项、甘肃省科技进步二等奖 1 项、中华神农科技奖 1 项。

席斌（1981—），硕士，助理研究员。主要从事畜产品质量及安全监测工作。主持项目2项，甘肃省星火计划及中央级公益性科研院所基本科研业务费专项资金项目。参加完成科研项目4项，参加国家标准6项，参与发明专利2项。以第一作者发表论文30余篇，其中核心期刊20余篇。参编出版专著1部（《动物纤维组织学彩色图谱》2007），并获第十六届2008年度中国西部地区优秀科技图书二等奖。

王玲（1969—），预防兽医学硕士，副研究员，研究方向为兽医微生物与免疫学，主要从事创新兽药的基础研究与应用。工作期间先后参加国家重点科技支撑项目、国家自然科学基金、农业部"跨越计划"、农业部和科技部专项资金项目、中央级公益性科研院所专项资金项目、甘肃省及兰州市科技攻关项目等课题40余项。主笔发表论文40余篇，其中核心期刊20余篇，主笔发表SCI论文1篇，署名发表SCI 2篇。工作以来作为主要完成人获得研究成果5项，发明专利4项，实用新型专利3项，甘肃省四类新兽药证书3项，转让成果1项。

曾玉峰（1979—），博士，副研究员。先后从事草食动物遗传繁育研究和农业科技管理工作，主持或参加完成了10余项国家、省部级项目，包括国家自然科学基金项目、863计划、公益性（农业）行业科研专项、国家肉牛牦牛现代农业产业技术体系、甘肃省科技重大专项、甘肃省国际合作计划、中国农业科学院科技经费等，组织申报各级各类项目400余项，作为主要完成人获得甘肃省科技进步一等奖、全国农牧渔业丰收二等奖等省部级科技奖励5项，参与制定行业标准2项，获得1项国家专利，副主编著作2部，参编著作10部，第1作者发表文章10余篇，其中SCI收录2篇。

李锦华（1963—），硕导、博士、副研究员、中国草学会教育专业委员会理事、主要从事牧草栽培和育种工作。"十五"期间，参与主持国家发改委"中兰1号和甘农系列苜蓿种子高技术产业化示范工程"项目。"十一五"期间主持完成国家科技支撑计划"优质牧草繁育及种子加工技术研究与示范"课题。"十二五"开始，主持西藏草业重点研究项目"西藏主要栽培牧草繁育研究"课题。发表论文60余篇；参编著作4部，主编1部；参加选育出国家审定苜蓿新品种1个，主持选育出甘肃省审定苜蓿新品种1个；作为主要完成人，获省、部、院级奖5项，其中省级一等奖2项。

王瑜（1974—），副研究员，1997年毕业于甘肃农业大学，硕士学位，助理研究员，中共党员。主持完成甘肃省科技厅中小企业项目1项，甘肃省农业创新项目等1项，基本科研业务费项目1项，参加并完成科研项目9项；获农业部农牧渔业丰收二等奖1项，甘肃省科技进步三等奖1项，甘肃省农牧渔业丰收一等奖2项，甘肃省农牧渔业丰收二等奖1项，兰州市科技进步二等奖1项；发表论文17篇，其中主笔7篇；作为主要撰稿人出版著作3部；参与并取得农业部新兽药注册证书1项、发明专利4项。

张小甫（1981—），硕士研究生、助理研究员、主持中央公益性科研院所基本科研业务费项目甘南州优质高效牧草新品种推广应用研究、甘南州优质草畜新品种推广与应用研究，甘南优质牧草及中草药新品种推广与应用。参加"十二五"农村领域国家科技计划课题甘南高寒草原牧区"生产生态生活"保障技术及适应性管理研究、公益性行业（农业）科研专项墨竹工卡社区天然草地保护与合理利用技术研究与示范。主笔发表西藏拉萨地区不同施氮量对鸭茅产量的影响、土壤微生物生态学研究进展、我国农业科技发展现状及趋势研究、苜蓿种子丰产因子研究进展，参与发表 2009—2014 年甘肃省高等院校获国家自然科学基金自主情况分析、2009—2013 年甘肃省获国家自然科学基金项目情况分析、我国湿地的退化原因及生态恢复措施研究、基于文献计量学的苜蓿研究领域进展分析。参与一种快速测定苜蓿品种抗旱性和筛选抗旱苜蓿品种的方法。参于研究的甘肃省旱生牧草种质资源整理整合及利用研究获得 2012 年度甘肃省科技进步奖二等奖。

程胜利（1971—），博士，副研究员。主要从事绵羊营养与饲料研究工作。先后参加完成的主要项目有：国家"863"项目"优质肉用小尾寒羊新品种培育与扩繁技术研究"、农业跨越计划项目"甘肃优质肉羊规模化生产配套技术试验示范"、国家基础专项"中国野牦牛种群动态调查及种质资源库建设"、科技部基础平台"西部稀缺偶奇蹄目野生动物信息资源库建设"等 10 余项。其中参与完成的"中国野牦牛种群动态调查及种质资源库建设"2005 年获中国农业科学院科技进步一等奖，"优质肉用绵羊产业化高新高效技术研究与应用"2008 年获甘肃省科技进步二等奖。共发表学术论文 50 余篇，其中 SCI 收录 1 篇，主编著作 1 部，副主编参编著作 5 部。

邓海平（1983—），硕士研究生，助理研究员、先后从事动物学、兽医微生物学方面科研工作，现从事科研条件建设项目管理。主持院、所级科研项目 2 项。以第一作者在国内核心期刊发表学术论文 10 篇，其中一级期刊 3 篇。参加作者发表文章 34 篇。参与出版著作 3 部，获得省部级集体奖 1 次，市级奖 2 次，中国农科院院级二等奖 1 次。

吴培星（1962—），日本山口大学兽医学博士，副研究员，硕士生导师。主要从事临床兽医学、兽医药理学、分子生物学、基因工程学等方面的研究工作。主持或作为主要参加人先后完成了"抗流感病毒感染性多肽的筛选及应用研究"、"利用转基因技术对癌细胞转移机理的研究及抗癌剂的的探讨"、"利用原代细胞培养法进行牛脂肪细胞分化机理的研究"、"乳中激素的酶联免疫法测定技术的建立及应用"、"新型高效农业生物药物载体和新制剂创制"、"新型抗动物病毒性疾病生物药的研究及开发"等十余项国内外前沿课题的研究工作。在国内外主要学术期刊及国际议发表科研论文 40 余篇，其中 SCI8 篇。获得发明专利 4 项，省部级各种奖项 10 项。参加编写《中兽医方剂大全（1995 年出版）》、《人与动物共患病（上、下册）（2013 年出版）》两部。